Hans Wehrli

Metaphysics

Hans Wehrli

Metaphysics

Chirality as the Basic Principle of Physics

wbg Academic

Die Deutsche Nationalbibliothek verzeichnet diese Publikation
in der Deutschen Nationalbibliographie; detaillierte bibliographische
Daten sind im Internet über http://dnd.d-nb.de abrufbar

wbg Academic ist ein Imprint der Verlag Herder GmbH

Satz und eBook: SatzWeise, Bad Wünnenberg
Gedruckt auf säurefreiem und
alterungsbeständigem Papier
Printed in Germany

Besuchen Sie uns im Internet: www.wbg-wissenverbindet.de

ISBN 978-3-534-40537-4

Elektronisch sind folgende Ausgaben erhältlich:
eBook (PDF): 978-3-534-40538-1
eBook (epub): 978-3-534-40539-8

Contents

Preface to the Second Edition 2020

This is a fundamentally revised version of the book's first English edition in 2008 (first German edition 2006).

In order to describe nature as it is observed, a strategy for the development of a physical theory is proposed that circumvents the contradictions between physical observation and mathematical formalism. Six paradigm shifts are required:

1. The laws of nature are as they must be for perception to be possible. The laws are not derived from nature, but from the way nature can be perceived. This way the theory does not simply work. It philosophically explains why it works, even if the formalism of the theory is not yet developed.
2. The theory does without the logical proposition $A \equiv A$, which is replaced by a new axiom of chirality, a prerequisite for any kind of order.
3. The theory does without the axiom of infinity because infinity is not perceivable.
4. The usual entities of space, time, substance (= mass, energy, information) and interaction are replaced by the new, mathematically defined term event, from which all other entities can then be derived.
5. Action without information transfer is non-local and instantaneous, but the information about any change of action travels with light speed.
6. The concept of the black hole is expanded: There are black mini-holes which are stable due to their high symmetry and black mini-holes within black mini-holes with specific action to the outside.

Already in the simple model of a space composed of only four points, terms such as neutrino, speed of light, mass, spin, fermion, boson, Planck's constant and black hole can be represented and the theory of relativity can be unified with quantum theory.

Following the above mentioned *paradigm shift number 1*, the metaphysical prerequisites of any physical theory are described, i. e. the limits of language (Chapter 1), the limits of knowledge and truth in mathematics (e. g. infinity and continuum according to *paradigm shift number 3*) and physics (Chapter 2), the distinction and relation between the ego as subject and nature with its objects (Chapter 3), the role of chirality as the basic principle of any kind of order, be it in thinking, in mathematics or in the perception of nature (Chapter 4), the problem of monism or dualism of body and soul regarding the flow of information from nature to the consciousness of the ego (Chapter 5), and finally the essence of measuring as a counting of events (Chapter 6).

In Chapter 7, a mathematics is developed which is as simple as possible and which does without any axioms that contradict the metaphysical prerequisites. Therefore, the axiom $A \equiv A$ is replaced by a new axiom of chirality according to the *paradigm shift number 2* mentioned above and the continuum with its infinity aspects is dispensed with according to *paradigm shift number 3*. It follows that irrational numbers such as π, e and $\sqrt{2}$ as well as continua are forbidden in mathematical models for physics. The term event is defined mathematically.

Surprisingly the resulting finite mathematics can be interpreted physically (Chapter 8). Space-time consists of single points which produce events by changing their relative configuration without any background space. The simplest such space is the 4-point space which is the model for the neutrino as the simplest physical particle. The counting of the mathematically defined events corresponds to the measurement of space, time, mass and interaction as postulated by *paradigm shift number 4*. According to the new model, actions always are non-local but information transfer between separate systems is always local as postulated by *paradigm shift number 5*.

In chapter 9, it is shown how gravitational interaction can be described as the simplest type of interaction in a many-point universe where action without information transfer is non-local and travels instantaneously according to *paradigm shift number 5*.

Chapter 10 outlines how the mathematical properties of gravitation lead to phenomena such as black holes, time dilation and length contraction of objects by their relative motion as well as by the action of black holes, red shift, deflection of light by gravitation, centrifugal force reversal in the vicinity of a black hole, the Big Bang and inflation of the universe thereafter. The duality of particle and wave and the Heisenberg uncertainty principle can be explained.

Chapter 11 describes electromagnetism by an electron model consisting of 16 points in 4 black mini-holes within a black mini-hole according to *paradigm shift number* 6. It exerts a specific action on the outside that is either attractive or repulsive depending on the arrows of time within these black mini-holes.

In Chapter 12, the concept of black mini-holes is extended to other particles consisting of 16 or more points such as quarks. A few simple rules for their construction and interaction are given. This way also the weak and strong interaction can be described and explained. It is shown how the four types of interaction were developed by symmetry breakings shortly after the Big Bang.

A comparison of the basic ideas and notions of this new chirality theory with other theories is given in chapter 13.

As is the case in all metaphysical and physical theories, there are numerous open questions, e.g. about the experimental verification, the usefulness of the mathematical model, the philosophical foundation of the theory and even about its theological consequences. Such questions are discussed in chapter 14.

Preface to the First Edition 2008

Alles in Allem
Kurt Guggenheim[1]

The question this book poses is: Why are the laws of nature the way they are? My answer is: The laws of nature are the way they have to be, so that perception is possible.[2],[3]

That is the metaphysical answer to a metaphysical question. The hope that natural laws can be deduced in this way from metaphysics motivates me as a scientist to practice metaphysics.[4]

The physical theory outlined in this book has not been fully perfected and is sometimes speculative. Even so, it forms a rounded whole, and as

[1] „Everything in everything“. With these three simple words, the writer and physical layman Guggenheim intuitively grasped the essence of modern physics. The words were uttered on the occasion of Albert Einstein's employment interview with the rector of the University of Zurich. The two were in agreement: Physics, everyday urban life and religion: Everything is contained in everything. Nature reveals to us such a governance of laws, such a superior reason, that that which one does, i. e., seek to recognize and grasp these laws, can be designated as nothing other than a service to God [Guggenheim (1952)].

[2] Kant formulated the same thought as follows: There are many laws of nature which we can only know through experience, but the conforming to these laws in conjunction with the appearances, i. e., nature in its totality, we cannot get to know through experience because experience itself requires laws which are *a priori* fundamental to the possibility of it. The possibility of experience at all is thus simultaneously the general law of nature, and the principles of the first are themselves the laws of the latter [Kant (1783/2001) § 36, 39].

[3] He who could, with sufficient intellectual capacity, analyze under which conditions experience is at all possible, would have to be able to show that all general laws of physics already follow from these conditions. The physics thus derivable would be precisely the uniform physics supposed [Weizsäcker (1999), pp. 17 and 344–348].

[4] Einstein (1918) also advocated the view that physicists must derive the laws of nature essentially via deduction from metaphysics and not by induction on an empirical basis.

far as I can see, there are no inconsistencies with any preceding empirical findings in physics. My theory means a paradigm shift for theoretical physics and thanks to this, it can uniformly and simply explain and interlink quite different physical phenomena, phenomena which up until now have been unexplainable or could only be described in separate theories. This is my legitimation for presenting the new theory in its current unfinished state. My basic approach is as follows: As a natural scientist, I am convinced that my consciousness exists and that it generates perception.[5] Other than that, I am convinced of nothing. Then I enquire into the basic philosophical conditions which must be met so that one can speak of perception, for instance, into the separation and connection of the perceiving subject and the object perceived.[6] I attempt to formulate these basic conditions mathematically, whereupon I forgo certain generally usual theorems of logic and axiomatic method. Examples would be the theorem $A \equiv A$ or the axiom of infinity, because they are not suited to the description of perception. Instead, I introduce a new axiom, the „axiom of chirality".[7] The finite mathematics which arises from this is probably akin to a finite group theory or a finite chiral topology. This, however, is not fully spelt out by me; it is a task which must be taken up by others. In any event, the simplicity of the theory is important to me. Yet the simpler a theory is, the more difficult it is to really understand it.

It appears that the resultant theory is nothing other than the basis of all natural laws, the so-called „theory of everything (TOE)". Weizsäcker (1971a, p. 24) considered such a „completion" of physics by derivation from meta-physics possible. Up until now, nature has always been described with the help of the four entities of space, time, substance (I regard the latter as synonymous with matter, energy and information) and

[5] It is difficult, perhaps even impossible, to define the term consciousness. Certainly, it involves subjectivity [Blackmore (2003) pp. 107, 159 and 198].

[6] A definition of the term object is difficult, since objects can never be completely clearly and individually distinguished, because impenetrable borders would make perception impossible. A definition could read: An object is a combination of quantities whose present values permit common predictions about these quantities (in the future) [Drieschner (1981)].

[7] The physical term chirality means handedness and expresses that although the right and the left hand are isometric, they are not properly congruent. That is, one hand cannot be aligned to cover the other.

interaction. All attempts in the last 2500 years to reduce the four entities to three or even fewer have failed so far. According to Plato (1961–1963a), perception always has the character of an event, and the attributes of perception are correlated with the current states of the perceiving humans, so that humans are the measure of all things. Plato tried to reduce physics to geometry, whereby Be-ing in nature is not physical, but rational, and masses and forces are merely psychological appearances which result from the interaction between the observer and the eternal, constant mathematical relations. For Aristotle, the becoming, the event, was real. Einstein tried to completely define space-time with his general theory of relativity with the help of matter. Eddington, on the other hand, sought the opposite; to define matter with the help of space-time [Northrop (1928)]. Also, Einstein, in the 1930s, considered describing elementary particles and space-time as a single entity where the particles would be a kind of knot in space-time [Musser (2004)]. For Descartes, space and matter were synonymous [Weizsäcker (1971b) pp. 9–24]. The most recent attempt to reduce the number of entities is string theory [Smolin (2001)]. In my chirality theory, a new and even more fundamental entity emerges from the axiom of chirality. I call it an event. Space, time, substance and interaction are then all simply different aspects of the event, which I will define mathematically. Whether something is perceived as space or time, as substance or interaction, as fermion or as boson, is relative and depends upon the state of the observer. Even perception itself is an event. According to Whitehead, perception comes about through perception events, a special kind of event. According to his theory, space and time are nothing other than relations between events. Matter is also an attribute of the event [Hampe (1998) pp. 61–73].

Events are countable, so in my theory only whole numbers, that is, values without a physical dimension occur. Resulting from this are the physical constants which are all reducible to the number one, since this is the smallest possible number of perceivable events. The physical constants play a fundamental role in the laws of physics, because they always form a bridge between entities which are completely different physically [Lévy-Leblond (1979)]. The three-dimensionality of apparent space, both theories of relativity and the most important theorems of quantum theory, for example the Heisenberg uncertainty principle, the equation $h\nu = mc^2$, the shortening of standard lengths and the slowing down of clocks at high velocities and in the vicinity of heavy masses, black holes and the reversal of centrifugal force in their proximity need no introduction. The elementary particles, with their remarkable properties and interactions from the neutrino to the 72 different quarks to the K^0-Meson with its time symmetry violation are derived from the theory. The four known interactions gravitation, electro-magnetism, strong and weak interaction are reconciled and accounted for by the new theory. The Big Bang gains a new significance and the inflation of the universe shortly thereafter becomes plausible. If the reader asks why all this is so, then the answer is ultimately always the same: if it were not so, perception would not be possible.

Central, in my view, are chirality and black holes, phenomena which receive rather little attention in contemporary physics. For example, to my knowledge no physicist has ever asked what happens when a black hole forms within a black hole, and most physicists erroneously believe that the spin has something to do with a direction of motion or with an axis.

The big questions of philosophy have new aspects and in part, answers as well. What is real? What is the difference between a perception and the perceived, between body and soul? New light is shed upon the controversies of Plato/Aristotle [Northrop (1928)], Leibniz/Newton [Leibniz (1715–1716/1990)], Einstein (1935)/Bohr (EPR)[8], Kant/Reidemeister concerning the difference of regions in space [Reidemeister (1957)] or they are harmoniously resolved. In Leonardo da Vinci's world view, space, time and bodies arise from an original movement of Platonic bodies, which is evident in the recurrence of a „symbolic form", the rotating spiral [Arasse (1997)]. Into the place of platonic solids, Leonardo da Vinci's spirals, the modern strings or Pauli's archetypal squares with the two diagonals and his *i*-Ring [Atmanspacher et al. (1995)], now steps the event and its simple mathematical description.

The reader may well ask, how necessary all these associations and digressions in Chapters 1 to 6 really are for the development of a new physical theory which is supposed to be a simple one. Such philosophical excursions may be all very interesting, but what have they got to do with

[8] EPR = Einstein, Podolsky, Rosen-thought experiment in quantum theory [Einstein (1935)].

the actual topic of the book? Yes, I believe that the excursions into theology, philosophy, psychology, biology, and literary history help not only to familiarize oneself with the new physical theory, but also to understand its rationale and nature. All too seldom are physical theories really understood; they are mostly applied because they are useful and because they work. A paradigm shift always calls for a rationale at a meta-level. Thus, a paradigm shift in physics must be accounted for in metaphysics. For this reason, the first half of this book is dedicated to metaphysics.

Chronologically my theory has, in essence, actually been developed as it is presented in this book, beginning with the philosophy of language and the development of a metaphysical approach. The discussion of chirality and measurement processes prompted me to formulate the axiom of chirality. Only then did I begin to concern myself with concrete physical theories and to sketch out a fundamentally new one. If the reader has no interest in understanding this theory, then he can begin by turning straight to Chapter 7. Conversely, a reader who is content to gain a broad appreciation of the theory can pick out what interests him from Chapters 1 to 6, simply take a look at the illustrations in Chapters 7 to 12 and then proceed to the open questions in Chapter 14. Those however, who aspire to truth and clarity, will find resonance in Friedrich Schiller's rendering of Confucius's words:

> Nur die Fülle führt zu Klarheit
> Und im Abgrund wohnt die Wahrheit.[9]

I have tried to make my theory generally understandable and where possible, to cite literature which is intelligible to all. In this way, the literature should serve to inspire, rather than form the actual basis of my theory. At times I have sought to summarize whole books in a few sentences, which are then inevitably not very sophisticated or exact. Without such streamlining however, my book would have become too long-winded and thereby unreadable. Those who wish to know „the fullness that makes us wise" will not escape having to read the books cited for themselves.

[9] „Naught but fullness makes us wise, Burried deep truth e'er lies." Proverbs of Confucius (Bowring's translation). The favorite verse of the physicist and philosopher Niels Bohr [Pauli (1955)].

Who ought to read this book? Every physics and philosophy student can learn something from it. In particular, he will learn to pose questions which are unfortunately not posed by their teachers, let alone answered by them. The book offers mathematicians tantalizing stimulation and I hope that it will give impetus to a mathematics of a Theory of Everything, just as Marcel Grossmann, Hermann Weyl and John von Neumann, together with many other mathematicians, have formulated the mathematics of the theories of relativity and quantum mechanics. Without a new qualitative prior grasp of space-time and its internal transformations, the search for a better mathematical model would become the search for a needle in a haystack [Saller (2003)]. My book delivers this qualitative prior understanding. Every natural scientist is certain to find stimulus in its pages. Those who enjoy interdisciplinary discourse, be they historians, psychologists, engineers or theologians, will also enjoy this book. When the interested lay reader understands three quarters of the book, I will be content. They can take consolation: I have not understood everything either. That is also not all that important. It is far more important that one notices when one has not understood something, for the questions, also those of the reader, are far more important than the answers. And even more important is amazement. Amazement indicates surprise, openness and also a little humility in the face of the amazing things being marvelled at. „Only those who see or experience something unexpected, or those who are able to ignite a sense of the sublime vis-à-vis the marvelous are able to be amazed“ [Janich (2000)].

I would like to thank John Bennett for the careful translation and his persistent efforts to grasp and convey the content and the language of the German original in equal measure. I express my thanks to Heinrich Baggenstos, Philipp Wehrli and Karl Wirth for their critical reading of my book and their numerous suggestions. My wife Christel has accompanied the formation of the present theory in countless conversations over the last 35 years and with her clever questions, stimulated me to ever new considerations and explanations.

The Tao[10, 11] that can be told is not the eternal Tao.
The name that can be named is not the eternal name.
The nameless is the beginning of heaven and earth.
The named is but the mother.

Therefore: He who looks calmly inward,
Experiences the wonder of unlimited being.
He who wishes to possess the world and holds onto names
Finds mundane limitation.

In origin these two are one, differing only in name.
Beyond comprehension is this unity.
The secret of secrets,
The gateway to the revelation of everything.
Lao Tse (ca. 600 BC/1978)

[10] Chinese „Tao" means way, path, course of nature, which underlies all appearances, the knowledge of which remains beyond comprehension, however. Tao is also often translated as sense [Wilhelm (1978)].
„Those who understand the sense (Tao), understand the laws of nature" writes Dschuang Dsi (ca. 340 BC/1972) p. 195.

[11] Dschuang Dsi, a contemporary of Plato, said this about the psychological precondition of the Tao (sense): „The state, where I and not-I (i. e., subject and object) no longer form a contrast." The distinction between subject and object only comes from the subjective viewpoint [Jung (1967) (pp. 542 ff.)].

Chapter 1

The Limits of Language

Language is the house of Be-ing[12]
Martin Heidegger (1946)

Writing a book, be it about philosophy or natural science, requires language. Even simply thinking about such things is not possible without language.[13] The philosopher and the natural scientist should therefore be aware of what language can and cannot do. All too easily a physicist can be led to believe that he has found a natural law, whereas a careful analysis reveals that his „law" is simply a consequence of the language used. The primary language of physics is mathematics, an abstract, strictly formalized and relatively precise language, which easily misleads the speaker about how inaccurate or even paradoxical his propositions are. Philosophers on the other hand are more concerned than physicists with the central significance of language. There are even those who simply define philosophy as the universal criticism of language and meaning [Heintel (1988)]. A language is a system of signs serving to convey information. The signs are signals that we are able to perceive with our senses, i.e., gestures, sounds, images, symbols or scripts. The signs or characters and words, are positioned in a particular relation to each other. In other

[12] Die Sprache ist das Haus des Seins.

[13] Thinking is more, and more complicated, than simply having a consciousness. For consciousness alone, humans do not yet require language. Einstein wrote: „The words or the language … do not seem to play any role in my mechanism of thought. The psychical entities which seem to serve as elements of thought are certain signs and more or less clear images which can be „voluntarily" reproduced and combined … The abovementioned elements are, in my case, of visual and some muscular type. Conventional words or other signs have to be sought for laboriously only in a second stage, when the mentioned associative play is sufficiently established and can be reproduced at will." [Penrose (1989) pp. 383 f., 423].

words, language has a structure. The signs themselves, as well as their relative arrangement, all have a meaning about which all the living creatures communicating with each other should be in agreement, lest misunderstandings arise. So, language needs not only to be conveyed, e.g., from a radio transmitter to a receiver, but needs also to be understood. For this, a receiver capable of understanding is necessary.

Accordingly, linguistics (or semiotics) is concerned with three relationships: firstly with the relationship of linguistic signs to each other, called syntax; secondly with the relationship between linguistics signs and the designated objects, called semantics; and thirdly with the relationship of language to speaker and listener, called pragmatics [Hinzen (2017)]. In the following we shall have a closer look at these three types of relationships.

1.1 Language and order

Irrespective of the kinds of signs composing a language, they always stand in a temporal or spatial relationship to each other. This applies even when the language is neither written nor spoken, but merely thought. A chaotic arrangement of characters or sounds would contain no semantic information and cannot be a language. A written text is ordered in space. When the text is read, it is ordered in time. Both language forms have, apart from script and sound characteristics, the same information content. This is familiar to us and has become completely self-evident. It is, however, a phenomenon which becomes all the more amazing, the longer one reflects upon it. Amazing in itself is the fact that a three-dimensional arrangement such as space can apparently be easily represented in a one-dimensional order such as time, without any significant loss of information. This gives us a first hint of a certain analogy between three-dimensional space and one-dimensional time. Moreover, this raises the following question: What are the structural interrelationships between language, time and space? Are their similarities coincidental, do they have a common basis or is one a consequence of the other? If philosophy does not exist without language, and if philosophy is, in a certain respect language criticism, could there even be philosophy without time or space? Answering the latter in the negative raises the question as to how free philosophers actually are when they philosophize *in* space and time *about* space and time. We can state: Language requires a certain framework and for its part, it sets boundaries.

1.2 Meaning

Scientists are used to explaining the meaning of a word with a definition, which in turn is made up of other words. Defining is an endless process, for which more and more new words are required. After all, in the ever-lengthening chain of definitions words appear which we have already met, perhaps in another context. Thus, we have already had the occasion to consider the meaning of these words. Complicating matters is the fact that most words have several different meanings and that they can stand for a concrete individual here and now, or for a whole class of things. „The dog“ could be my dog Barri, or an abstract class of elements, which all have a certain number of dog-like attributes. Even a toy dog is a dog. In the first case the dog is an individual, in the second an idea. However, the dog can also be something quite different, e.g., a constellation of stars. Moreover, there are different words for one and the same object, all meaning the same, e.g., „chien“ or „Hund“ or „собака“. Incidentally, there is nothing to prevent me from writing as of now „blabla“ or „MÜ MÜ“ instead of „dog“. In light of all this, can language be precise at all?

Let us take an example from mathematics, an exact science. How, for instance, can one define the term „point“? „A point is an infinitesimal place in space“ could be one definition, or „A point is a thing whose only core attribute is that it exists“. These are quite sophisticated definitions and one could discuss them for hours without becoming any wiser. After all, every child knows what a point is. The relationship between the word „point“ and the speaker or hearer of this word, i.e., the child, will be discussed in the next section. For now, we are looking for a definition of „point“ which is valid, whether it is communicated or not. For this we have used words such as „space“, „place“, „is“, „exists“ and „infinitesimal“, words which are themselves not easy to define. Space could be defined as the sum of all points and place as the sum of neighboring points. In so doing, we see immediately that the definition goes in a circle: „point“ is defined with the help of „space“ and „space“ is defined with the help of

„point". Such circular definitions are unavoidable, even in strictly formulated logic. It is clear, that in the same way, no language can be constructed precisely. Language is thus fundamentally imprecise from a scientific point of view, i.e., we can never know precisely what a text means. Here, also, language has limitations [Weizsäcker (1972)]. Nevertheless, there are a few important terms, as they are to be understood in this work, which should be defined.

Physics describes observed nature and formulates by means of mathematical methods laws of nature, which permit predictions (about the future). *Nature* is the entirety of all those things which can, in principle, be empirically – directly or indirectly – perceived (Whitehead 1925/1939). In this sense, nature is *real* or *material. Existence* in chirality theory is an ontological property of anything that is either real or – if not real – must have a Be-ing in nature. For example, a point or a natural number is not real but it exists. A boson is directly, a fermion indirectly perceivable; therefore, both are real. Also, black holes can be indirectly observed and are real. Space and time *per se*, all numbers except the natural numbers, and infinity are neither existing nor real but only ideas or mathematical models. The examples mentioned for this definition of reality and existence will be discussed later in the book. I am aware that depending on the philosophical and epistemological point of view there might be other definitions. *Mathematics* studies patterns in abstraction of the individual things which are patterned (Hampe 1998). *Empirical perception* is a flow of information from the outside into the conscious mind of a subject. The subject is an entity, which can take up, store and consciously process information. If the observer is transcendental, he is called a *presumed observer* and he is not part of reality. It is left open as to whether a subject can itself be part of nature. For the considerations in the present work it is sufficient to proceed from only one subject, the *ego* [Descartes (1641/1960) 1 (18–19), 16 and 2 (26–27), 23]. Possible further subjects and intersubjective communication are not brought up for discussion, because such further subjects can never be unequivocally differentiated from objects. An *object* is a summary of mathematical quantities or patterns, whose current values permit predictions about these very values (in the future) [Drieschner (1981)]. *Information* can be defined as answers to potential questions which can be reduced to a countable number of so-called *binary choices*, i.e. to alternatives which can be decided with a simple yes/no answer. The binary choices are computable as bits or qubits [Weizsäcker (1986), pp. 163–173].

We shall now examine the extent to which today's usual methods and terms of physics contradict the definitions above. Subsequently, those basic metaphysical conditions for a physics which conforms better to these definitions are to be described.

1.3 Understanding

If the reader has not given up so far, he is probably confident that he has understood at least part of the text and that, by and large, the author has the same understanding. This is by no means self-evident, considering all we have just discussed about the limitations of language. On what then, do we base the belief that despite occasional misunderstandings, we actually understand each other quite well? In order to answer this question, it is helpful to examine how a child learns to understand something.

Even prior to birth, a child receives stimuli. The sense of touch reveals spatial boundaries, the baby becomes accustomed to the constant comfortable temperature or it hears sounds. The child receives these stimuli unconsciously, though perhaps also with an increasing consciousness. I am reasonably certain that even today I can still recall that comfortable, simple and orderly pre-birth feeling, although it would of course be impossible for me to name a concrete event. There is no question of a language at this early stage of development, not even a primitive one. However, experiences are already taking place, in which information is transmitted, processed and stored as memory. In this way, the first foundations of a later language are laid.

After birth, the number of stimuli increases at a bewildering rate. The infant begins to compare certain perceptional experiences with his still unconscious memory and relies at first on the relatively simple experiences of taste, smell and hunger. Between perception and memory an order begins to crystallize, which is always ordered in time, since the memory precedes the perception with which it is being compared. It can remain open here as to whether the order is established solely in the brain or in a transcendental consciousness as well. It is equally irrelevant whether the perception is sensual, psychic or mental in nature. Of sole importance is that the order is chronological and that the infant develops a sense of time, albeit a rudimentary one, quite early.

Next come the light bulb moments of „if – then“: If I cry (or more precisely: if there is crying), then the mother will come. Even if this does not always work right away, the infant will unconsciously recognize that there is such a thing as cause and effect. That he, as an independent being can himself be a cause, is outside the infant's awareness in the beginning. He is not yet capable of distinguishing between I and you. However, his sense of order and time develops, since the cause always precedes the effect. The child learns that it is often worthwhile to differentiate between the signals perceived and to store them in his memory.

The visual sense, sight, develops more slowly. This is primarily because sight is more complicated than all the other senses. The child must learn on the one hand how to reproduce a three-dimensional space in the brain via a two-dimensional visual field, on the other hand to simultaneously distinguish between colours, light and dark, and left and right. That is difficult and takes years. That said, we now know from experiments carried out with small children, the theory that language is a precondition of spatial perception is wrong. Most children are able to manage spatial perception, at least the perception of topological arrangements, before they start speaking [Piaget (1970)].

Thus, the four conditions necessary for understanding a simple language are met:

- An initial speech signal can be received.
- The information of the signal is stored and is associated with the context of the experience or stream of experience in which the signal is received.
- The stored information or memory can be compared with the new perceptions, i. e., with newly received signals as well.
- Should this comparison reveal that the new signal is similar to the first signal, the child will then associate the new signal with the experiential context of the first signal. If the child does not err in doing so, it will have understood the second signal.

Communication is not dependent on the capacity to understand language alone. Equally important is that the language expresses what the speaker wants to say. The mechanism by which this occurs is, in principle, the mirror image of the mechanism of understanding described above:

- The child has a psychic or mental experience in his brain or consciousness.
- The child compares this experience with images of memorized earlier experiences and looks for a similar one.
- The child associates the earlier image with its signal.
- The child sends an appropriate speech signal.

This reverse process is more difficult for the child to learn than the receiving and understanding of signals, so it takes quite some time before children learn to differentiate in their speech in the way they do in understanding what they hear. The ability to understand a language is not limited to humans. In principle, animals are able to do so as well. Some animals are better equipped to receive certain signals than humans and have superior memories. However, in associating different signals with different contexts of experience, human beings surpass all animals.

What about computers? Are computers only able to receive, store and process, or can they understand as well? Understanding requires a stream of experience, which is more than the registration of stimuli. Of course, a computer is also able to connect the word „red“ with the wavelength of the colour red. Even so, it will not experience the sensation of red in the way humans do when they see red, hear the word „red“ or visualize red. (Some humans can even *smell* red!) It may well be possible, in principle, to program such a sensation, however this will always be completely different for a computer from how it is for a human, since being a machine it can never have precisely the same experience and store it in a stream of experience as humans can. Thus, the computer cannot understand humans, mainly because it is not human itself. Since animals are also not human, they will also – independent of their limited mental potential – only ever partly understand a human being.

Ultimately, this also applies to different human beings. Each person lives in its own stream of experience and accordingly, understands language as an individual and differently from everyone else. Ask for instance ten different people what „God“ means and you will receive ten different answers, although „God“ is a pretty familiar concept, about which most children have to learn quite a lot. God is the father in heaven, he is the creator of the world, or the law of nature; he is the manifestation of creation, he is spirit, he is love, he is the trinity, he is the final cause, he is the almighty, he is the substance of Be-ing. Two people who have known each other for a long time and have shared many experiences and discussions, will certainly better understand what the other means by saying God than two people who are complete strangers to each other. Of course, this is not to say that common understanding equates to a common opinion. Indeed, even discussions over years will never lead to complete understanding between two people. Most philosophers are aware of this and thus seek to express themselves in a way that cannot be misunderstood. As a result, they construct sentences in which all important words and relationships are explained or relativized in subordinate clauses, so that the language often gets extremely complicated and barely comprehensible.

We conclude therefore, that not only language itself has limitations. This is especially so for the understanding of language. I may well be convinced of having understood everything, but I can never know, whether the persons I'm talking to have understood the same. They will never think exactly the same about a given word as I do. Language, therefore, is always subjective. With and within language we are on shaky ground, or better, in the swell of an ocean [Han (1999) p. 103]. All understanding is simultaneously a little misunderstanding [Humboldt (1903)].

Chapter 2

The Limits of Knowledge

> Awareness of ignorance is the highest.
> Not knowing what knowledge is, is suffering.
> Only if one suffers from this suffering,
> does one become free of suffering.
> The fact that the appointee does not suffer,
> is because he suffers from this suffering;
> that is why he does not suffer.
>
> Laotse, (ca. 600 BC/1978)

Humans crave knowledge. It is a deeply rooted instinct in man. Evolution has given humanity an urge to know that is more pronounced than in other creatures. Clearly, knowledge helps humanity to survive.

Now, what do I know, when I know something? Thinkers have sought the answer to this question for thousands of years and have not reached agreement to this day. The answers change from century to century; they also differ fundamentally in the cultures of east and west. I don't presume to offer a final answer. It is my belief that this is simply not possible, because every answer is language-related. As we have seen, language is subject to strict limitations. In the following chapter I am concerned with something else. I intend to demonstrate that there are different kinds or degrees of knowledge and to explain which terms I use for each type. Even in these matters academics are not in agreement, which causes ongoing confusion. I don't expect the reader to understand me from the outset. Yet the longer he continues to read in the later chapters, the clearer my way of thinking will become and the closer he will come to my understanding of what it is to know.

Let's begin with a series of examples. I know that the sun rose yesterday. I know that the sun will rise tomorrow. I know that the sun is shining today in Zurich. I know that there's a 20 % chance of rain tomorrow. I know that I get wet when it rains. I know why I get wet. I know why I learn. I know that $a = a$. I know that $ab = ba$. I know that $2 + 2 = 4$. I know that the probability that $2 + 2 = 5$ is infinitesimal. I know that the atom is made up of electrons and a nucleus. I know that the speed of light is constant. I know that space is three-dimensional. I know that I think. I know that I am. I know that I have a temper. I know that you have a temper. I know that the world will end on August 13. I know there is a God. I know that Michelangelo's David is beautiful. I know that is unjust. I know that I know. I know that you know. I know that I know nothing. I don't know.

They are all simply statements with the phrase „I know“ in which it remains open as to whether „know“ has the same significance in all statements. (The „I“ in „I know“ will be discussed in Chapter 3.) The statements could be true, they could be false, they could be contradictory, they could be vague, and they could have a certain probability of being true. This second statement or „meta-statement“ regarding the degree of truth of the first statement can in turn be true, false, contradictory, vague or true with a certain degree of probability. A meta-meta-statement about this meta-statement is also possible, and so on. Furthermore, different people can have different opinions about the statements' degree of truth.

It is already apparent that knowledge is not absolute. Perhaps we should be pleased when we can, little by little, increase the degree of truth of our knowledge and when we can, to some extent, agree with others about the degree of truth.

In each of the different research disciplines – I intentionally avoid the use of the term „sciences“ – knowledge has a certain and different significance. Since this paper essentially deals with metaphysics and physics, and because physics operates with mathematical methods, I will limit myself in the following to explaining what relationship these three disciplines, as I like to interpret them, have to knowledge. I don't wish to speak about disciplines such as theology, paraphysics, art and other humanities or psychology. Nevertheless, I will take the liberty of pointing out that the latter two can, at least to some extent, be ascribed to the natural sciences and that many natural scientists today are of the opinion, that the natural sciences should be based on physics.

2.1 Metaphysics

Metaphysics is a philosophical discipline, to which the three following statements apply [Kamitz (1980)]:

- Metaphysics seeks to reach conclusions which inform us about reality or a certain aspect of reality. Thus, it makes propositions which are synthetic and not analytic – at least that is the intent.[14] That said, there are philosophers for whom this distinction is irrelevant.[15]
- The theses and conclusions of metaphysics cannot be tested empirically, i.e., by observation or experiment. It's a case of synthetic statements *a priori*. This means that there are aspects of reality which cannot be perceived directly but are preconditional to perception.
- The results of metaphysics should take us beyond what can in principle be reached by way of a single scientific approach. It concerns things which are „beyond physics". It strives for outcomes which, with respect to certainty or universality, are vastly superior to the findings of science.

So, metaphysics begins where scientific research reaches a dead end. This is a proud claim, which has made metaphysics (and to some extent mathematics as well, as a sub-science of metaphysics) the queen of the sciences for over 2000 years. Yet the advance of the natural sciences has

[14] The distinction between synthetic and analytic goes back at least as far as Plato. If the degree of truth of a statement depends on the sense of the expressions used as well as on the facts to which it refers, then one calls the statement synthetic. If it only depends on the sense of the expressions used, one calls the statement analytic. Whereas synthetic truths are propositions about the world and can be assessed empirically, analytic truths are not empirically examinable and are merely „by-products" of the language [Kant (1783/2001) § 2].

[15] Willard Van Orman Quine, in his famous lecture „Two dogmas of empiricism" to the Eastern Division of the American Philosophical Association in December 1950, expressed the view that there is no test for determining, where the assignment of empirical reality stops and where the assignment of word meaning begins. There is no safe method of separating necessary and contingent truths. He suggested replacing this old dualism of synthetic and analytic judgements with sliding transitions – with a spectrum which would range from convictions, which we hold uncircumventable to convictions which we consider easily revisable by future observations [Rorty (2001)].

pushed metaphysics further and further into the background. Fewer and fewer questions which were still „beyond physics“ remained to be asked. Natural scientists and increasingly philosophers as well, asked themselves whether metaphysics was anything more than blind speculation and whether it had any use at all [Kuhlmann (2017)]. In biology class I was told that metaphysics was not a science and that metaphysicists may well carry out very clever or seemingly very clever discussions, but never about tangible things. The most radical rejection of metaphysics arose in the twentieth century as a result of „neo-positivism“. Proponents of neo-positivism consider a proposition or thesis meaningful only if it is empirically verifiable. This renders all metaphysical questions meaningless by definition. Philosophers coined the term „post metaphysical thought“. This was intended as an endeavor which rejected the prevailing scientific hierarchy with its underpinning philosophy (and with it, metaphysics) at its peak. A universally applicable basic science or a universal tenet, a search for the final „why and wherefore?“ was supposed to be impossible or at least unscientific. Each individual scientific discipline was to formulate its own theory as to what it would consider true or real. This naturally pleased in particular the psychologists, sociologists and political scientists.

However, criticism of this attitude was not long in coming [Langthaler (1997); Mittelstaedt (1972) pp. 166–207]. The very question as to whether metaphysics has a use is a typical metaphysical question which can only be answeres within metaphysics. Science must distinguish between the existing things (one could also say nature) and existence in itself (the Be-ing), between the sensory and the extra-sensory, and between the empirical and the transcendental. Nature is not comprehensible when one has no concept of what being, what existence, means. Closely connected to this is the concept of perception.[16] What is true? What exists? What is possible? The genius of great physicists such as Einstein and Bohr is characterized precisely by their courage in all simplicity to continually pose such questions. In Section 2.3, which concerns

[16] The German *Wahrnehmung* is stronger in this context. It is a nominalization of the verb *wahrnehmen* which has a variety of meanings including to perceive, observe, appreciate, sense, etc. This in turn is made up of *wahr* (true) and *nehmen* (to take).

physics, I will use several examples to show how physicists have been variously led astray, because they have not learnt to distinguish between existence and that which exists, between form and content, and between an object and a model[17] of that object.

Admittedly, there is always a connection between existence and that which exists, and a rather complicated one at that. [Heidegger (1949); Wenzel (1998)]. It is examined in ontology.[18] Perhaps the traditional polarity of „empirical“ and „*a priori*“ is actually part of the same continuum. We never philosophize in a vacuum. When we speak of „existence“ or „Be-ing“, we always have an existing something in mind. Empirical knowledge can thus contribute greatly to answering metaphysical questions. For my further thoughts, however, it is not necessary to go into details about this.

Like all scientists, metaphysicists also want to make true propositions. When is a proposition true? What is meant by truth?[19] It is doubtful that the term truth is definable at all. Even so, two possible definitions may be cited [Gloy (2004)]: For Tarski, a statement is true when it corresponds to reality. This is rather trivial and isn't much help. For Habermas something is true when we agree that it is. This modern definition has been strongly criticized by other philosophers [Langthaler (1997)], since even an apparently rational collective can be greatly mistaken, and who is qualified to determine whether all members of the collective have really understood each other and are in agreement? What is required are criteria of truth, against which the truth of something can be judged. According to Popper (1979), there can be no such criteria. Somebody has to determine that these criteria are true, and for that, they require at a meta-level criterion of truth as well. These in turn must be judged on a meta-meta-level whether they are true, and so on. There is no absolute

[17] Models are only approximations (of reality), however not always due to an inaccuracy, but because they only consider certain aspects [Kreisel (1980)].

[18] Ontology is a multi-faceted discipline. It is concerned with five questions: What is the existent? Are there different forms and meanings of existent and are these connected? What constitution must something have, if it is to have Be-ing? What structure must something have in order to form a possible world? What is the constitution and structure of the real world? [Burger (1998)]

[19] Popper has given a great deal of thought to the notion of truth from a metaphysical and physical perspective [Popper (1972)].

certainty. Every theory of truth is beyond true and false. Not even the falsification of a theory by a concrete experiment proves with certainty that this theory is false, since the falsification could itself be in error [Carter (1993) pp. 51 f.]. At best a theory is obvious or immediately certain, but certainly never provable. Actually, one can only believe.

The three classic preoccupations of metaphysics are: whether there is a God, how the cosmos is composed as a whole, and how and as what am I to understand „me". The first question I leave to theologians. The second is the purpose of my book. The third will be discussed in the next chapter and in Chapter 5.

2.2 Mathematics

Humanity's desire to know the truth is irresistible. Where are we most likely to find such a thing as absolute truth? For a long time, most thinkers agreed that mathematics was best placed to give true answers. This impression was given in schools as well, when the teacher wielded his red pen in unmistakable judgement on true and false answers in a maths test. There's nothing more to discuss. The slogan was: That which is proven, is correct! The essential character of mathematics is that it studies models in abstraction of the individual things which are modelled [Hampe (1998) p. 31]. Mathematics appeared to be based on a small number of axioms which originated directly from God. The axioms of logic appeared simple, direct and plausible. These form the initial foundation of set theory and a theory of relations and structures, and in turn set theory is the foundation of the whole numbering system, from natural, rational, surd to the complex. This is then the basis of algebra, geometry, topology and differential calculus. From the latter came the development of complex analysis, functional analysis and probability theory.[20] With the introduction of the concept of „infinite", mathematics indeed became more difficult and more abstract, yet at the same time, in some way all-embracing and eternal as well. It became „higher mathematics" [Hilbert (1926)]. The task of mathematicians appeared to be to find the right

[20] This hierarchy of the branches of mathematics is sometimes questioned today (see Section 14.7) [Reinhardt et al. (1974) p. 12].

axioms, to formulate theorems and then to prove these according to the rules of logic. But how does one find the „right" axioms and is conventional logic the only possible one?

We now know that it is fundamentally impossible to prove that an axiomatic theory is without any contradictions and that it is impossible to determine which of the known theories of logic is the true one [Quine (1963)]. Since mathematics as a whole depends on both these questions, it is now a bit up in the air [Wilholt (2017)]. It is not simply that we don't yet exactly know the answers to the questions relating to the axioms and logic. We know that we can basically never know the answers. 2700 years ago, there were already hints that it could be so. Even then theorems were known which were simultaneously logically true and logically false. These were the paradoxes. In the following I will summarize our present-day knowledge about paradoxes, axioms and theories of logic. In the „Physics" section, I will also discuss the question of infinity.

2.2.1 Paradoxes

„I know that I know nothing", said Socrates 2400 years ago and 300 years earlier still, the Cretan Epimenides asserted, „I am a liar". Are these statements true or false? First, let's observe what happens when we accept that Socrates' statement is true. It means that Socrates knows nothing, yet he begins with the assumption that he does know something. Therefore, the statement cannot be true. Now let's see what occurs when we assume that Socrates' statement is false. That would mean that he doesn't know that he doesn't know. He knows precisely that however, as he has said himself. Therefore, his statement cannot be false either. Some paradoxes are much more complicated, at times lengthy anecdotes. Ultimately however, all paradoxes can be reduced to one statement which makes a statement about itself, a so-called self-referential statement.

Accordingly, there are two different avenues for solving the paradox problem. Firstly, one can postulate a logic in which self-referential statements are simply forbidden [(Wechsler (1999) p. 116]. Or one could introduce time into mathematics. In this way, a statement which is uttered later is always on a higher meta-level than the statement which has preceded it. The logic must then stipulate that statements from two different levels cannot be compared with each other, since they belong to two different classes. This can be justified in that the speaker of the second statement had another, broader knowledge than the – possibly same – speaker of the first statement; namely he knows, unlike the speaker of the first statement, that this first statement has been spoken. Thanks to this knowledge, he is able to form a more precise judgement than the speaker of the first statement. He will establish that the statement is false, because it is contradictory. Basically, paradoxes stem from an abuse of language. No distinction is made between ordinary statements and statements about statements, and ordinary characteristics and characteristics of characteristics.

Applied to set theory, into which all syntactic antinomies or paradoxes fall, this means that the relationship of elements in set theory must be so ordered that a distinction is made between the elements of a basic set, sets of elements of the basic set, sets of such sets and so on, and that the variables for elements of different levels are identified. x can then only be an element of y, if the level to which y belongs is one level higher than the level of x [Russell (1919)]. Since the number of levels is essentially unlimited, mathematicians assume that „The intolerable state with respect to paradoxes can only be overcome without a betrayal of science, if the nature of infinity, a concept which is not to be found in reality anywhere, is fully understood" [Hilbert (1926) p. 170]. Infinity is, in principle, not empirically perceptible and therefore impermissible as the basis of thinking in natural science. In the opinion of most mathematicians however, it could prove quite useful as a theoretical construct.

It is evident that in mathematics, it is easy to reach the wrong conclusions when one ignores time. Mathematics also is nothing more than a language, albeit a special one. It follows that it is subject to the same limitations as every other language too, and they are, above all, limitations placed by the structure of time. Most mathematicians are convinced however, that their science is independent of time. Such a thing as an „arrow of time" is completely lacking in mathematics. From a mathematical perspective, time belongs to the sphere of Be-ing, i. e., which exists, to nature and is quite foreign to mathematics, almost beneath its dignity. That has now come back to haunt mathematicians. There is no timeless mathematics.

2.2.2 Axioms

It is usual to construct mathematics on the basis of principles which do not derive from other principles, i. e., cannot be proven. In this sense, axioms form the foundation of mathematics. The axioms are, however, not simply unfounded assumptions, but are immediately evident. From these, concepts and theorems can then be derived. I do not intend within the framework of this book to give an overview of the axiomatic systems of the different branches of mathematics. However, as an illustration, I would like to highlight a couple of examples of axioms which are controversial or could probably prompt some discussion.

Taken from logic, the following axioms will both play a special role in my deliberations:

$$A \equiv A.$$

The axiom will be examined in Section 6.4.

$$\text{Not not } A = A$$

means that either A or not A is valid. There is no third option: „*tertium non datur*" or „a statement is either true or it is false". The axiom does not apply in quantum theory and in the so-called quantum logic, sometimes also called three-value logic [Mittelstaedt (1972) pp. 166–207)].

From the area where logic meets set theory, of primary interest to us is George Boole's second law:

$$AB = BA,$$

in which A and B are mental activities, through which an element of a set is selected. This axiom means that the act of selecting an element A leaves the elements A and B unaffected, so that the order of selection is immaterial and produces the same result [Boole (1847)]. This axiom does not apply in quantum theory [Mittelstaedt (1972) pp. 166–207; Finkelstein (1996) pp. 3 ff.]. Since the axiomatic systems of set theory described in the literature are largely incompatible and since none of them can be clearly designated as the standard system, we must speak of a whole panorama of equally valid alternatives. The requirement that axioms must be theorems which are immediately evident in no way precludes the devising of different axiomatic systems for the same field, all of which are equally suitable. There are no „true" axioms [Quine (1963) pp. 15 ff.)].

Despite all the relativisation of the axiomatic method, it came as a shock [Carnap (1963)] to many mathematicians and philosophers when Gödel showed in 1931 that all consistent axiomatic systems of number theory contain indeterminable propositions, i. e., propositions which can neither be proven nor disproven. He went on to generalize his conclusion to include any axiomatic theory: „There is no constructive procedure with which one could prove that the axiomatic theory is consistent". The proof [Gödel (1931)] is complicated and difficult to understand, yet Hofstadter has written a lively, generally understandable account of the courageous and surprising line of argument with its philosophical background and implications [Hofstadter (1979)]. A good and more easily understandable account, especially of the work of Turing concerning the boundaries of mathematics has been written by Chaitin (1993). The reasoning is comparable with the analysis of Epimenides' liar paradox, „I am a liar". In its place Gödel formulated the selfreferential statement: „This statement cannot be proven". If this statement were false, then it would be provable, yet this would contradict itself. To avoid this contradiction, the statement must be true. However, although it is true, it cannot be proven because the statement (which we know is true) establishes just that. Gödel was able to translate the above statement into mathematics and thereby show that there are statements in mathematics which are true, but can never be proven, the so-called undecidable statements [Singh (1997)]. The surprising conclusion is: provability is a weaker concept than truth, irrespective of the axiomatic system concerned. If we know the truth, then it is not because we have proven it.

2.2.3 Infinity?

Of course, there are also axiomatic systems for number theory [Dawson (1997) pp. 37–52]. Zermelo has produced evidence based on set theory according to which order is possible without spatial or temporal configuration, solely by means of sets and subsets with only a single first element [Zermelo (1908a)]. Therein plays the axiom of infinity a particularly important part. According to Zermelo, set theory involves a domain of objects which he calls things, among which the sets form a part. The axiom of infinity then says: „There is at least one set within the domain which contains the empty set and one which contains the set *a*, whenever it contains *a* itself" [Zermelo (1908b)]. However, it is quite legitimate to question the extent to which infinity can be something „immediately evident." In any event, mathematics can also be conducted without the concept of infinity. This is finite „discrete mathematics" [Biggs (1996)]. Since infinity is never perceptible, not even – as sometimes is claimed – an „approximate infinity" a physical theory should completely do without the term infinity and thus without „higher mathematics" [Hilbert (1926); Weizsäcker (1971b, pp. 53–87)]. Thus, irrational numbers (surds) such as π, e, $\sqrt{2}$ e.g., or integrating and differentiating are dropped and mathematics becomes substantially simpler without the axiom of infinity. Zermelo's axiom of infinity contradicts physical perception because there are no empty sets in physics. Events cannot be empty. They happen or they don't happen. Yes or no. The physicist as a real observer counts the events and the counted numbers always are finite. It is always clear where one event ends and where a new event begins and the counting is done by natural numbers. π, e and $\sqrt{2}$ are not aspects of nature. An ideal circle, e.g., never can empirically be observed in nature.

The axiom of infinity changes mathematics considerably. As every axiom, it leads to less freedom for the mathematical model which is intended to describe physical perception. Suppes (1972) pp. 252f., has shown that dispensing with the axiom of infinity has a similar effect to dispensing with the parallel postulate in Euclidean geometry. So, this opens up similar perspectives to the introduction of curved spaces in the general theory of relativity.

2.2.4 Continuum?

The first attempt to construct a geometry on the basis of axioms originated with Euclid. There also, it was found that several contradictory axiomatic systems were possible. Alongside Euclidean geometry, in which the axiom of parallels applies, other geometries with curved spaces are possible. They are, in fact, sometimes more appropriate for describing nature [Tarski (1959); Brandes (1995); Luminet et al. (1999)]. For Kant, the axioms of geometry, i.e., „the theorems which are perfectly certain", are yielded by the construction prescription for the geometric figure with which the axiom dealt. However, he never specified any such structural specifications, evidently because he had no reason to doubt Euclidean axioms.

The renunciation of infinity means that no continuum is allowed, be it time or space of the model. A continuum is anything that goes through a gradual transition from one condition to a different condition, without any abrupt changes. In set theory, the continuum is the real line, i.e. the line whose points are the real numbers $\mathbb{R}$.[21]

Now, the point is that the physical universe is finite by definition because otherwise it would in principle not be perceivable and measurable by a finite observer during his finite life time. Therefore, physical measurements always are a finite counting of events (see Section 7.6 and 8.13.5). The counting is done by finite natural numbers $\mathbb{N}$ and not by real numbers $\mathbb{R}$. In $\mathbb{R}$ there is a gradual transition from one number to the next without any abrupt changes, in $\mathbb{N}$ there is not. In the real world there is neither a „gradual transition" from one event to the next nor a path between the structures of the observed object before and that after the event. Therefore, it makes sense and simplifies the mathematical model if it is finite too. This excludes continua.

[21] In the mathematical field of point-set topology, a continuum is a nonempty compact connected metric space. A topological space is said to be connected, if it is not the union of two disjoint nonempty open sets. A set is open if it contains no point lying on its boundary. The space does not have to be Euclidean. In general topology, compactness is a property that generalizes the notion of a subset of space being closed (that is, containing all its limit points) and bounded (that is, having all its points lie within some fixed distance of each other). There may be different notions of connectedness that are intuitively similar, but different as formally defined concepts. We might wish to call a topological space connected if each pair of points in it is joined by a path. However, this concept turns out to be different from standard topological connectedness; in particular, there are connected topological spaces for which this property does not hold. Because of this, a different terminology is used: spaces with this property are said to be path connected. While not all connected spaces are path connected, all path connected spaces are connected.

If all this can be accomplished, the artificial renormalization and perturbation techniques can be dropped although they might still be useful to simplify certain calculations. Feynman (1985 and 1966a) emphasized that with such techniques he can describe nature mathematically but he does not understand it because nobody can understand nature.

Following all these paradigm shifts, not a lot remains of the mathematics we learned in school. For this reason, it is not yet possible to formulate the proposed new theory in a precise mathematical language as it is usually done with the description of physical theories. The formalisms of quantum and general relativity theory as well as that of set theory cannot simply be applied, even if modified. The proposed new physical theory is at a stage similar to that of classical mechanics in 1715 when Newton and Leibniz discussed the reality of space [Leibniz et al. (1715–16/1990)]. Independently of one another they both discovered infinitesimal theory.

2.2.5 Logic

Just as axioms form the foundation of mathematics, logic is the mortar which holds the structure together. Logic is a theory which addresses the connections between different propositions which are valid, independent of the content of the propositions concerned [Russell et al. (1932)]. Since Aristotle, logic has been regarded as the prototype of a science, which is valid independent of experience, whilst nevertheless being applicable to experience. In the previous section we got to know axioms of logic which don't apply in quantum mechanics. Without going into quantum mechanics for the moment, we need to be aware what that means for logic itself. Can the axioms of logic be „immediately evident“ or „obvious“ at all, when afterwards they are no longer valid in physics? Putnam (1998a) was convinced that there is no *a priori* plausibility. This means that different logical systems had to be formulated according to the questions posed, and that perhaps logic might one day be modified, similar to how the Euclidean *a priori* geometry of Kant was modified in the theory of relativity.

It is clear that the axioms of logic must be subject to an additional requirement other than the evidence: The propositions, which logic addresses, must be unlimited in their application. This is the so-called „protological condition“, i. e., a requirement that presupposes that logic can be applied at all. In everyday life and in mathematics, this additional condition is generally met. In contrast, with respect to propositions concerning quantum mechanical systems, the applicability of the proposition is limited, whereby classical logic can no longer be applied *tel quel*. Now there are two possible ways of dealing with this situation. Either one can steadfastly maintain that classical logic applies generally. Logical propositions about quantum mechanics are then forbidden, since the protological condition of unlimited applicability has not been met. Or one forms the axioms of logic in such a way that they also hold for propositions with limited applicability. This logic is called „quantum logic“. It is a logic of ambiguous language [Mittelstaedt (1972)].

What do logical theorems actually reveal? A proposition A may concern a given proposition taken from daily life, natural science or mathematics, which can in principle be proven, i. e., for example, „It's raining“, „The speed of light is constant“ or „The sum of the angles of a triangle is 180°“. Using such simple statements, with the help of links such as $\rightarrow$ (implication), $\equiv$ (identity), $\wedge$ (conjunction, and) or $\vee$ (disjunction, or) compound statements can be made, namely the very axioms or the theorem is derived from them.

Let us consider for example the theorem $A \rightarrow (B \rightarrow A)$. It means that if A is true, it follows that aslong as B is true, A is also true. At first sight, that appears to be quite trivial: If A is true, A is true independently of B. But the theorem is then only correct when statement A, or more precisely, when the proof that A is true, is universally applicable. First the truth of A must be proven. Then follows the proof of the truth of B. This proof of B must have no influence on the already adduced proof of A, otherwise the proof of A is no longer universally applicable and can no longer be used to verify that from B follows the truth of A. Proofs of the truth in physics are measurements. In quantum theory, two measurements of the same system are never independent of each other. As a result, the allegedly trivial logical theorem $A \rightarrow (B \rightarrow A)$ does not hold in quantum theory. One says that A and B are not commensurable statements, i. e., they are not independent of each other and it is not irrelevant in which chronological order A and B are proven or measured.

Often it is not known at all, whether the different statements are commensurable or not. In these cases, one needs to establish which logical theorems are rendered invalid by not meeting the protological condition of universal applicability. Thus, a once-proven proposition *A* is in a later stage of the discussion, i. e., when it has possibly been altered by the proof of truth of *B*, no longer available and can no longer be cited. A closer examination shows that in quantum logic, the law of the excluded middle (*tertium non datur*)[22]

$$\neg\neg A = A \text{ (not not } A = A)$$

in particular becomes invalid. A statement in quantum logic is not either true or false, it can also be indeterminate. An important result of this is that in all cases where quantum logic is employed, the law of causality no longer applies [Mittelstaedt (1979) pp. 264 ff.]. As early as 70 years before quantum theory was known, George Boole (1847) formulated his second Boolean law $BA = AB$ and pointed out that „It is, for example, true that the result of two successive acts is unaffected by the order in which they are performed. This law will perhaps appear so obvious to some as to be ranked among necessary truths, and so little important as to be undeserving of special notice. And probably it is noticed for the first time in this essay. Yet it may with confidence be asserted, that if it were other than it is, the entire mechanism of reasoning, nay the very laws of constitution of human intellect, would be vitally changed. A logic might indeed exist, but it would no longer be the logic we possess." This new logic is today called quantum logic. *A* and *B* are for Boole neither propositions nor measurements. Rather, they are mental activities of the logician, acts whereby a subpopulation of individuals is selected from a general population. *A* and *B* define classes of elements which are selected from a set. „*AB*" means that first the elements belonging to class *A* are selected, and then in a second act, those belonging to class *B* are selected from these. The result of „*AB*" is thus the set of elements, which in addition to class *A*, also belong to class *B*. This product should, according to the Boolean law, be independent of the sequence of both acts of selection. In quantum theory, one deals not with mental acts, but rather with physical acts or measure-

[22] Literally, „a third is not given".

ments. In quantum theory, not only is information gained about the object by taking a measurement, but the object itself is always influenced thereby, such that the information gained from a second measurement could be different from what it would have been, had the first measurement not taken place.

Kant (1786/1957) p. 16, considered that in psychology „even the observation itself alters and distorts the object being observed" and that therefore, a fundamental difference would have to exist between psychology and natural science in which the observation wouldn't inherently alter the state of the object being observed.[23] This insight wouldn't directly result in quantum theory, yet nevertheless, quantum logic can and must be derived from it, but to my knowledge nobody has ever attempted this. Those who assume, like Kant, that the laws of nature are ultimately a consequence of metaphysics, would then have to infer quantum theory and the uncertainty principle from quantum logic.

2.3 Physics

In the first lecture of his studies, the physics student learns that there is three-dimensional space and one-dimensional, directed time. When the path can be presented as a function of time, velocity is revealed by a differentiation according to time. Physicists thereby usually assume that space and time actually exist and that both represent a continuum. Is that really physics?[24, 25]

[23] Plato (1957) argued similarly: „If to know is active, to be known must be passive. Now Be-ing, since it is, according to this theory, known by the intelligence, in so far as it is known, is moved, since it is acted upon, which we say cannot be the case with that which is in a state of rest."

[24] The conception of an existing (absolute) space and an existing (absolute) time goes back to Newton, who for his part relies on Christian-theological roots, i. e., the observer role of an omnipresent and constant God. For Newton this was not physics however, but „philosophia naturalis" (natural philosophy) and „mathematica". The distinction between physics and philosophy only became usual later, and today's scientist is generally no longer conscious of where the boundary between the two sciences lies.

A physicist is a scientist who observes nature,[26] describes it with the help of mathematics, derives laws from this and based upon these laws, makes predictions about the future [Scheibe (1994)]. So, in the beginning, there is always empirical observation. The language of mathematics or the mathematical model which is used to describe this observation should in each case be so chosen that the description is distorted as little as possible. The physicist should always remain aware to what extent his model deviates from the observation. If he does not do so, then he must accept that his laws do not describe nature but are only an expression of the chosen mathematical model. That could then make his predictions accordingly false or uncertain.[27]

Can space and time be perceived at all [Grünbaum (1973) pp. 3–65]? It is obvious that this is not the case. Einstein (1979) is quoted: „… time and space are modes by which we think and not conditions in which we live." According to the opinion prevailing today, space and time do not, as things in themselves, actually exist [Mehlberg (1980) pp. 225–234]. A space, in which there is no object, cannot be perceived. (In this context, a field is considered an object.) Only objects are ever perceived, never the space between them [Leibniz (1704/1981) and (1715–16/1971); Mach (1889)]. The impression of space arises in that objects are not simply a chaotic mass but form a certain order relative to one another. It is analogous with time: a time, in which nothing occurs, cannot be perceived. It stands still or doesn't exist at all. Only the events, which form a certain order relative to one another, convey to us an impression of time [Maxwell (1881)]. Space and time, in themselves don't exist. They cannot be perceived and are only ever introduced as an artificial construct by means

25 According to Kant (1786/1957) pp. 25 f., absolute space is not a thing in itself, but only the background against which the subject can conceive of motion.

26 I define the concept of nature, like Kant (1783/2001, § 16), as the epitome of all objects of experience.

27 The connection between physics (laws of nature) and metaphysics, language and meta-language is discussed by Mehlberg (1980).

of the model which is designed to describe the perception [Cushing (1990) p. 33].[28, 29]

In principle, the same holds for the continuum. The precondition of a continuum is that space and time can be divided into an infinite number of infinitely small segments (see Section 2.2.4). Such a thing is easily possible in mathematics. Yet because a physicist or a measuring instrument, as finite entities, will never be able to perceive infinitely small, or an infinite number of objects, infinity is not an aspect of observed nature which the physicist wants to describe, but always only a part of the mathematical model which he uses for the description [Weizsäcker (1971b) pp. 53–87; Leibniz (1904)][30]. Both Schrödinger and Einstein also held the opinion that the continuum of mathematics was not suited to the description of the reality [Penrose (2004) p. 62)]. However, Penrose himself believes that without infinity, physics is scarcely possible at present. If $\mathbb{N}$ is the set of positive, whole numbers and $\mathbb{N}_0$ is the cardinality of $\mathbb{N}$ and is thus the smallest mathematically possible infinite number, then the smallest possible infinite number for the description of physics is $2^{\mathbb{N}_0}$

[28] Einstein turned prominently against the Kantian *a priori* of space and time and decreed it „one of the most wretched acts of the philosophers that they shifted certain conceptual bases of natural science from the areas of the empirical-practical, which are verifiable, into the untouchable heights of the *a priori* … This particularly applies to our concepts of space and time, which physicists – forced by the facts – had to bring down from the Olympus of the *a priori* in order to repair them and put them back into a useful state“, whereby „the concepts cannot be derived from the experiences by logic (or in any other way), but in a certain sense, are free creations of the human spirit … nevertheless with just as little independence of the nature of the experiences, as for instance clothes have of the shape of human bodies“ [Janich (2000) p. 66].

[29] Space and time continuums are abstractions of counted events, a theoretical framework for the order in natural occurrences. Space and time in themselves are not perceivable as such [Einstein et al. (1950)].

[30] When Weizsäcker was studying physics in Leipzig with Heisenberg, the latter asked him what he was currently doing in mathematics. „I'm studying set theory“. Heisenberg: „You shouldn't study that.“ Weizsäcker: „But set theory is the basis, and it interests me philosophically as well.“ Heisenberg: „No, it is sheer nonsense. Don't believe the mathematicians when they want you to believe that there's such a thing as an actual infinite point set! Could anyone observe such a thing?“ [Weizsäcker (1999), p. 305].

[Penrose (2004) pp. 363–367]. Often, such „unnatural models" lead to immense difficulties. How does one correct, for example, the theoretically infinite energy within every single electron which is generated because all the negative electrical charge is in an infinitesimally small location, so that the internal repulsion becomes infinite? Infinite energy however, cannot exist in reality. Therefore, the theory must be modified so that wherever an infinite energy appears, it is adjusted with a theoretical, equally infinite counter-energy and made finite again, thus conforming to real energy. The problem may well be solvable mathematically, but the solution no longer has much to do with reality [Georgi (1989a) pp. 446 ff.; Cushing (1990) pp. 25–39]. Einstein stated that a finite system with finite energy should be describable in quantum theory by a finite set of quantum numbers. That ruled out a continuum theory and had to be grounds for seeking a purely algebraic theory. Unfortunately, nobody has yet conceived how the basis of such a theory could look [Einstein (1956a); Weizsäcker (1971b) pp. 53–87]. However, the lack of such a theory can in no way mean that the conventional continuum theories are true or at least particularly functional. „One would indeed like to insist that only fundamentally observable quantities be introduced into physics. Are we supposed to be completely on the wrong track with the continuum theories for the field in the electron's interior?" [Pauli (1919)]. After all, the complete continuum theories were only formulated at the end of the 19th century [Feyerabend (1961)]. „Infinity was raised to preeminence by Cantor (with enormous contributions from Frege and Dedekind) and enjoyed a period of the highest triumph. The infinite is not realized anywhere; it is present neither in nature, nor permissible as the basis of our intellectual processes – a remarkable harmony between being and thinking. The infinite only retains the role of an idea, i. e., an intellectual concept which exceeds every experience and with which the concrete, in the sense of totality, is supplemented [Hilbert (1926)].

All physical laws, in which space, time or a continuum are found, are therefore at best an approximate description of the observed nature. The question arises as to whether there are no simpler and better laws and models which distort the perceptions less and make predictions for the future which are accordingly more accurate. It is the purpose of this book to provide an indication of how such natural laws can be found without the concepts of space, time and continuum.

All human sensory perception results from the information conveyed by photons, according to today's recognized theories. Other physical interactions, such as gravity for instance, play no part in human perception. Humans cannot perceive gravity directly, but only ever via their different senses, i. e., via photons. The photon has a remarkable series of properties. It moves at the speed of light. Since, according to the special theory of relativity, time stands still for objects which move at the speed of light, the photon, from the photon's perspective, exists for zero seconds, even then, when from the observer's perspective, it has been in motion for many light-years, perhaps even since the Big Bang. Interference experiments further show, that although the photon comes into being at a certain instant in time in a certain location, and likewise, at a certain point in time in a certain other location is annihilated, it nevertheless exists simultaneously in different locations between these two places and points in time. One and the same photon can move through two different slits from sender to receiver, interfere with itself and nevertheless hit a particular single point in the detector [Jauch (1973)]. This state of affairs is so difficult to conceive of, that even a specialist such as Stephen Hawking has illustrated photon interference incorrectly, that is, although a single photon indeed splits in two, after passing through both slits, it does not strike a single point in the detector, but two different ones [Hawking (1988)]. In later editions of his book Hawking has corrected his erroneous drawing. Also, the quantum theoretician Brian Greene (1999) pp. 97–103, describes the interference of the photon in the double slit experiment incorrectly when he says that photons separated in time can annihilate one another by interference at the double slit. In fact, the interference doesn't require multiple photons, but every individual photon annihilates itself during interference past the double slit on certain points of the detector.

Moreover, the photon as such is not perceptible at all.[31] What we perceive when a ray of sunlight, i. e., a photon, hits our eye, is strictly speaking not the sunray, but the annihilation of the sunray. Once per-

[31] As far back as the fifth century BC, Parmenides (ca. 515 to 450 BC) was aware that light must be a non-thing and thus a mere illusion of the observing subject [Popper (1998) pp. 87 f.].

ceived, a sunray no longer exists. All observations of the physicist, then, rely upon „particles" which in a certain sense, exist for zero seconds, which during this „existence" are simultaneously in completely different locations and which cannot as such be observed at all. This remarkable state of affairs should give pause for thought as to how a physical perception actually comes about precisely, and how it can most appropriately be described mathematically. The best generally understandable account of photons has, to my knowledge, been composed by Richard Feynman (1985). Yet also he stresses that quantum electrodynamics may well describe photons very precisely, however nobody has ever understood the wherefore of this theory.

A further difficulty of physics is that physical laws cannot be proven in the way that mathematical laws can. Naturally we may assume that the sun, having reliably risen every morning for several billion years, will also rise tomorrow. One could with the help of mathematical or quantum theoretical probability models, presumably even calculate how small the probability is that the sun will not rise tomorrow, but proof, that it will definitely rise, is impossible. Every physical law is only valid until an experiment shows that it is false. (One should bear in mind that an experiment can never produce an absolutely certain result, but only ever one, the correctness of which is more or less probable. Thus, it can also never be shown that a physical law is false, but only that it is perhaps very likely to be false.) Every prediction based on a physical law is therefore risky. Experience doesn't provide a certain basis for our knowledge; it is simply the designated test of our assumptions. To learn from experience means then, to determine on the basis of observations that we have erred with an assumption. It is characteristic of empirical-scientific hypotheses that they can conflict with theorems of observation. According to Karl Popper (1959) they are not verifiable but are falsifiable. An observation can show us what the case is, but not that it must be so. Yet as I have already explained in Section 2.1 about metaphysics, a falsification is also never one hundred percent correct, since the theory behind the measurement process used for the falsification could be false. In summary, we conclude that physicists, too, have knowledge much less precise than is generally assumed, and that a substantial part of this knowledge is not even knowledge about natural laws, but only the result of the mathematical models used [Wüthrich (2017)]. John Maddox (1998) p. 120, speaks of an actual crisis in the fundamental research of physics, which has not yet grasped the nature of space, time, matter and spin.

2.4 Science and truth

After all the sceptical and qualificatory remarks with respect to metaphysics, mathematics and physics, the impression might arise that I am concerned with a total dismantling of science. On top of all this, there are also the limitations of language commented upon in Chapter 1, without which there can be no science.

Is serious science at all possible under these circumstances? Or can we do no more than believe? Can there be anything, which is true, real or provable? Or is there something fundamentally wrong with my assessment?

The answer to this question depends upon the demands made of the terms science, reality and truth. In the event that knowledge should be absolute knowledge and truth should be absolute truth, then one can actually only believe. The impossibility of absolute knowledge was first formulated by the medical writer Alkmaion (ca. 500 BC), a pupil of Pythagoras [Richardson (1998) p. 201; Popper (1972); Poser (2001)]. Xenophanes (ca. 570–470 BC) wrote: „But as for certain truth, no man has known it, nor will he know it; neither of the gods, nor yet of all the things of which I speak. And even if by chance he were to utter the perfect truth, he would himself not know it; for all is but a woven web of guesses" [Popper (1998) p. 115]. Knowledge is faith mediated by symbols [Santayana (1940)]. Our experience of life and good common sense tells us, however, that a certain degree of knowledge is possible in addition to belief. How do we gain this knowledge and to what extent is it true? This is a fundamental question of metaphysics. Natural science can provide no fundamental answer to it.

There is a whole series of criteria according to which one can determine the degree of truth of a proposition. The criteria depend on the one hand, on the nature of the statement, in particular on whether it concerns a statement which is empirically checkable or a transcendent one, and on the other hand, different people will weight these criteria quite differently. A psychologist has confidence in psychological criteria, a doctor in medical criteria, a natural scientist in criteria open to empirical scrutiny, a religious person in theological criteria, a philosopher in intellectual criteria, an artist in aesthetic criteria and a pragmatist in criteria of usefulness. „Objectivity and rationality, humanly speaking, are what we have; they are better than nothing [Putnam (1981)]". Truth is thus always *human* truth, truth related to *human* experience, without validity about things beyond the human, human understanding, human language [Rogler (1996)]. Accordingly, different are the opinions with respect to the degree of truth. The following list of criteria of the truth is arranged according to my subjective weighting, i. e., the first criterion is the most important for me, but none of the following is irrelevant.

1. The degree of a statement's truth is best ascertained when it pertains directly to the self, i. e., has a direct connection to the „ego", to the self. When Descartes says „cogito" then he is firmly convinced that he thinks, that there is a something or someone thinking, since he himself is a model of this statement. Saint Augustine puts it thus: „Truly within me, within, in the chamber of my thoughts, Truth, … without organs of voice or tongue, or sound of syllables, would say, ‚It is truth'." [Augustine (396–400/1946)].
2. The degree of an explanation's truth is all the greater, the simpler the explanation is. The postulate of simplicity is based on the assumption that the laws of nature are themselves simple. „Nature itself is complex. Simple conditions result from the relations of complex objects, and not the opposite, i. e., of simple objects forming complex ones. The search for the simple is therefore not a search for material basic elements, but for ideal abstractions" [Hampe (1998) p. 76]. From Aristotle (1987) book II (192bff), to Kepler (1937) and Newton [Weizsäcker (1999) pp. 167ff.] to Mach (1889), Einstein (1991a) and Heisenberg (1971) pp. 312 f. most of the great natural scientists worked according to this principle and had success with it. Einstein (1956b, p. 72) made the following remark about the concept of simplicity: „What actually interests me is whether God could have made the world differently; that is, whether the requirement for logical simplicity leaves any freedom at all". Nature should be described with the simplest theories possible. A theory is simple when it is expressed in simple words, when it is brief and when its predictions can be falsified with a few yes/no statements. The simplicity of the laws of

nature is a subjective assessment, not different in principle from assessments in ethics [Putnam (2003)].

A simple theory is often not simple or easy to understand, since the degree of abstraction sometimes increases with the simplicity. Similarly, a theory such as quantum theory, which is simple in itself, can necessitate a demanding range of mathematical measures. As a methodological postulate, simplicity requires minimal explanations: What are the fewest theoretical assumptions necessary to explain the observed phenomenon completely and consistently? This principle of parsimony is referred to in the literature as „Ockham's razor" after the philosopher Wilhelm von Ockham, who operated in Munich in the 14th century [Carter (1993) pp. 33 ff.]. A theory cannot be simplified *ad libitum*. If the theory is based on only one theorem, then the limit has been reached. The discovery of simplicity is a genuine discovery and not just a methodical measure, nor merely a means of economy of thought. But what is being discovered here? What is discovered when it appears that precisely the fundamental laws, when penetrating right down to the atoms and elementary particles, are really simple? Weizsäcker (1999) p. 52, although convinced of the success of the simplicity principle, believes that the scientific theories of the 20th century have no answer to this question.

Related to simplicity is beauty. It is consistently apparent in the biographies of great mathematicians and natural scientists how beauty and simplicity have intuitively led to the correct theorems, elegant proofs and successful theories, which were then only later verified by logical deliberation and scientific experiments. A good example of this is Newton's theory of gravity. Newton (1973) was so convinced of the beauty and therefore truth of his theory, that he even faked his data in order to prove this theory. Einstein later wrote in his autobiography about the beauty of quantum theory: „But unfortunately, nobody has ever succeeded in finding solid ground under the feet in this field. And it is the highest musicality in the sphere of thought that Niels Bohr has succeeded in erecting the wonderful structure of his theory of the atom on such shaky ground [Weizsäcker (1999) p. 319]." With respect to theoretical physics, Dirac (1977), a founder of quantum mechanics, represented a similar viewpoint: „Today it seems to me that the best starting point which one can have in physics lies in the assumption that physical laws are based on beautiful equations. The only really significant requirement is that the underlying equations should be of pronounced mathematical beauty." However, beauty can also be beguiling. Many physicists today believe in string theory, simply because it is mathematically beautiful and not because there is an empirical basis for it [Penrose (2004) pp. 869–933].

3. Propositions appear to be true when they are „immediately evident", i.e., in line with common sense and are consistent. The simplest of these propositions are the axioms of logic. I have already indicated in Section 2.2.2 that caution is advisable here. As a further example, I will question the logical theorem $A \equiv A$. In so doing, space and time will acquire quite a different significance.
4. Propositions which can be proven are considered valid. One must make a distinction here between logical and empirical proof. Logical proof carries more weight, since it is independent of time and is not subject to the numerous deceptions and mistakes which can occur in all empirical observations and measurements. But neither are mathematical proofs absolutely certain, since they are based on axioms and principles of logic, which for their part can be called into question.
5. When a proposition is consistent with an already proven theorem or law, that increases its degree of truth. As a rule, we turn a blind eye to the fact that every theorem and every law can also be wrong.
6. If we have perceived something ourselves with our own senses, we regard it as true. In so doing, we don't consider possible sensory deception. Such rather certain propositions are only possible for the past, i.e., when a perception has really taken place. Concerning future perceptions, there are never certain propositions, but only those with a greater or lesser degree of probability.
7. Should there be a document for a past observation which we can consult in addition to our memory, for example a temperature chart or a meteorite, that will increase the memory's degree of truth. Yet even the document can be faked, it can be deceptive, it can be arbitrarily collected, and it must for its part, in turn be perceived with all the mistakes which can occur in so doing.

8. If the same observation is repeated or indeed made periodically, it will increase the probability that it concerns a genuine occurrence. Sunrise is such a case. Also, belief is often based on habituation. We believe that the sun will rise tomorrow, because up until now, it has risen every morning. But we cannot know this with absolute certainty. Only repeated equal observations permit predictions about the future. The predictions can, however, never achieve the same degree of truth as statements about the past.
9. Of great philosophical significance today is the criterion of objectivity. A statement is objective when different observers or subjects, as independently as possible from one another, make the same statement or observation and are in „intersubjective“ agreement about their truth. There is agreement between the subjects, that it is *a priori* clear, that a disagreement about true or false means that one is mistaken. The term „objective“ can mean also „independent of the subject“. The contents of objective knowledge however, are always connected with the quality of the subject, because there is no knowledge without the knowing subject. The more precisely we are able to determine what the input of the subject into the knowledge of the object is, the more precisely we know how the object is actually constituted [Schaff (1980)]. The criterion is overrated in my opinion. The catch is that to agree, the different subjects have to communicate. In so doing, two things occur: First, every subject becomes an object for all the other subjects, so the subject can never be certain whether it is facing a second subject or an object without consciousness. Second, the subjects influence each other, which can lead to mass psychological excesses. Objectivity is actually always a fiction [Dürr (1997) pp. 166f.]. Truth is never objective, but always subjective [Kölbel (2002) pp. XV and 69]. On top of all that, there are all the language-related limitations and misunderstandings described in Chapter 1. I will go into detail on these matters in Chapter 3.
10. The certainty of a statement increases when the subject has been previously examined with respect to credibility and sensory health by psychologists and doctors. The risk of hallucinations, illusory dreams or misperceptions can thus be reduced. In the case of statements coming about intersubjectively, then naturally all subjects should be vetted. This process is sometimes taken quite far with

> psychiatric reports for criminal proceedings. How much these contribute to establishing the truth is something for the experts to decide.

At the beginning of this chapter, I made a whole list of sentences beginning with, „I know“. The reader is now in a position to assess the degree of truth of these sentences with the help of the preceding criteria of truth.

Scientists and thinkers disagree as to which criteria have to be met so that one can actually speak of science. It is usual that the criterion of objectivity or intersubjectivity, which appears in second last position on my list, is required as *conditio sine qua non* for every science. Many scientists further demand the reproducibility of the observation, i. e., criterion 2.4.8. Personally, I regard it as appropriate to understand the notion of science as broadly as possible, yet to always critically scrutinise the degree of truth of all statements according to the criteria presented. If the line between natural science and the other sciences becomes blurred thereby, that is deliberate and probably even advantageous. This line is rather artificial in any case and it can be dangerous to act as though natural science is much truer and more accurate than the intellectual sciences. In this sense, I regard philosophy and theology as sciences as well.

2.5 The aim and methods of science

The objective of every science is to know more, i. e., to be able to make statements with the highest validity possible, and when possible also those which concern the future and – if possible – to explain the matters of fact under scrutiny. The statements are always expressed in a language and are subject to the limitations of language described in Chapter 1.

In natural science, there are three additional features: First, observations and experiments are monitored intersubjectively. Second, the observations must be repeatable, whereby laws can be derived which make predictions about the future possible. Third, the theory of a field of research can vary in form and content from the pre-systematic representation of the empirical observation, that is, from the description in simple terms of the observation. In this sense, the theory is a model or representation of reality. The model is a simplified representation of the observed state of affairs. The scientist deliberately does not include in his model all structures which he has observed. Natural science is always based on certain questions *not* being asked. In fact: were we to ask all questions at the same time, we wouldn't answer a single one, since the truth is that all questions are connected [Weizsäcker (1999) p. 183]. The natural scientist therefore, is interested only in those aspects which are presumably relevant for possible predictions about the future. Thus, the model becomes simpler than the reality; its validity becomes more general and, in this sense becomes truer. The simplification can, however, lead to a later observation not fitting the chosen model, be it because the model lacks one or more degrees of freedom, or be it that the observation is simply too complicated. This is the principal motivation for the natural scientist to carry out his experiments under strictly controlled conditions with as few parameters as possible. If, in spite of this, the contradictions between observation and model do not disappear, then the model must be expanded or altered. Either new degrees of freedom are introduced into the model or the relations between the parameters in the model are altered.

This portrayal of procedure in natural science shows that it is the close connection of empiricism and transcendence which makes natural science at all possible. Observation and measurement are empirical; the pre-systematic description of it (i.e., a simple description without measurements) lies on the border between sensory experience and transcendence, that is, the non-sensory realm. Models and theories are products of the intellect. It seems that natural science can only exist at all, because it has an open door to metaphysics [Weizsäcker (1986) p. 634; Atmannspacher et al. (1995)]. If there can be no natural science without transcendence, are then at least the humanities without empiricism possible, as for example, Kant (1781/2003) presupposed with his *Critique of Pure Reason*? Is there an intellect *a priori* which sets the laws for nature as an object of the senses? Is there a faculty of reason *a priori* which sets the laws for freedom and its causality? This cannot be proven, because a metaphysical theorem can never be proven. It is equally impossible to prove the opposite. What is certain is that barely anyone uses his intellect or faculty of reason without being indirectly influenced by empirical experience. Even Kant himself would scarcely have deliberated over judgements a *priori* had he not, as a flesh and blood human, had quite concrete sensory experiences. Even so, he considered it important to clearly separate the study of the transcendent and that of the natural science, the study of the material world. The former was for him „proper science“ which was considered apodictically true and which in principle could not be falsified; the latter was merely empirical certainty or „improper knowledge“, which only applied until it was falsified by an experiment. According to Kant, a theorem was generally valid experientially if it articulated the preconditions of every possible experience. Such preconditions according to Kant are space, time and causality. We will have explained natural laws if we have traced them back to the preconditions of experience. Thanks to these uniform preconditions, in individual cases, we always, in a certain way, perceive the general. It is primarily these which give nature its unity. Science is then successful when it finds the hidden simplicity of the form and with it, the unity in the experience. Weizsäcker (1971a) supposes that on the whole, physics is essentially nothing more than the totality of those laws which must already be valid simply because we objectify and are also able to objectify that which physics studies, i. e., that the laws of physics are none other than the laws which formulate the conditions which makes it possible to objectify the event. To „objectify“ is thereby defined as the reduction to empirically determinable questions.

I'm inclined to accept the view that a strict separation between the transcendent and the material is not possible, that there is possibly even a continuum linking the two sciences and that they complement each other. In this way, science obtains a sort of cyclical structure: The experimental physicist measures. The measuring instrument is thereby an object of his experience and is accordingly subject to the empirical laws of physics. These laws, in turn, have an effect on the structure of the experience the physicist can have with the aid of these instruments. Physics thus becomes a theory of observable quantities which describes nature as it is revealed when one examines it with material measuring instruments. In short, the theory defines, what can be measured, and what is measured, must, within the framework of the theory, be possible. When an experiment turns out differently from what the theory being tested predicts, then this only implies anything of importance if I can be shure that the equipment „is in order“ and that it works according to a true theory. Yet since there can be no true, i.e., verified theories, the thesis of a falsification doesn't bring us any certainty [Görnitz (1999) pp. 33]. Science becomes an interplay between the transcendental and the material without a clean separation between the two being possible. Since this loop is a short one, science should actually reach a conclusion quite quickly, that is, produce a theory which explains all the natural laws. I have even considered whether I should design my book in loose-leaf format with actual 360° rings. The reader could then begin reading anywhere in the book he likes. Once he has read all the pages he would be in possession of the complete and consistent laws of physics. That is unfortunately impossible, namely due to the limitations of language and of knowledge which I have portrayed at length in the preceding sections. They show that science can never be finalised. It is, however, possible to come a little closer to the truth with every loop. The loop is – strictly speaking – not a loop but an endless spiral! With every rotation of the spiral we make a small scientific advance.[32] The spiral argument is an argument of infinite length, in which the same, never quite clear terms, e.g., „measure" and „theory", periodically crop up time and again, but with every rotation of the spiral, are specified a little more precisely.[33] The beginning lies at a bottom-most loop in a primitive language, which is simply a sign language. The transition from this metalanguage to everyday speech to scientific formalised language is not an abrupt one but is rather a gradual progression.

Is the spiral regular, or does it make leaps from time to time, namely always then, when a so-called paradigm shift takes place? According to Heisenberg, a finalized theory is one that can no longer be improved with small changes. Only fundamental modifications are then still able to open up new perspectives and they are changes in the meta-language which describe the theory. Such paradigm shifts include the introduction of classical mechanics, the special and general theories of relativity and quantum theory. These discoveries were supposedly not about normal

[32] The metaphor of the spiral was also used in this context by Gadamer (1965).

[33] When Niels Bohr had washed glasses under the not very clean conditions of a ski hut, he opined: „Fancy being able to make dirty glasses clean with dirty water and a dirty cloth – if one were to tell a philosopher that, he would not believe it." But that's how science really works! [Heisenberg (1969) p. 190].

science but concerned conceptual revolutions. This view is contrary to my conception of science. Naturally, certain discoveries are „more revolutionary" than others, because their repercussions are more wide-reaching. But it must be stressed that the thought processes which make such discoveries possible are indistinguishable from those which are used in solving a small physical problem of detail. Paradigm shifts are encountered all over the place in science; there are simply bigger and smaller shifts at different levels [Kuhn (1967)]. To a far greater extent than is commonly supposed, the ways of the big shifts as well have been prepared in almost all cases by various researchers who are barely known today [Hofstadter (1979) pp. 659 ff.].

To have found a simple theory, which permits the correct predictions, actually ought not satisfy the scientist yet, even if the theory is still useful. What I am looking for is a scientific explanation. A good theory can, after all, only recognize natural laws and formulate the rules and methods, by which our intellect creates sensual ideas. But „after all we do not want merely to describe the world as we find it, but to explain to the greatest possible extent why it has to be the way it is [Weinberg (1977)]." Natural science can indeed record incidences of regularity, but not explain them or make them comprehensible [Feynman (1988) pp. 9–12]. That applies, even if two originally independent theories can be unified by a common mathematical model. Only the unity of ontology at best would be able to deliver an explanation of the unity of nature [Morrison (2000)]. To science, only, so to speak, is the How of the event accessible, but not the Why [Duhem (1914)]. We have no explanation as to why space has three and not more dimensions, why the speed of light is constant, why there must be a Planck's constant or why there are black holes. The question as to Why is not a physical one, but rather a metaphysical or if need be, a theological one. The answer therefore is always dependent on a metaphysical frame of reference and the scientist can decide for himself which metaphysical frame of reference he wishes to recognize as the true one. I will avail myself of this freedom and in so doing, arrive at my own striking explanations of natural laws.

Chapter 3

My Consciousness Exists

The non-existent, verily, was here in the beginning.
presumably Parshva, ca. 800 BC[34]

cogito ergo sum.
René Descartes (1619/2006), Neuburg an der Donau

Yea, this ego, with its contradiction and perplexity, speaketh most uprightly of its being – this creating, willing, evaluating ego, which is the measure and value of things.
Friedrich Nietzsche, Also sprach Zarathustra[35]

3.1 Metaphysical foundations

Where should we begin now with our science? Actually, we are already in the middle of the spiral of argumentation without noticing it. In the first chapter about the limitations of language, the spiral already turned 360°, in the second chapter about the limits of knowledge a second time. However, the reader's science had begun, long before he started to read this book, i. e., at that time when he collected his very first experiences, perhaps even before he was born.

We have used terms such as God, Be-ing, reality, soul, mind, think, consciousness, memory, observation, perception, phenomenon, model,

[34] Zimmer, Heinrich (1994).

[35] Nietzsche, F. (1930a): *Ja, dies Ich und des Ichs Widerspruch und Wirrsal redet noch am redlichsten von seinem Sein, dieses schaffende, wollende, wertende Ich, welches das Mass und der Wert der Dinge ist.* English translation: Project Gutenberg etext of Even so, we have reached an albeit not decisive, but yet important point, where a course is set.

subject, object, ego, you, communication, information, nature, world, matter, person, brain, and computer. How do these terms relate to one another? Is there a hierarchy of these terms? Can one begin with one, all-embracing or particularly true term and then derive the others from it?

3.1.1 God

Since Babylonian times[36] it was common to place God as „the final cause of the world" above or beyond all other terms [Dürr (1997); St. Thomas Aquinas (1267–1273/1947)]. Or can, and should one, like Descartes discover God in oneself and conclude that this conception is only possible, if he who is being conceived of, i. e., God, exists? Accordingly, this conception would be primary; from this the existence of God would be inferred. For people who did not succeed in finding God within themselves, this derivation would be invalid, however. For Spinoza and Einstein, God revealed himself in the regulated harmony of what exists, i. e., of nature; however, that does not mean that he caused and created this harmony of nature, which would not be possible in any event without a superordinate concept of time. Four weeks before his death, Einstein wrote to the surviving son and sister of his life-long friend Michael Besso: „And now he has preceded me briefly in bidding farewell to this strange world. This signifies nothing. For us believing physicists the distinction between past, present, and future is only an illusion, even if a stubborn one." Since God is outside time, and therefore cannot cause anything in the earthly sense, I will forgo making God the foundation of my science.

3.1.2 Thought

Another possible introduction to science proceeds from the soul or mind, two concepts which are synonymous for many scientists. The aged Goethe in one of his epigrams[37] has another person ask him,

[36] As far back as 1800 BC, the Sumerians believed that their chief god Marduk had created the laws of nature and thereby determinism [Stent (2002) p. 109].

[37] Goethe (1823–28/1921): „Wie hast Du's denn so weit gebracht? Sie sagen, Du

„How have you then achieved so much?
They say you did it with great touch."

To which he answers:

„My child, I did it as I ought,
I wasted not a thought on thought."

That is an expression of Goethe's great attachment to the world. Can the organ of thought we are endowed with for use in dealing with reality, suffer no harm when we use it on itself [Weizsäcker (1971b) p. 106]? Rudolf Steiner (1894/1992) considers thinking about thinking not only illegitimate, but actually impossible. „I can never observe my present thinking; I can only make my experiences of my thinking process the object of subsequent thinking. I would have to split myself into two persons, into one who thinks and another who observes this thinking. But this I cannot do." Accordingly, to Rudolf Steiner, thinking comes before the ego, before consciousness. Thinking requires no consciousness, no intellect, no subject, and no ego as carrier. To start with, it is not so that „I" think, but that „there is thinking". If I desire enlightenment about the kind of relationship which exists between my thinking and my consciousness, then I have to think about it. Consequently, I presuppose thinking. Most philosophers are of the view, however, that thinking without a thinker is impossible [Frege (1918/1999)].

Nevertheless, I don't wish to make thinking the foundation of my philosophy, the main purpose of which is to derive elementary laws for natural science. In natural science, observation, perception and with them the subject as observer, all play such a central role, that I prefer as the basis of my science, concepts which lie closer to natural science than purely thinking.

hättest es gut vollbracht.", „Mein Kind, ich habe es klug gemacht, ich habe nie über das Denken gedacht."

3.1.3 The ego

Philosophy can also begin with the ego. That leads to the so-called solipsism, which some philosophers consider to be not very productive [Rae (2004); Nagel (2003); Conrath (1973)]. Nevertheless, in constructing my metaphysical approach in Section 3.2, I will proceed from a moderate solipsism because for the natural scientist, that is the most honest path. Beforehand however, I want to sketch out a few further possible introductions to philosophy, which all have their advantages and disadvantages.

3.1.4 The paradigm of mutual understanding

Solipsism was criticised as „subject philosophy“ by Habermas (1988) among others. He suggested replacing the consciousness paradigm, whereby an individual subject contemplates and decides what is true or real, with a „paradigm of mutual understanding“, whereupon several subjects were to agree about truth and reality. The communication in the scientific community thus gives the „world-developing language primacy over the world-engendering subjectivity“. The beginnings of such a philosophy can already be found in Kant's (1910) „Rules for completely avoiding error“, which require that one „thinks for oneself, to imagine oneself in the place of another, and to be at all times at one with oneself in one's thoughts.“ With Kant, however, it was still clear that ultimately somebody, viewed empirically, can only think for himself alone – in solitude and freedom. Hegel went somewhat further in the direction of intersubjectivity with the remark, „The living spirit is the medium which creates the kind of community where the subject knows itself to be at one with the other subjects and can nevertheless remain itself“ [Langthaler (1997) p. 221]. Fifty years later Peirce (1903/1997) defined truth and reality as follows: The opinion which is destined to be finally agreed to by all investigators is the truth, and the subject which is represented by this opinion is real. In short: truth is that about which we agree.

The paradigm of mutual understanding was criticised at length by Langthaler (1997) pp. 308 ff.:

1. Since agreement about the truth can only, if at all, be reached by means of a language, the truth according to the paradigm of understanding is always subject to the limitations of language and is correspondingly blurred.
2. A subject can never determine empirically whether the objects he observes have the character of a subject, i.e., their own consciousness, or whether they are, for example, simply dreams, hallucinations or robots.
3. In the event that there really are several subjects which can communicate with each other, in attempting understanding they will influence each other. It could happen that a single dominant subject, consciously or unconsciously, imposes a false opinion about the truth on the other subjects.

For these reasons, I prefer to forgo making intersubjectivity the basis of knowledge and truth.

3.1.5 Extrasensory perception

One approach which partially clears away the objections to the paradigm of mutual understanding is the communication between different subjects with the aid of extrasensory perception. Such a communication requires no language and no object which the subject can perceive with its senses. Also, the mutual influencing of the subjects could have a different character from that of the influence during spoken communication. Experiments with telepathy have allegedly, according to Joseph Rhine et al. (1957), yielded results of great significance, showing that the correct transference of information was not attributable to mere coincidence. However, these results have been criticised or refuted by other writers [Blackmore (2003) pp. 293 ff.]. There are essentially four possible explanations for extrasensory perception [Rýzl (1992)]:

1. There is no extrasensory perception; it is illusion or mere coincidence.
2. Extrasensory perception is not genuine perception, since it takes place unconsciously.

3. Extrasensory perception occurs through a new, still unknown sense.
4. Extrasensory perception essentially cannot be explained by the known laws of nature.

For C. G. Jung (1952), the last explanation is the most plausible. According to his observations, parapsychic phenomena are independent of space and time and accordingly do not fit the context of causality with respect to the processes in the universe. He therefore introduces alongside the traditional categories of space, time and causality an additional category, synchronicity, which is intended to constitute the key to understanding parapsychological phenomena. Whilst according to Jung the principle of causality makes the connection between cause and effect necessary, the theory of synchronicity implies that the elements of a significant synchronicity are bound to each other by their concomitance and their sense [Jung (1952) and (1967) pp. 475–591].

Since I have scarcely made any extrasensory observations personally and since these are scientifically controversial, for me they cannot be considered as the foundation of metaphysics. Even C. G. Jung, as a professor at the Swiss Federal Institute of Technology in Zurich (ETH), hesitated some twenty years before daring to publish his findings about synchronicity, since he feared that afterwards he would not be taken quite seriously as a scientist. Wolfgang Pauli, the Nobel laureate for physics, who as a quantum theorist contributed to similar paradigm shifts, was first able to persuade him to publish his still somewhat vague theory [Atmanspacher et al. (1995)]. According to Thomas Görnitz (1999) pp. 290 ff., there's not necessarily a contradiction between extrasensory perception and quantum theory. Fritjof Capra (1991), as a nuclear physicist, is even convinced that the eastern philosophies, with their teachings of an anima mundi[38], of acausal enlightenment, of a holistic universe as a network of timeless changes and of a living void, could serve very well as a basis for most modern physical theories.

[38] Latin: soul of the world.

3.1.6 Nature

The five potential foundations of science itemised thus far in Sections 3.1.1 to 3.1.5 essentially all lie in the transcendental realm. Nature, by contrast, can be experienced empirically. The distinction between transcendental and empirical, put simply, is the difference between Be-ing and that which exists. The term nature is formally defined by Kant (1786/1957, p. 11) as „the first internal principle of all that belongs to the existence of a thing." So, the things of nature are „there", they are present, they can be perceived empirically, so they exist. Nature is the totality of that which is accessible to our senses, including those senses armed with technical instruments of observation [Whitehead (1925/1939)]. The transcendent, observing and thinking subject, however, is not part of nature according to this definition. What is more obvious, particularly for the natural scientist, than to make that which actually exists, i. e., nature, the foundation of science? The problem is threefold:

1. The division between the transcendental subject and nature is an artificial one, since a perception directly involves the perceived nature as well as the perceiving subject. Indeed, the perception consists precisely of the gap between nature and subject being overcome and the flow of information from nature to the subject.
2. The subject can never know with certainty whether nature also exists when it is not being observed. The true derivation of nature from sensations is impossible [Hume (1739/1978)]. Nature behaves in this respect similarly to a subject, of which a second subject cannot say with certainty, whether it really exists or whether it is possibly just a robot. Many philosophers therefore distinguish between a subjective nature, being in itself, and the perceived objective nature. Both together constitute nature [Langthaler (1997) pp. 84 f.].
3. But if one limits oneself to observed nature, it then remains unclear to what extent observation changes nature. According to quantum theory, there is no observation which doesn't change the object, and the change cannot be measured nor precisely calculated.

The relationship between nature and subject is complex, insomuch that it becomes very difficult to make nature the foundation of science. The same applies for the same reasons to the similar concepts of world, cosmos and universe.

3.1.7 Brain and computer

Some scientists are convinced that there is no difference in principle between thought and cerebral activity, that both are essentially the same material process. The capacity for knowledge and thought, in Konrad Lorenz's view, is linked to the structures and functions of the human brain which has developed in the evolutionary process to the extent that it is able to produce these peak performances. Lorenz doesn't exactly say: „Mind is nothing other than matter", rather he opines that spirit or consciousness are a „newly emerged system unity, a specific function of the brain" in evolution. All consciousness phenomena can be regarded as the result of specific patterns of integration between material elements in the central nervous system – especially in the brain. Thus, the material, physical sphere is declared to be the absolute basis of consciousness, without making the latter identical with cerebral activity [Wuketits (1983)].

Hofstadter goes a step further still. He investigates the possibilities – in principle – of theoretically constructible computers and comes to the conclusion that it is essentially possible to invent computers which imitate cerebral activity so perfectly that they can not only think, but that they even have a consciousness and a soul in the following sense: As a subsystem within the computer it continually communicates with the other subsystems of the computer, it registers which symbols are active and which are not. That means that it must possess symbols for mental activity, in other words, symbols for symbols, and symbols for the activities of symbols. „To be aware" is, according to this theory, a direct effect of the hardware and software complex, which is designed so that every theoretical activity of a subsystem is monitored by another subsystem in the computer itself [Hofstadter (1979) pp. 337–390; Turing (1948/1992)]. Actually, Turing writes not of a consciousness of self, however, but of thought, the capacity to learn and the processing of symbols. Minsky (1968) advocates the view that such a computer must also possess a free will and with it, could make a free choice, which arises from an intention or an intuition. However, there are also well-known scientists, such as

Penrose (1989) and (1994) for example, who consider the human intellect something qualitatively different from a computer, since the brain, with strokes of mathematical genius, can produce performances which no computer working by algorithms can simulate.

According to Gödel, such theories about a material consciousness are, however, always self-contradictory and mathematically inconsistent. For mathematically logical reasons, the human intellect must be equivalent to a finite machine, which is not in a position to fully understand its own workings. Even so, Gödel doesn't claim that his incompleteness theorems disprove the mechanistic view of the intellect. On the contrary, he believes in a disjunction of philosophical alternatives: „Either the working of the human mind cannot be reduced to the working of the brain, which to all appearances is a finite machine," or else „mathematical objects and facts … exist objectively and independently of our mental acts and decisions." Those alternatives are not, of course, mutually exclusive. Indeed, he was firmly convinced that both were true [Dawson (1997) p. 170].

I don't know whether or how the reader perceives his own soul, but in my personal experience, the soul is not satisfactorily explained or modelled by this theory. It is impressive what a theoretically constructible computer might possibly be capable of, my soul however is something else and that of my wife as well, otherwise I would scarcely have married her. In psychological, emotional, ethical, and aesthetic respects I am more than, and different from, a computer. I can know that about myself, but not prove it. Other people can believe me or not.

The theories described, attractive in themselves, are based on the assumption that a transcendental realm neither exists nor is required, or if it does in fact exist, is no more than a copy of the real world. The scientists who develop or advocate these theories, as a rule do not explicitly claim that consequently all questions can be reduced to ones of natural science, but they do give rise to such hopes. The decision about the acceptance or rejection of such theories is always a metaphysical one and I believe it cannot be made by the free will of a computer, but only by me. I therefore do not consider such theories as a basis for my science, even when they are inspired by natural science.

3.1.8 Matter, substance and information

Since the natural scientist seeks to describe things which exist, it would actually be self-evident to base all science on such things, i.e., tangible objects. For this there are different designations, such as matter or substance. „For us Germans, matter is more or less dirt" Bertold Brecht is meant to have once paraphrased his philosophical take on materialism. I see it, I feel it, I can wash it off. It would never occur to anybody that dirt doesn't exist. For the materialists, matter is the origin, and consciousness is dependent on it, or derived from it. Matter is real. Consciousness can reflect, model and recognize reality. As an optimist, the materialist considers the world fundamentally discernible. From the earthy, speculative materialism of Thales around 600 BC to the dialectic materialism of Marx and Engels in the 19th century, materialism may well have continued to increasingly liberate the concept of matter from its attachment to the recognition of the material characteristics of reality. Despite this, according to Brecht, matter has remained dirt for the average person.

In the philosophical tradition, substance is that which survives over time. Substance could be divided into a spatial and a thinking one, i.e., into matter and consciousness. According to the philosophy of quantum theory, such a division is not admissible, however, since matter and consciousness can never be cleanly separated during the act of observation. That which survives over time is the quantity of the form, expressed in more modern terms, information. Occurrences over time are then nothing more than an information stream. Since the energy in closed space remains constant, i.e., since it correspondingly „survives over time", there must also be a simple correlation between information and energy [Weizsäcker (1986) pp. 163ff.]. So why not begin with the information, from which then concepts such as substance, matter, energy, probably consciousness as well, could be quite easily derived? Such a path would probably be practicable. If I nevertheless favour the path of solipsism, then that's because the information stream doesn't make sense scientifically until a consciousness which registers the information is present. The difficult relationship between information and consciousness is undoubtedly central to the understanding of natural science. The relationship in German everyday language is referred to as „*Wahrnehmung*", a formation from the verb , „*wahrnehmen*", i.e., , „*wahr*" (true) + , „*neh-*

men" (to take), which can be understood as, „I take something and consider it true". For that, however, I first need an „I"!

3.2 Solipsism

Solipsism is based on the notion of the „I", the ego. If I want to structure my science around the ego, then that has nothing to do with egoistic conceit, rather I do it for the following reason: I want to begin with a concept which is immediately evident to me and with a degree of truth which is therefore very high. Since science seeks true statements, it should begin with a concept which has the highest possible degree of truth. In my opinion, this is the ego. Einstein reached quite a similar conclusion: „The, as it were, religious attitude of the scientist to the truth is not without influence on the overall personality. Since, except for the results of experience and deliberation, for the researcher there is in principle no authority whose decisions and pronouncements can lay a claim to „truth". From this arises the paradox that a person who dedicates his energies to objective things becomes regarded socially as an extreme individualist, who relies – at least in principle – on nothing other than his own judgement. One can even very probably hold the opinion that historically, intellectual individualism and scientific striving first arose together and have remained inseparable [Einstein (1950/1991)]." I would like to stress however, that that is my personal judgement and my free choice. Other people can easily come to another conclusion and their science – by other paths – ultimately still yield the same result as mine does. This is because knowledge in science advances in a spiral, as it was explained in Section 2.5. The more revolutions the spiral makes, the deeper my knowledge becomes. The more comprehensive the science should be, the more wide-reaching a turn of the spiral must be, i.e., the longer the circumference of the spiral becomes. Since in a comprehensive spiral all fundamental concepts occur over and over again and are specified further in the new context, sooner or later I still arrive at each of the concepts which I initially rejected in Section 3.1 as foundation of my science. Despite this, it is relevant where I begin my deliberations in the spiral because the ensuing argumentation can become simpler or more involved, more plausible or more questionable. What a scientist accepts as plausible is subjective however and depends both on his previous experiences and on his aims.

Cogito, ergo sum. I think, therefore I am. Descartes arrived at this thesis when an officer for Maximilian von Bayern in the middle of the Thirty Years' War. Perhaps it's no coincidence that precisely a soldier, who – voluntarily in the case of Descartes – puts his life at risk, comes to such a conclusion. A soldier in war is more intensively aware that he is, and he is confronted with the danger that tomorrow, he might no longer be. Thus, it can happen that on a quiet, cold winter's day he ponders the question of what that actually means, „I am" [Spierling (1990)]. Descartes wants to build his philosophy on a hard and fast, absolutely certain foundation. He therefore begins from the position, that he doubts everything which comes into his mind. The more the doubt includes, the clearer it becomes that it cannot include itself. There is no doubt about doubt itself. Everything can be denied, only the act of denial cannot. In doubting, the ego becomes aware of itself. Doubting is a thinking act, which cannot be separated from the ego. Therefore, the ego must exist. This truth is so unshakeable for Descartes, that he makes it the first principle of his philosophy. It is impossible to think without existing at the same time. I know myself as a thinking consciousness. I exist thinking. Thinking cannot be separated from me. „Thus, I am precisely one thinking being" [Descartes (1619/2006)].

This special transcendental-logical position of the ego is also well recognized by Plato (1961–1963a) Theaitetos 159e, and Kant (1783/2001) § 46 f. It cannot and must not be replaced with „we". For always, when somebody says „we", he appoints himself as the spokesperson of the group to which the personal pronoun „we" refers. But then another member of the group could object, „Who says that we say that? Rather, we say something quite different." Then, however, the first speaker would respond, „I say we say". This consideration shows that every „We say" sentence actually has the structure „I say we say". Only the „I say" sentences would regress in infinite perpetuity, in that purely theoretically, one could formulate them „I say I say I say …" *ad infinitum*. This infinite possible repetition yields no new meaning at all, however, and can therefore be eliminated [Klein (1984)].

In spite of this, every use of the word „I" creates a problem: „I" is the subject. But now, when thinking, every notion in face of the thinking subject must become an object. Also, the „I", when it is thought, becomes such an object. It must, then, be an object whose essence exclusively consists of being a subject. Humboldt (1903) pp. 204–207, therefore makes a distinction between the „true and real feeling of „I-ness" which defines the subject, and the „I" in the sense of Self, which is actually an object. According to Humboldt, the semantic relation of language always has three strands: First it refers to the speaker, second to the dialogue partner and third to the object being discussed. When I think „I", then the „I" is both speaker (or thinker) and partner, as well as the topic under discussion, thus it covers all three strands [Langthaler (1997) pp. 308 ff.]. No other term has this characteristic. That is reason enough to take the ego as the foundation of science. Here a distinction is being made between the *totality* of the Self and the conscious, or thinking & feeling part of the Self, what Jung calls the „*ego*".

The simplest statement I can make about a thing is that it exists. If I am convinced that my ego, my Self, exists, does it then have further characteristics? What is this ego? Is it simply a thinking consciousness, an awareness of self or is it a person? Does the content of the thinking also belong to the ego or is it a third something? Is the ego divisible? Is it free? Can there be an ego outside time? The answers of philosophers to these ancient questions, unsurprisingly, are widely divergent.[39], [40] It is not necessary for the purposes of this book to set out, contrast and exhaustively

[39] The highest and most distinctive achievement of Indian philosophy is the discovery of the self (atman). Atman, etymologically related to the German word Atem (the breathing) and the more poetic Odem, is also translated as breath or soul. The self was generally conceived of as eternal, beyond space, time, causality, logic and all visible things. Indian philosophy is a 4000 years old story of the conflict between dualism and monism, of body and atman on the one hand, god and atman on the other hand [Zimmer (1994)].

[40] Parmenides (515–450 BC) can also be viewed in a certain sense as a solipsist. Being, according to Parmenides, was a conscious Be-ing, and the individual yet complete content of its consciousness forms the fact of its existence. I am, and I am the ‚One Being'. Apart from me, there is nothing. There is no place for a second someone to stand by and acknowledge this Be-ing [Fränkel (1951); Popper (1998) pp. 111–127].

discuss the possible answers in great detail.[41] It is sufficient to briefly say what I choose to understand by the word ego.

The ego is not a flesh and blood person, and also not a brain. I don't differentiate between a transcendental ego, which faces the whole world and an empirical ego that finds itself in a world as one amongst many others. The ego is a subject, which can perceive and think and in so doing, influences the perception and thinking. That is only possible when a temporal relation exists between perception and thinking, in that a perception must chronologically precede the thought influenced by it. The perception is the connection between the world and the ego. The ego is thus not part of the world, but it encounters it, the object, as a subject. This is not to pass comment as to what extent the world really exists.

Thoughts belong to the ego, likewise the memory of earlier thoughts and earlier perceptions. In this sense the ego may indeed be broken down into different memories, but it is at the same time also an indivisible whole in the sense of an independent consciousness. With respect to perception, the ego is not free. It can only perceive things which are present and is limited in the process by certain basic conditions, which philosophical tradition has labelled space, time and causality. Without a certain measure of self-perception, there is no sense of space and time [Kather (2003) pp. 139–181]. With respect to thought, the ego may well be dependent of stored memory, apart from that, however, it is free and can also develop a free will. So, the ego is part unfree, part free. It finds itself unfree in the world of perceptions and realises in itself the free spirit. Due to its freedom, the ego is responsible and the subject of moral law. The anchor role of subjectivity cannot be lifted.

Not only our sensations and feelings are individual; our thinking is as well. That need not necessarily exclude an *anima mundi*, in whose stream different subjects become connected with each other and can exchange information via telepathic means. It could even be that in such a „stream of the world soul" neither temporal nor spatial restrictions apply.

[41] A good overview of the current controversial perspectives concerning concepts such as consciousness, awareness of self, subject, free will, etc. was given by Blackmore (2003).

Consciousness perceives the world and reflects reality. It encompasses all focuses of attention and experience, such as for example understanding, wanting, feeling or doubting. Accompanying consciousness in humans is the awareness of self, which goes with these modes of experience. Consciousness and awareness of self are often very clearly separated, so that animal consciousness can be distinguished from the differentiated human consciousness and the capacity to experience oneself as an individual. We are conscious of our own existence, we know that it is we ourselves who think, experience, feel and have a will [Dürr (1997) pp. 111–156]. Research into animal behaviour, e.g., the observation of chimpanzees, which can recognize themselves in a mirror and laugh at their own face-pulling, indicate however that the distinction between human and animal, also with regard to a consciousness of self, is not a fundamental one, but rather one that is gradual [Gould (1994)].

The Be-ing of the existent is therefore in one way or another always subjective, always an existence imagined by a subject [Heidegger (1961), Santayana (1940), Metzinger (2003), Searle (1997)]. Also, Frege (1993) was convinced that there cannot be a conception without a bearer of this conception: „Not everything is conception, which can be an object of my recognition. I, as the bearer of conceptions, am not a conception myself … The grasping of thoughts presupposes a comprehending entity which thinks. This, then, is the bearer of the thinking, not however the bearer of the thought." Whitehead's (1936) pp. 252 ff., subjective principle says that the entire universe is composed of elements which are revealed by subjects in the analysis of experience: „… apart from the experiences of subjects there is nothing, nothing, nothing, bare nothingness."

Heidegger (1997) called this situation anthropomorphism, that is „the conviction that the existent on the whole, is what it is and how it is, due to and in accordance with the imagining, which in humans, i.e., in the animal endowed with reason, proceeds as one process of life among others". Thereby the nature of Be-ing remains quite unexplained. The only thing this theory explains is that Be-ing without a subject is impossible. Anthropomorphism means that the laws of metaphysics must be so configured that thinking by a human-like subject is possible. Since it is very difficult to think of human-like subjects independent of human persons, i.e., humans of flesh and blood who are part of nature, Heidegger's theory has been expanded to the „anthropic principle", which amounts to: The conditions of the world are as they have to be, because we exist as humans. Of all the conceivable universes, we can only perceive ones which make our existence as observers possible. The anthropic principle has an effect on cosmology, philosophy and theology and places astonishingly strict limits on the laws of nature, natural constants and initial conditions of the universe [Bertola (1993)]. For example, only dimensionless numbers are permitted as natural constants [Penrose (2004) pp. 1030–1033]. Since both the universe and the humans in it can only be perceived as phenomena by subjects, the question arises as to whether a general subject-principle couldn't be formulated which is the basis for the laws of nature, e.g., for the structure of space and time. Such a subject-principle has in fact been postulated by Michel Bitbol (1993): „The topological structure of the world must be such that the world can be perceived as an object by a subject."

Can the ego, as subject, judge objectively, or are such judgements arbitrary and thereby unscientific [Kölbel (2002)]? The Greek philosopher Protagoras of Abdera (485–415 BC) was the first to assert that truth cannot be objective. „That which appears true to you, is true for you; that which appears true to me, is true for me" [Richardson (1998) p. 204]. A judgement which expresses an awareness only makes scientific sense if it can't be arbitrarily, i.e., not at my convenience, disposed of. Furthermore, objectivity also means independence from the judging subject and validity for the other subjects as well. Precisely this intersubjectivity is a fundamental characteristic of reality which assists me in distinguishing it from dreams and illusions. For scientific research in particular, intersubjectivity is one of the most important requirements. First of all, I refer the reader to my provisos with respect to all intersubjectivity in Section 2.4.9. However if one accepts the idea that multiple independent subjects exist side by side and can freely communicate with each other, then the contradiction between the subjective perception, which is always dependent on the individual stream of experience, and the perceptions of other subjects, which always differ from it, can, according to Carnap – his blueprint for a „constructional system" based on propositional logic proceeds from solipsism [Carnap (1928/1998) pp. 84–91] – be solved as follows: Certainly, the material of the streams of experience in different subjects is always completely different and not at all comparable; but certain structural characteristics are consistent in all streams of experience. Science must limit itself to statements about such structural characteristics, since science should be objective. And it *can* limit itself to statements about structure, since all objects of experience are form, not content and can be portrayed as structural entities. Science is thus by its nature always structural science and there is a way to construct objectives based on the individual stream of experience [Carnap (1928/1998) p. 208].

The latest brain research has yielded that the three most important characteristics of consciousness: unity, the ability to differentiate between a mental image and the external world, and its experience of authorship or a self, for instance in feelings, thoughts or perceptions, are all functions in the human brain and can be localised there as well [Lyre (2017)]. They can each be separately and distinctly disrupted, e. g., in the case of schizophrenics or of the terminally ill. The ego is then possibly a form in the brain and thereby subject to natural laws [Roth (2003)]. I will come back to the connections between brain and ego, and between body and soul in Chapter 5.

Arguably the most difficult question which comes up in solipsism concerns the relationship between the ego and the world. Reality could be defined as that which exists independently of the ego and independently of every perception. Is there any such thing at all and in the event that there is, can we know that there is? How must consciousness be constructed, so that it can judge such questions? I will discuss this problem in Chapter 5. Yet in order to be able to understand the connections between ego-consciousness, perception and the world, I must first introduce the concept of chirality.

Chapter 4

Without Chirality no Order

> The distinctions between things are not based in their existence. They come only from the subjective point of view. The bases of distinction are: left and right …
>
> Dschuang Ddsi, 340 BC (1912/1972), p. 47

4.1 Chirality as the prerequisite of order

Usually in the constitution of a philosophy time is introduced at this point [Grünbaum (1973) pp. 179–208]. It has the function of bringing order to consciousness. Perceptions and thoughts follow one another in time. Without time, chaos would reign. Perceptions have a cause which occurs before the effect in the ego. This process is not circular, so it is impossible that the cause in turn is dependent on the effect. It follows from this that time must have a direction. Time is a one-dimensional, directed continuum. I have already indicated in Section 2.3, however, that time *per se* can never be perceived and that every continuum is at best a mathematical, artificial construct which deviates from the reality perceived. Therefore, I look for a more basic approach to introduce a concept of order into theoretical physics.

No information transfer, no language and therefore no physical theory can exist without a certain order. Zero order is chaos. This precondition for every physics is often neglected. It is seldom precisely defined what order means, where it comes from and how it is represented. The minimal condition for every order is that there is a duality of things that can be differentiated although they are characterized the same way, e. g. before and after, left and right, clockwise and counter-clockwise, positive and negative, yes and no.

The metaphysical origin of order in physics is the direction of the information flow from object to subject. This is best expressed by an arrow („if …, then …") as in category theory. The information flow is ordered in the direction from the object to the subject, otherwise it would become chaos. In order to define an order, one needs a duality of two distinguishable orientations. Every type of physical order is based on such a duality, be it yes/no-answers, left-handed/right-handed spin, past/future or positive/negative charge.

The duality of yes/no answers leads to an information theory which defines information as answers to potential questions which can be reduced to a countable number of so-called binary choices or bits, i. e. to alternatives which can be decided by a simple yes/no answer. The order or entropy of a system can be calculated from the probability of any answers to single questions (Weizsäcker 1986, pp. 163–173). The other types of duality, spin, time and charge are mathematically represented by the signature plus or minus or by the symbols < and >.

The duality of two distinguishable orientations is the core of the phenomenon of chirality. Chirality (Greek *cheir* = hand) means handedness. The right hand is the mirror image of the left. Although both hands are isometric, they cannot be brought to coincidence with one another. So, they are different from each other. However, a third hand, which is likewise isometric to the right and to the left hand, which nevertheless cannot be brought to coincide with either of them, does not exist. For each hand, there is one, and only one, counterpart with opposite handedness [Kant (1768/1960)]. An object is called chiral if it has a mirror image which is not identical to the object. Lord Kelvin (1893/1904) was the first to define chirality: „I call any geometric figure, or group of points, chiral, and say that it has chirality, if its image in a plain mirror, ideally realized, cannot be brought to coincidence with itself."

Such a duality of mirror images can be defined for mathematical or physical spaces with as many dimensions as desired. Every chiral object has an orientation. Its mirror image has the opposite orientation. Orientation is not an absolute characteristic; it only can be defined relative to the orientation of another, likewise chiral object or subject.

Both types of order, bits as well as mirror images, are based on chirality and they can be combined by means of the bits formalism. The imaginary qubit is a unit of quantum information – the quantum analogue of the classical bit. A qubit is a two-state quantum-mechanical system described by imaginary numbers. A metaphysical prerequisite of qubits is chirality because the orientation of imaginary numbers can only be defined by a clockwise or counter-clockwise chiral orientation. Therefore, chirality is more basic than information and the latter should be derived from the former. The mathematical advantage of qubit theory is that quantum theory is easily derivable from it [Weizsäcker (1986) pp. 300–412; Wehrli (2017)]. But since the quantum of qubit theory is not introduced as a metaphysical, but simply as a physical axiom, the qubit is no metaphysical help for understanding quantum theory. It is beyond the scope of this book to discuss qubit theory.

In set theory order as the duality of orientation is best expressed by the axiom of inequality of relations $(a,b) \neq (b,a)$ [Russell (1903/1956) pp. 199–217]. A real proposition is to be symbolized by *aRb*, whereby *a* and *b* are the terms and *R* is the relation between the terms. *aRb* is then always different from *bRa*. That is, it is characteristic of a relation of two terms that it, as it were, goes *from* one *to* another. That is the source of order. When reading the axiom, the physicist as an observer must know that there is a basic difference between left and right or between before and after respectively. For that the physicist himself has to be chiral – even if he is a transcendental subject.

Mathematically, chirality plays a fundamental role in category theory [Mac Lane (1969/1998); McLarty (1992); Dümont (1999)]. Here chirality is represented by arrows ($\Rightarrow$) which are meant to symbolize a change, a morphism: each morphism has an origin and a goal. Mathematical morphism is therefore a good model for representing physical perceptions as a causal information flow from an object to a subject. A mathematical structure changes, whilst something else remains unchanged. How the change is interpreted mathematically or physically, as a time, location or structure of an object or as a relation, is left open by category theory. In category theory, each morphism can also apply in the reverse direction, even if it is not an isomorphism [Mac Lane (1969/1998) pp. 31 f.].

4.2 The chirality in mathematics

The concept of chirality has to date barely featured in philosophy although it is a very simple principle of arrangement, without which perception is not possible. The result of my philosophical approach will be that the concepts of space and time will receive a new, less absolute meaning and the discrepancy between perception and the mathematical description of this perception will diminish. Philosophically thinking physicists such as Einstein (1949/1983) pp. 233–249, and Weizsäcker (1986), pp. 379–412, have sought and hoped for such a theory.

The following thoughts on the subject of chirality go back to 73 discussions which from March 1973 until July 1975 were conducted by a group led by André Dreiding with Alex Häussler, Martin Huber, Dimitri Pazis, Karl Wirth and myself, initially in Dreiding's office in Zurich and later in his house in Herrliberg.[42] The meetings always took place on a Monday morning. Norma Dreiding made the coffee and I donated the *Gipfeli* and kept the minutes. Accordingly, the more or less symmetrical coffee pot was a prominent object of our attention, on which we critically and thoroughly exercised our intersubjective observations. However, there were also gentlemen, who for a bit of joviality, were quite happy to substitute the object coffee pot with the image of an attractive lady, which did no harm at all to the seriousness of the discussions. The discussions were recorded on cassettes. Since then, the question for the foundation of the laws of nature has never lost its fascination for me.

Every chiral object as defined by Lord Kelvin (1904) has an orientation. Its mirror image has the opposing orientation. Orientation (in n-dimensional space) is mathematically more than the mere change of the sign. Rather, it is a change in the direction of rotation. For a mathematical description of orientation, see Russell (1903/1956) pp. 417 f. The orientation is not an absolute characteristic, i.e., it can only be defined on the basis of, or relative to, the orientation of another, likewise chiral object. So, one can agree by telephone with a person in another galaxy that he and we have a right and a left hand, which are both chiral but it is impossible to explain to him, which is his right and which is his left hand.

[42] The discussions were part of a government funded research project and are mentioned in the corresponding reports.

This applies at least as long as the person in the other galaxy is unable to conduct physical experiments with weak interactions, which would violate the parity of space.

Chirality is an attribute of symmetry. A figure is called symmetrical when there exists a non-identical congruent isomorphism of itself [Nakahara (2003) p. 469]. If it involves a mirror reflection, point reflection, or a rotation, then one speaks of line (or bilateral)-, point- or rotational symmetry. A square reflected along the diagonal, along a central perpendicular bisector, from the central point, or at a rotation of 90°, 180° or 270° will be an image of itself. A rotation of 360° in geometry always leads to an identical congruent image. In physics, that is no longer the case, however. A fermion with spin $\frac{1}{2}$ must be rotated 720° in order to turn back into itself [Sternberg (1997)], p. 23. The higher the number of symmetrical operations that are possible with an object, the greater its symmetry is. The three-dimensional object with the greatest symmetry is the sphere. It permits an infinite number of congruent images. An object without any non-identical congruent image is chiral.

The reason why I prefer the term chiral to the term asymmetrical is that the concept of symmetry in physics carries baggage and no longer has all that much to do with symmetry in colloquial terms. Physicists always speak of symmetry when something changes whilst something else remains unchanged. A train in motion changes its location, but the train itself remains unchanged, thus the process, in the physicists' vernacular, is symmetrical. Symmetrical in this process is not the train however, but the space and the time in which the train moves.

Chirality, as I choose to understand the term, is possible in spaces with any number of dimensions. In one-dimensional space, a line with a red point at one end and a yellow point at the other is chiral, since it cannot be an image of itself by reflection. Colour is naturally not a geometrical object. It serves here only to mark out the endpoints and thus make them distinguishable. In Chapter 7, I will outline a procedure whereby the points can be differentiated from each other simply on the basis of geometrical characteristics, indeed without having to use in addition the geometrical term distance. Even if the one-dimensional space is supposed to be closed, e. g., as a circle, it is impossible to reflect and subsequently move the chiral lines within the one-dimensional space in such a way that it can be brought to coincidence with itself. This would only be possible if one could rotate the reflected line on a plane, i.e., in two-dimensional space. However, we wanted to confine ourselves to one-dimensional space (Figure 1). Time is also a one-dimensional chiral space. In a reflection of time, the past becomes the future. Since the past and the future have fundamentally different characteristics, similar to the red and yellow endpoints on the line, time must be chiral.

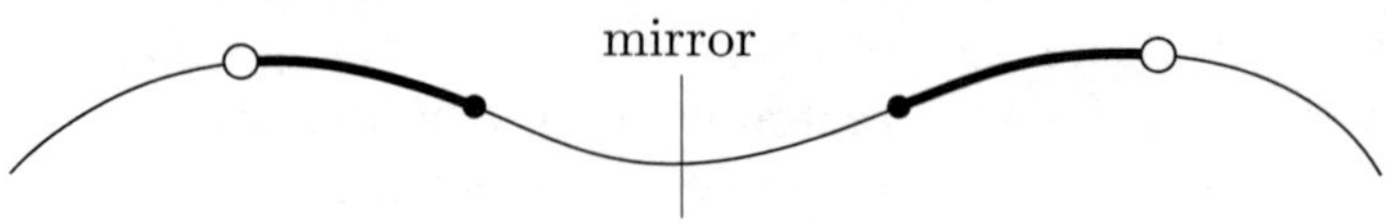

Figure 1. Chirality in one-dimensional space: *One-dimensional space with a* chiral line *which cannot be brought to coincidence with its mirror image.*

In two-dimensional space, the line with the two differently-coloured endpoints is no longer chiral, since its mirror image can always be moved about on the plane, so that it can be brought to coincidence with the original: image and mirror image are called to be „properly“ congruent. The simplest chiral planar object is the scalene triangle. It is not properly congruent with its mirror image (Figure 2).

If it were permitted however, to turn the reflected triangle around in three-dimensional space, then it could be brought to coincidence with the original; but we have ruled out such rotations. Things look different when the scalene triangle is reflected not on a line, but on a point. In this case, the reflected triangle will be congruent with the original (Figure 3).

Moreover, in two-dimensional space there is the case of the Möbius strip, a continuous closed band, formed by twisting one end of a rectangular strip through 180° and joining it to the other end. Here a normally reflected scalene triangle can also be aligned with the original by pushing it around the Möbius strip, since the front and back, inside and outside cannot be distinguished on a Möbius strip (Figure 4).

The Möbius strip itself is a chiral two-dimensional space with a mirror image counterpart [Nakahara (2003) pp. 11 and 204 f.].

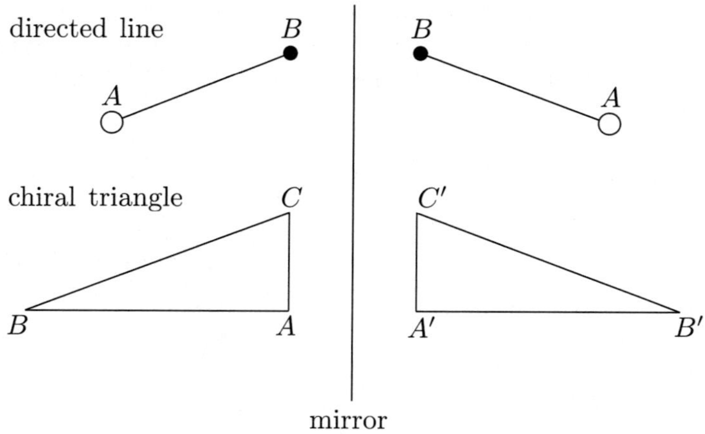

***Figure 2. Chirality in two-dimensional space**: In two-dimensional space, the directed line can be moved and rotated until it can be brought to coincidence with its mirror image. This is impossible with a scalene triangle; it is chiral.*

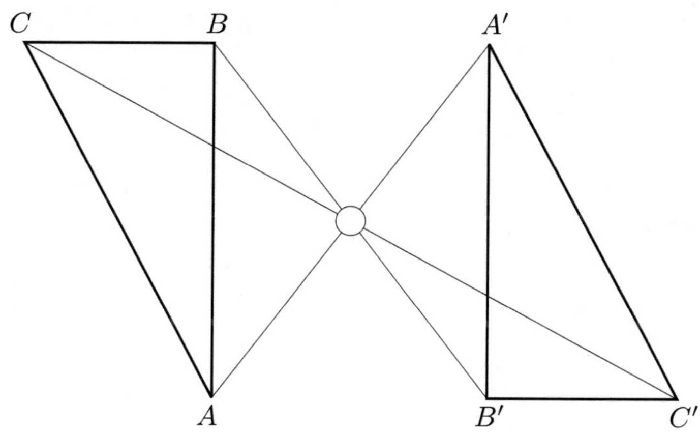

***Figure 3. Point reflection:** When a* chiral triangle *is reflected on a point, the mirror image is properly congruent with the original.*

In three-dimensional space we have already discussed the example of the hand, which is chiral. The scalene triangle is in three-dimensional space, as we have just seen, not chiral. Chiral by contrast would be a tetrahedron with all edges of unequal length or with four differently coloured corners: When such a tetrahedron is mirrored on a plane, then that mirror image is not properly congruent with the original (Figure 5) [Nakahara (2003) pp. 98 ff.].

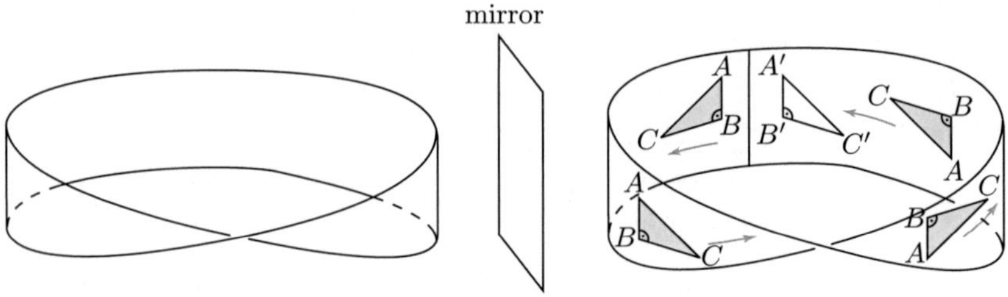

Figure 4. Möbius strip: *The* Möbius strip *is a two-dimensional chiral space which cannot be brought to coincidence with its mirror image. The triangle ABC is properly congruent with its mirror image A'B'C' in the Möbius space, which can be demonstrated by pushing the triangle around the Möbius strip.*

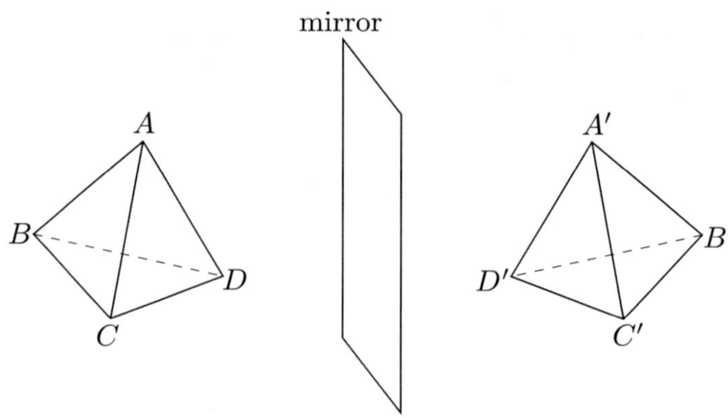

Figure 5. Chiral tetrahedron: *A* chiral tetrahedron *cannot be aligned with its mirror image in three-dimensional space.*

In three-dimensional space, helixes are also chiral (Figure 6), as well as all elementary particles with non-zero spin. The reason we have the impression that our space is three-dimensional is that we can distinguish between our right and left hand. We can never bring both hands to coincidence with each other by any number of rotations. For that we would require a space of at least four dimensions. This also applies to tiny elementary particles, for example, the neutrino. A neutrino can be distinguished from its mirror image, the antineutrino, since neutrinos and antineutrinos have completely different physical properties. For details (distinction between Majorana- and Dirac-Neutrinos) see Section 8.2.

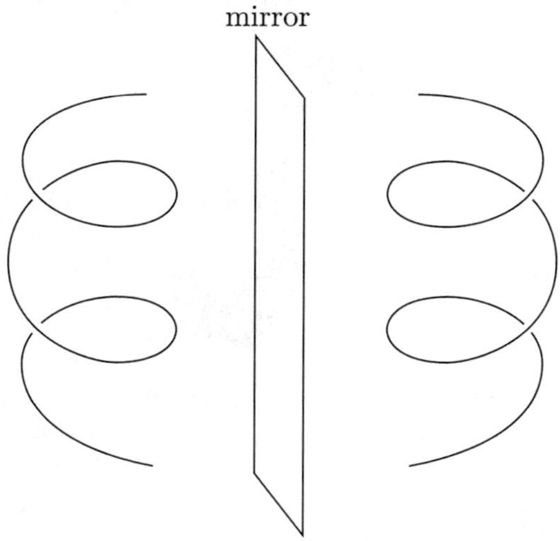

Figure 6. Helix: *The left-helix and its mirror image, the right-helix, are chiral and not properly congruent.*

If space had fewer than three dimensions, then there could be no three-dimensional hand in this space. If the space had more than three dimensions however, then we could bring the right and left hand to coincidence in it. Therefore, physical space must perforce be three-dimensional.

Another chiral two-dimensional space which corresponds to the Möbius strip is the so-called Klein bottle. Since it is not possible to give a four-dimensional representation of it, in the three-dimensional depiction it inverts and appears to pass through itself and cannot be turned into its mirror image (Figure 7). If one is able to imagine this in four-dimensional space, however, then it could be turned into its mirror image. Also, in the case of the Klein bottle, the outside and inside cannot be distinguished. For a topological description of the Klein bottle see [Nakahara (2003), p. 116]. The capacity for orientation is a topological invariant of space. Chirality is therefore not an absolute, intrinsic attribute of an object; rather it depends on the number of dimensions and on the topology of the space, in which we view the object.

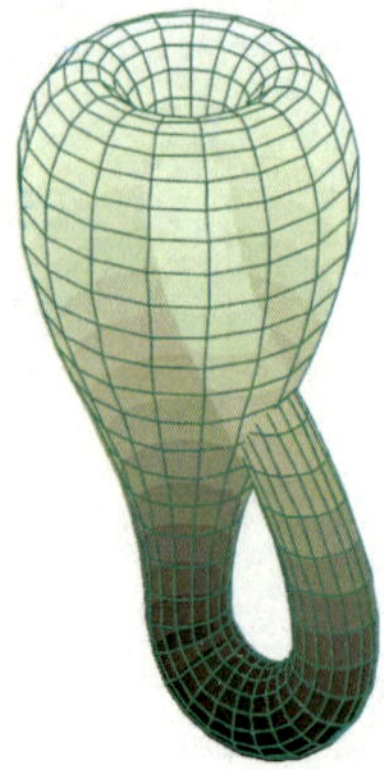

Figure 7. Klein bottle: *The* Klein bottle *is a chiral, two-dimensional space. Inside cannot be distinguished from outside.*

Generalized to spaces with any given number of dimensions one can say: If an (n – 1)-dimensional, orientable space is mirrored in an (n – 2)-dimensional space, then congruence between the object and its mirror image can only be produced by rotation in the n-dimensional space [Möbius (1827)].

An important indication is that chirality is not reliant on a space continuum. If, for example, the corners of a chiral object are marked with different colours, the object remains chiral, even though the distances between the corners may change continuously. The only condition for chirality is that such corners exist, that they can be distinguished from each other and that no corner coincides with the space which is stretched by all the other corners. Concerning the distinction between different points, a suggestion is made in Chapter 7 for which neither a marking by colours, nor a numbering of the points, nor a term for distance are needed. A good mathematical description of the concept of chirality has been made by Nakahara (2003), pp. 98 ff., 204, 469.

4.3 The chirality of Be-ing and of the existent

Chirality plays a fundamental role in nature, as well as in metaphysics, since without chirality, there is no conveyance of information, and thereby no perception, no memory and no thought. A graphic vivid and lively description of the interesting implications of chirality in nature is given by Martin Gardner (1964), and what is more, without even using the word chiral a single time.

Even the simple question why a mirror actually only reverses the right and left sides, but not up and down, confuses most people. The answer is that the mirror in reality reverses neither left and right, nor up and down, but rather back and front. Humans have only one plane of symmetry, which passes vertically through the centre of the body and divides the body into two more or less equal mirror-image halves. There is no similarity between its front and back, nor between the upper and lower halves of the body. For this reason and because gravity pulls all things downwards equally, we construct thousands of objects which all have a two-sided (bilateral) symmetry: tables, chairs, rooms, houses, cars, aeroplanes and so on. We speak therefore of a reversal of right and left when we look into a mirror, because that is the most convenient mode of expression for distinguishing a bilaterally symmetrical figure from its counterpart. It is only because we imagine standing behind the reflecting glass ourselves and looking in the reverse direction, that we speak of a left-right reversal. A mirror, however, only effectively reverses the axis which runs perpendicular to its surface. This human method of observation also influences their aesthetic taste. Whereas strictly symmetrical pictures are mostly seen as rather boring, partial deviation from symmetry often generates a stimulating tension, as for example in Leonardo da Vinci's Last Supper. In almost all such pictures, the axis of symmetry runs vertically, like that of humans themselves. If right and left in a picture are reversed, the aesthetic value of the picture is also changed. This is connected to the fact that we don't simply see a picture, but that we unconsciously „read" it from left to right. Those people who are used to reading from right to left however, usually prefer the mirror images to the originals.

In nature, one finds chirality from the most enormous galaxies with their magnetic field through to living and dead matter and down to the smallest elementary particle, to the neutrino with its spin [Gardner

(1964) p. 45]. The image and mirror image of space, time, spin and electrical charge have inherently different physical properties, an example being the weak interaction [Wolschin (1999)]. However, the *CPT* theorem (*CPT* symmetry) nevertheless shows that a connection between the orientation in space, time and the sign of the electrical charge must exist. Only if the orientations of space, time and electrical charge are reversed together the physical laws are preserved. This corresponds to the situation in category theory where the binary counterpart to each mathematical expression, to each theorem is obtained by turning around all arrows in that theorem [Mac Lane (1969/1998) pp. 31 f.].

There are few other phenomena with such a universal general validity as chirality, probably none at all. Not even time with its direction from the past into the future is so generally valid. The photons for example, so important for physicists, are in a certain sense timeless, but chiral. This is the case also for the instantaneous non-local interaction between entangled particles.

On the surface of every astronomical body there develop all kinds of interesting asymmetries, which are left-handed in one hemisphere and right-handed in the other. When, for example, a pilot in the northern hemisphere flies his aircraft directly towards the North Pole, then he must compensate for a pronounced tendency of the aircraft to stray to the right. Conversely, if he flies southwards in the southern hemisphere, then the deviation is to the left. This deviation is an example of the Coriolis force. The effect comes about because a body on the earth's surface, owing to the earth's rotation, has varying velocities in space, according to the position it has on the earth. Due to inertia, every body strives to maintain the original velocity. The closer the aircraft comes to the North Pole, the less its movement eastwards becomes as result of the earth's rotation. So, an aircraft, when it flies towards the pole, drifts towards east, or right, as the case may be.

When the bathwater is drained, a vortex is formed at the plughole. The directional rotation of the vortex conforms on the one hand to the circulation of the water when the bath was filled, which remains the same for many hours, even when the water has long since settled, and conforms on the other hand to the earth's rotation and the Coriolis effect. The bathwater is chiral, even though the individual water molecules are symmetrical. Depending on the position and direction of the tap and depending on the tub's location on the earth, the water drains with a left or right orientation [Gardner (1964) pp. 58 ff.]. The chirality of a hurricane has an analogous explanation.

It is similar with many crystals which in the crystallisation process are formed from essentially symmetrical ions or molecules. The most well-known are the quartz crystals, composed of silicon dioxide. These crystals rotate, either clockwise or counter-clockwise, the plane of polarisation of the polarised beams of light which penetrate the quartz. So, all quartz crystals are chiral, but their orientation, similar to bathwater, is not always the same.

A great many, above all organic compounds, are chiral and therefore also rotate the plane of polarisation of the light when they can move freely in a solution. To these asymmetrical molecules belong all sugars, the amino acids (with the exception of glycine) and accordingly, all macromolecules formed from them, i. e., polyglycosides, proteins and nucleic acids, all important building blocks of life. The asymmetries of these compounds are based, as a rule, on the fact that they have one or several carbon atoms with four different ligands, which form the four different corners of a tetrahedron. The phenomenon was discovered in 1848 by Louis Pasteur as he experimented with the chiral natural tartaric acid and with achiral artificial tartaric acid. The latter separated when crystallising into dextrorotatory and levorotatory crystals. The tetrahedral structure of tartaric acid was first postulated in 1874, independently from one another by Jacobus van't Hoff and Joseph Le Bel (Figure 8).

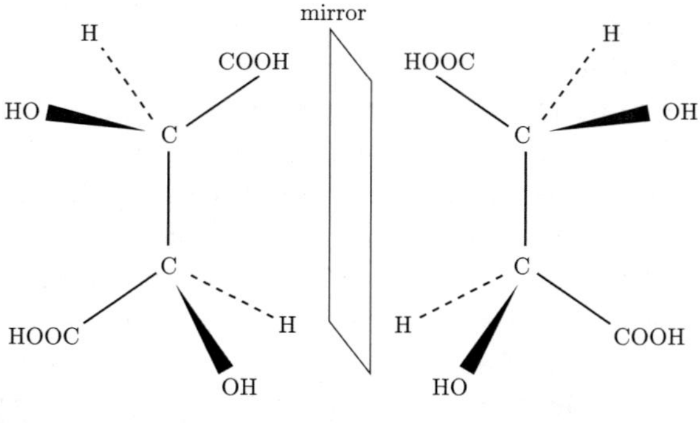

Figure 8. Tartaric acid: *(+)-tartaric acid and (–)-tartaric acid are different from their mirror image and therefore chiral.*

4.4 The chirality of life

Can there theoretically be life without chiral macromolecules? One speaks of life when cumulatively, four processes are possible: replication (copying of information), mutation (alteration of information), metabolism (exchange of information) and death (irreversible destruction of information) [Eigen (1987)]. If these requirements are fulfilled, evolution is the automatic outcome [Dawkins (1976)]. Replication is necessary for the reproduction of competitive creatures, mutation for the improvement of competitive capacity, metabolism supplies the energy which is necessary to sustain the order of the organism during the simultaneous increase in the entropy of the total system [Weizsäcker (1986) pp. 163–189], and death destroys the less competitive creatures and creates space for the new generation.

All of this already exists in inanimate nature, yet only macromolecules can fulfil all conditions at the same time. It may well be possible that a salt crystal which is thrown into a saturated solution triggers the formation of many further salt crystals, yet this is not a case of replication, since the new crystals assume different sizes and shapes. By contrast, the replication of a macromolecule, provided it proceeds error-free, creates an exact double of the original. Mutation and death are basically nothing other than an irreversible change in the order. This is possible with macromolecules without changing their primary structure, i. e., without the covalent bonds in the molecule having to be broken. It is sufficient to destroy the spatial configuration (the chemists call this the „conformation“) of the molecule strands, as it occurs for example with the gelatinization of starch during cooking or by the coagulation of egg-white when frying eggs. In mutations, however, covalent bonds are mostly changed as well. (Exceptions are the prions. With these, a change in the secondary structure without the breaking of covalent bonds is possible and can, for example, lead to mad cow disease.) Macromolecules, under the right conditions, are also able to absorb substances from outside and with the help of catalysts, make specific chemical changes; that is, they are able to metabolise.

One may assume that a single macromolecule, for example, something like a short ribonucleic polymer originating in a „primeval soup“ already had a small amount of all four characteristics necessary for life and reproduced itself in a suitable environment, whereby it possibly catalytically accelerated its own duplication [Shapiro (2007)]. Such molecules have already been manufactured synthetically. Later, perhaps various such molecules consolidated in a kind of symbiosis into bimolecular proto-viruses, which in the course of evolution, developed into viruses. Different viruses and other molecules could in turn have combined in symbiosis to form a very simple proto-cell, which with yet more cells developed into a more advanced cell.

The emergence of multicellular organisms required a gigantic catastrophe, such as the impact of meteorites, volcanic eruptions and enormous temperature fluctuations [Ward (2006)] and finally a relatively stable climate, which first became possible with the longer coastlines and smaller continents which resulted from tectonic plate movement. So, life didn't emerge on day X, but rather the four characteristics vital for it consolidated gradually by degrees and transformed an inanimate nature through evolution into a living one. The process took a very long time and certainly did not proceed continuously, but rather in irregular leaps of development over short periods: Every time many species suddenly died out as result of catastrophes such as meteorite strikes, volcanic eruptions or temperature fluctuations and new, free ecological niches were created, these were quickly filled by the evolution of the surviving species. Evolutionary leaps of this kind, can give an impression of creation. Once the niches were filled, it was very difficult for the newly-emerging species to compete with the existing ones which had already adapted.

However, that is not to say that macromolecules necessary for life also have to be chiral. There appears to be no theoretical reason for why this should be so. One could quite easily imagine achiral macromolecules, which replicate themselves, mutate, metabolise and can die. The existence of such symmetrical molecules, however, is just as unlikely as symmetrical balls of wool. Balls of wool are never exactly symmetrical. Since macromolecules are not symmetrical anyway, it is advantageous to life when they nevertheless have a clear structure. They should at least always be similarly asymmetrical, so that they can reliably play their part in replication and metabolism, since in these processes the spatial arrangement of the molecules is important, otherwise they would fit together as badly as a key pushed the wrong way round into a lock. Therefore, the orientation of the chiral molecules necessary for life is in most cases always the same. Many chiral agents, such as for example vitamin C, adrenalin or nicotine don't function if they have the wrong orientation.

Life on earth would probably function just as well if all organisms and molecules were suddenly turned into their mirror images. Why the orientation is set up the way it is, we don't know. In the event that life, as is currently presumed, emerged spontaneously in an achiral primeval soup under specifically suitable conditions on earth or on another planet in space, the orientation of asymmetry found today can be explained by the following theories: It is possible that all life goes back to a single chiral self-replicating molecule, the „Adam molecule" and accordingly took on its orientation for the whole living world. Another theory says that life initially began in only one hemisphere, where the Coriolis force somehow ensured the necessary rotation. Or elliptically polarised light, which is produced when light is reflected from a surface, could have combined with the earth's magnetic field in order to produce the rotation.

Perhaps molecules of both orientations emerged in the primeval soup. Then, each molecule could only subsist on molecules of its own handedness, until the mutation of a left-handed molecule imparted the capacity to consume both left and right-handed compounds and perhaps even the living right-handed competitor. Over the course of reproduction its progeny had a massive competitive advantage. Finally, it would be conceivable that weak interactions, which violate space parity, play an as yet unknown part in living matter whereby the left-orientation would have certain advantages. In this case, organisms in a possible galaxy made of antimatter would have to be right-orientated. For example, the isotope carbon 14 is created during neutron bombardment of atmospheric nitrogen by cosmic radiation, which is a natural source of beta rays and can be incorporated in organic molecules like normal carbon. These molecules are then constantly exposed to a bombardment of electrons with predominantly left-orientation. Laboratory tests, during which levorotatory and dextrorotatory leucine was irradiated with left and right polarised electrons, showed that levorotatory electrons mainly destroyed the dextrorotatory leucine and vice versa. This could explain why in nature, practically only left-oriented amino acids are found [CERN (1979)]. Most plausible at present is the theory that the chirality of nature is the result of a strong, extraterrestrial magnetic field, which causes the dextrorotatory amino acids to be disintegrated by light more rapidly than the levorotatory ones [Rikken (2000)].

4.5 The chirality of thought

Would a symmetrical creature be able to think and communicate? Apart from the fact that the existence of such a creature, as we have seen, is extremely unlikely, it's capacity would be at least so severely limited that compared to chiral organisms, it would be at a huge competitive disadvantage. A symmetrical animal is incapable of distinguishing between left and right. It could not consciously and purposefully move in a specific direction. Locomotion would probably only be possible by snail-like crawling or rolling. Such a symmetrical animal, which resembled a billiard ball and moved by rolling, would not have the option of deciding whether it wanted to roll past an obstacle on the left or the right, since it cannot determine left or right at all, neither in its body, nor in the environment.

Sooner or later it would probably trip over an obstacle and fall on its symmetrical nose. So, the symmetrical creature would be either plant-like or a microorganism. In both instances, intellectual abilities apparently bring no great competitive advantage. Even so, one can justifiably ask whether thought would be fundamentally possible for symmetrical creatures, or whether for that they have to be chiral. So contrary to Goethe's advice, I want to think about thinking, however not for the moment about transcendental thinking, but about the function the brain has in thinking. Even so, I undertake this thinking in the expectation that human thought corresponds to a material structure which satisfies the laws of physics.

During thinking, information stored in the memory is processed such that additional new information is created which is available to us as knowledge and which is in turn stored as memory. Pieces of information can be defined as answers to potential questions, which can be reduced to a countable amount of so-called final alternatives. Every alternative has to meet the following three conditions:

- The alternative is determinable, i. e., a question can be posed to which one of the alternative possible answers is correct and therewith becomes a fact.

- If an answer is correct, then all the other possible answers are false.
- If the alternative has been determined and all answers except one are false, then this is the correct one.

The final alternative is an alternative which can be answered by a simple yes/no answer. Weizsäcker (1986) pp. 379 ff., also called the final alternative „*Ur*“ and postulated that the world is constructed entirely of „*Ur* 's“.

Now it is important to distinguish between the information content of a structure and our knowledge about this information. The more complicated a structure is, the more questions I have to ask until I have gathered all the information about it, so the greater its information content is. The more ordered a structure is, the less its (potential) information content is. With a few questions I can then learn a high percentage of the information contained in the simple structure. Chaos has the highest information content, where I have to know all the parameters of every single point, i. e., all the relations to all the other points, in order to know everything about it. Despite the high information content – or precisely because of it – I know practically nothing about chaos, whereas my knowledge about simple, ordered structures is much better as a rule. So, information in this sense is merely potential knowledge, not actual knowledge itself.

This now raises the following question about the thought processes in the brain: How is the information stored? How does it become accessible to us? How can we process it so that our knowledge is increased?

Pieces of information, as I have defined it above, should be stored as answers to the final alternatives as simple yes/no answers. This is possible most simply through chiral units, which are accordingly orientated left or right. Nothing is yet said thereby about the number of dimensions which the space must have for this left/right orientation. The essential thing here is solely the duality of the possible answers: yes or no, left or right. Accordingly, in the brain, the chiral topology of certain molecules or even only parts of molecules would have to be such, that an external stimulus, namely the question, produces a specific reaction, a signal as the „answer“. In so doing, the reaction must not change the physical configuration of the molecule, so that the information is not lost.

The spatial configuration of these information molecules thus has exactly two possibilities, which can be distinguished from each other in that the spatial orientation of certain parts of molecules is different. In order to change a yes-molecule into a no-molecule, covalent bonds don't necessarily have to be broken and newly formed. The change can also occur via simple rotations within the molecule, by which the spatial configuration of single atoms with varying degrees of separation in the molecule changes, and with it, the chiral unit postulated above changes as well (Figure 9).

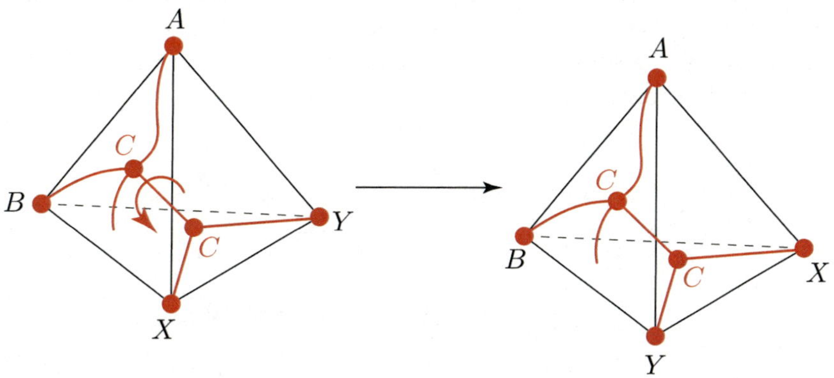

Figure 9. yes/no molecule: *By rotating the red molecule about the C-C axis, the tetrahedron of the ligands ABXY is transformed into its mirror image.*

In order to be accessible to the questioner, the information molecule must be connected with the environment by one or several channels, through which the stimulus, or the question, can be sent to the molecule. The brain cell which contains the information molecule must be able to send its answer, i. e., yes or no. This in turn transpires most simply in that a uniform signal is either emitted which means „yes", or not emitted, which means „no". In order to increase knowledge, the organ as a whole must also be capable of learning, i. e., the information must be subject to change and improvement. That means that certain signals should change a yes-molecule into a no-molecule and vice versa. Since thinking takes place without external stimuli, otherwise it would be rather a case of perception than one of thinking, these learning signals must be emitted from the brain itself. Most simply, one takes the previously postulated yes-and-no signals, which must be received somewhere: „Yes" means that the orientation of the receiving molecule must be changed and „no" means that the orientation remains as it is.

Alan Turing (1948/1992) probably had such thoughts in mind when he pondered over how a computer capable of learning would have to be constructed. Since nobody at that time took his playful idea seriously, his paper (Figure 10.) wasn't published until 1968, fourteen years after his suicide. Today we know that essentially, our brain also functions in exactly this way: 10^{12} to 10^{14} neurons can stop or let through a signal transmitted along nerve fibres by other neurons, according to how each of their 1'000 to 10'000 synapses are programmed by signals, which for their part are transmitted or not transmitted by a third group of neurons.

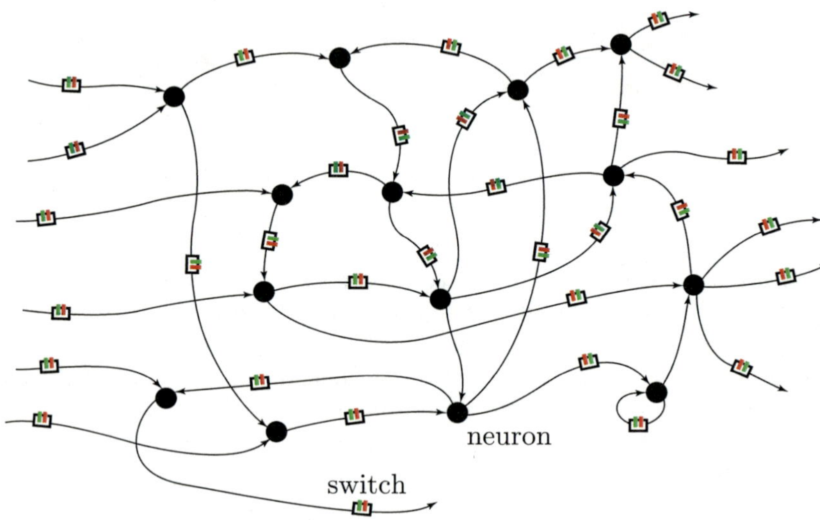

Figure 10. Model of a learning machine: *Alan Turing's unorganized learning machine consists of a network of artificial neurons. Every connection passes through a switch which either allows the signal to pass (green filament) or stops it (red filament). With the help of the switches, the system can be trained. Every neuron has two inputs: When both signals are 1, then the output is zero; in all other cases the output is 1. Turing demonstrated that the switch could also be constructed of neurons.*

The neurons are formed in the emerging brain and are initially unprogrammed. Learning takes place little by little via external signals, perception, and internally, via thought. The signals programme the neurons and create order in the brain. The signals themselves are also chiral. Either a stream flows through the nerve fibres, or nothing flows. So, the yes is not simply the opposite of the no, but something fundamentally different.

In order for thought to function, the postulated chiral units, information molecules or neurons must be networked with each other via connections or channels through which signals can circulate unhindered and uninfluenced. This is only possible if the different channels do not intersect. So, a space with at least three dimensions is necessary. In two-dimensional space, thought would be impossible. This is undoubtedly one of the reasons we have the impression that a space with three dimensions exists. However, this doesn't explain why the space apparently has no more than three dimensions. The networking must not change in the course of life, otherwise the information is lost or corrupted. However, no particular arrangement of the network is required.

Asymmetrical information, for example the picture of a levorotatory spiral, can by means of yes/no signals only be clearly modelled in a chiral memory. A symmetrical memory, if there was such a thing, may well be able to establish that the spiral is chiral, but it could not determine the orientation of levorotatory and dextrorotatory spirals. Useful information, as a rule, also includes information about when, in which sequence and in which context the information was received. For this it is not sufficient to store the correct sequence of information, which in principle would also be possible in a symmetrical memory; rather one needs to know which information arrived first and which came afterwards. Since the passage of time is itself chiral, it can only be correctly and clearly mapped and read in a chiral memory. Only here, not simply the coexistence is clear, but also the sequence. A mathematical alternative to the definition of the chronological orientation of information would be to mark each piece of information individually, for instance with numbers, so ordered that the information with the higher number is always the more recent. As is generally known, nature doesn't function that way, presumably because the fixing of chronological orientation by means of the memory's chirality is simpler.

The conditions listed, which must be met for thought to be possible at all, apply not only to the brain or to a computer, they probably also apply in principle to transcendental, non-empirical thought, to consciousness. The question whether a computer can, in principal, have a consciousness, is answered in the affirmative by Douglas Hofstadter (1996). His criterion for the existence of a consciousness is that the subject knows that it knows. That is programmable in a computer, at least in a primitive form.

Only when consciousness is chiral, can it distinguish left and right, before and after. Since every measurement, probably even every perception, is ultimately based on such left/right or before/after distinctions, chirality is not only a prerequisite for thought, but also for all perception, i.e., for bridging the gap between the empirical and the transcendental world. The nature of perception will be discussed in the next chapter.

4.6 The chirality of elementary particles: Spin and angular momentum

First of all, we ought to consider the role which chirality plays in elementary particles. The concept of elementary particles is unclear. The further experimental physics advances and is able to produce in giant accelerator machines ever higher energetic collisions between subatomic particles, the more elemental the elementary particles become, but we are never quite sure whether we have now really arrived at the smallest and most elementary building block of matter. There are also physicists who have come to the conclusion that it has become absolutely meaningless to speak of elementary particles, since all elementary particles somehow appear complex and primitive at the same time. They are all composed of themselves and blend into each other [Dürr (1968)]. Even so, I wish to adhere to the term elementary particle whilst acknowledging that it is a fuzzy concept. Elementary particles have internal and external properties. Internal, or intrinsic properties, are those which are perceived identically by all observers, irrespective of their state, whereas external properties are dependent on the state of the observer. Intrinsic properties are rest mass, electrical charge, spin, parity and a series of quantum numbers such as lepton number, strangeness or isotopic spin. External characteristics are location in space, time, momentum, velocity, energy and orientation of the spin. It is usually assumed that the intrinsic properties are independent of the external, and vice versa. However, connections exist between the intrinsic properties, as well as between the external properties, amongst themselves. Spin, parity and perhaps also the electrical charge have to do with chirality and therefore are a subject of this chapter.

All elementary particles have a spin, which is not zero. Particles with spin zero are always combined. There are, for example, mesons with spin zero. They consist of one quark with spin $+\frac{1}{2}$ and one antiquark with spin $-\frac{1}{2}$ or vice versa, so that the total spin is exactly zero. Distinctions are made between fermions with spin $\pm\frac{1}{2}$ $\pm1\frac{1}{2}$, $\pm2\frac{1}{2}$ etc. and bosons with spin 0, $\pm$ 1, ±2 etc. Fermions conform to the Pauli Exclusion Principle, which states that no two particles can exist in the same state at the same time in the same place. This principal was formulated long ago by Aristotle (1984). The Pauli exclusion principle is based on the conclusion that there are essentially different possibilities for statistically ordering a population of objects („spaces of states") which stand in relation to each other, i. e., which can be converted into each other by symmetry groups [Finkelstein (1996) pp. 208–230]. Bosons do not obey this rule. The spin is a quantum number which belongs to the particle, i. e., a purely mathematical quantity, whose actual physical significance is rather unclear. Mathematically, spin 1 means that a particle must be turned 360° once to return to the same state. A particle with spin $\frac{1}{2}$ must be turned 720°, i. e., twice and a particle with spin 2 (e. g., a graviton, if it actually exists) must only be turned 180° to reach the original state [Sternberg (1997) p. 23; Penrose (2004) pp. 198–208]. If the reader has difficulty imagining an object which one must turn on itself twice for it to come back to its starting position, he can take a glass of water in his right hand and then turn the glass inwards on itself without moving the legs, without letting go of the glass and without spilling a drop of water. He will find that with rather difficult contortions he must rotate the glass twice before it is back in its original position. In so doing, he will have to move the glass up and down in a spiral at the same time. Spin can be mathematically positive or negative, which is to say, spin is a dual property, a property which can come in two varieties that can only be distinguished by one being the opposite of the other. In three-dimensional space, that means that the spin is chiral, that it must have something like a three-dimensional rotational direction. In physics and chemistry books spin is often portrayed as a vertical arrow, ↑ or ↓, something like a three-dimensional rotational direction, whereby one is given to imagine that the arrow forms the axis for a clockwise or anticlockwise rotation. In this portrayal, spin appears to have an axis and a rotational direction.

In reality however, spin does not have an axis. If an elementary particle has an axis, then that is the consequence of its magnetic field with a north and south pole, or motion at light speed and not the consequence of spin. The neutrino, which has no electrical charge and which, in the event that it should have any mass, cannot move at the speed of light, doesn't have an axis. (In Chapter 8.7 and Figure 25a I make a proposal for a mathematical model for a chiral neutrino structure without any axis.) To describe spin in quantum theory an axis is laid in an arbitrary direction through the particle being described, which then turns on this axis. This yields an attribute of the particle which must be conserved and that can only occur in two opposing variations. This attribute is then called spin. Indeed, the theory of spin in quantum mechanics presupposes a theoretical axis. Since its direction is completely free however, for an external observer it is not an intrinsic feature of the particle. The intrinsic feature is solely the chirality in three-dimensional space. With the transition to the large systems of classical mechanics with rotations of a radius, the spin then becomes angular momentum, which must also be conserved and which has an axis perpendicular to the radius [Feynman (1966a) pp. 10–7 and 17–3].

In the 1920's, Uhlenbeck and Goudsmit (1926) pp. 264 f., discovered that the electron behaves like a gyroscope. It turns on an axis and what is more, with an angular momentum which comes to exactly half of the Planck's constant divided by π, that is, $\frac{h}{2\pi} = \hbar$. Thus, spin is no longer a pure number, rather it aquires the dimension of an angular momentum ar an action, namely mass · distance^2 · time^{-1} or energy · time. Also, in the case of particles such as photons, which have no mass, the dimension of mass appears as a characteristic of their spin. That doesn't sound very plausible and raises the question once more whether we have really understood the nature of the photon.

4.7 The symmetry of space, time and charge, and their violation

The second chiral property of the elementary particle is its parity. The relativity of left and right expresses itself physically in that every process which exists in nature must have a mirror image counterpart, and what is more, with the same probability. Mathematically this theorem entails a well-defined reflection behaviour of the quantum-mechanical wave amplitudes: The wave amplitudes of quantum physics are mathematical quantities without direct physical significance. One can, however, by means of suitable calculation specifications (so-called operators) compute physically measurable quantities from them, such as energy for example. The square (positive or negative) of the amplitude is the probability of finding with a possible measurement the particle in the location of this amplitude. If we count the radius vectors x from the system's centre of gravity, then the amplitude at location x which is the mirror image of location x, must agree either with the original amplitude or change its sign. According to whether the first or second case is present, one speaks of even or odd parity of amplitude. If all physical processes were on equal footing with their mirror image counterparts, i. e., the natural forces were mirror invariant, then the parity of a system during all possible processes would not change. The parity P would be, like energy, momentum and rotational direction, a strictly maintained quantum-physical quantity: equalling either +1 or −1. In the 1950s, a study of the K-meson found that the K^0-mesons can decay into two as well as into three pions:

$$K^0 \rightarrow \pi^0 + \pi^0 \qquad \text{or} \qquad K^0 \rightarrow \pi^0 + \pi^+ + \pi^-.$$

However, that can only be explained if the parity is either violated during this process, or if the two K^0 concerned are two different elementary particles, one with even and one with odd parity. Lee and Yang (1956) investigated all the experiments with decay modes resulting from weak interactions and found that in not a single case had parity conservation been proven. They accordingly proposed several experiments with which possible parity violations in weak interactions could be concretely demonstrated. In early 1957 Chien Shiung Wu (1957) found during the decay of polarised cobalt 60 that parity was indeed violated (Figure 11). The electrons in this β-decay are longitudinally polarised to a high degree, and further, their spin, more precisely their angular momentum, is the opposite: So, the two together define a left-handed rotation. There are no right-handed electrons in β-decay! Also remarkable is the behaviour of the associated antiparticles, the positrons: They proved to have a right-handed polarisation. So, in β-decay, the symmetry between particles and antiparticles is also destroyed.

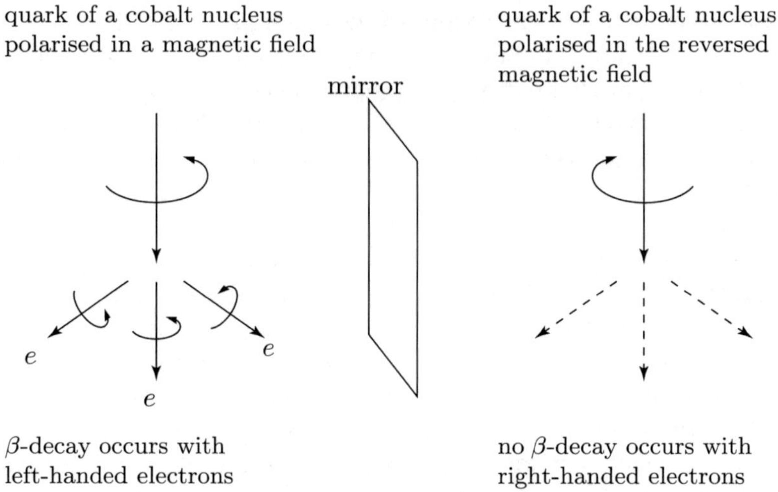

Figure 11. Decay of K^0-mesons

This means: The invariance under charge conversion (or „charge conjugation *C*") is violated (Figure 12). One also sees here how despite the violation of parity *P* and the charge conjugation *C*, a symmetry operation can nevertheless be found which is respected in β-decay: One needs only also to exchange right and left during the transition from particle to antiparticle, i.e., employ operations *C* and *P* together. Then the left-handed electrons are converted into right-handed positrons, and these do actually exist. This *CP*-symmetry significantly relativises the difference between left and right. To be able to make an objective decision about left and right, it's essential to know whether one is dealing with matter or antimatter. However, these also are relative terms – and so in a deeper sense, the relativity of left and right is re-established [Faissner (1968) pp. 141–164]. These new insights came as a shock for physicists, and Lee and Yang were awarded the Nobel Prize in recognition of their

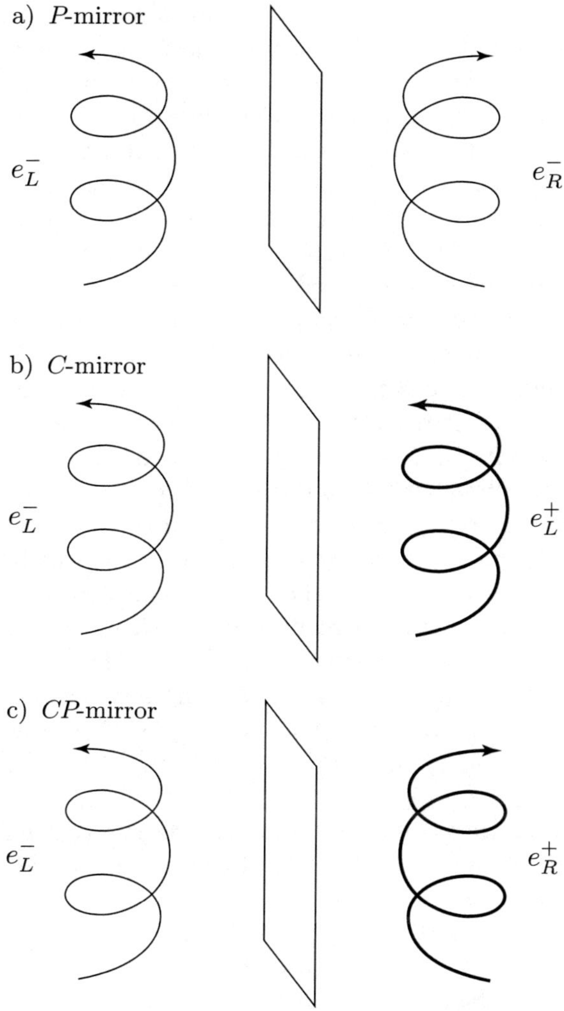

Figure 12. CP-symmetry: *A left-handed electron:*
a. is seen as a right-handed electron in the P-mirror (space mirror)
b. becomes a left-handed positron by a charge conjugation C
c. becomes a right-handed positron by simultaneous mirror reflection and conjugation (CP)

discovery. To postulate an objective distinction between left and right meant, as Pauli put it „to allow space an absolute rotational direction". Still, there would be quite another explanation if one were to allow not space, but the electrical charge an absolute rotational direction. This would have to occur in such a way however that didn't alter the spin of the charged particle. To my knowledge, nobody has seriously pursued this idea, but I will come back to it in Chapter 12.10. Spaces with a rotational direction would be, for example, the two-dimensional Möbius strip or Klein bottle, if they are viewed from the perspective of three-dimensional space. The space could, however, only have a rotational direction if it existed. Yet I have already indicated in Chapter 2 that it is highly questionable whether one can claim that space exists as an entity *per se*.

One must add that further studies of *K*-meson decays have shown that possibly even the *CP*-invariance is violated. That would mean that matter and its antimatter mirror image are different. In this way, it would be possible to determine objectively whether one was dealing with matter or antimatter, and one could perhaps also establish why the universe is composed primarily of matter and not antimatter. However, if one inverses not only the spatial orientation and the charge in these radioactive decays but the time axis *T* as well, then one would reverse the nuclear reaction by coalescing the pions into a kaon, thus, the symmetry would be conserved and we would at least still have a *CPT*-invariance [Faissner (1968) pp. 165–191; Schmutzer (1972); Mehlberg (1980) vol. II, pp. 174–188].

This invariance had theoretically already been postulated before 1956 and generally proven by Pauli (1955). Recent studies of the spontaneous transformation of uncharged kaons into uncharged antikaons and vice versa indeed showed that violation of the *CP*-invariance during this transformation is compensated by a simultaneous violation of the *T*-invariance, so that the *CPT*-invariance is conserved [Wolschin (1999)].

All these insights with respect to parity violation are based on experiments with so-called weak interactions. In Stanford, in 1978 however, scatter experiments on protons with electrons, whose rotational direction was either purely left-handed or purely right-handed, showed that the probability of an electromagnetic interaction between an electron and a proton is 0.01 % greater with levorotatory electrons than with dextrorotatory electrons. It seems therefore the perceived invariance of the parity in the electromagnetic interaction is only an apparent one. There are indeed dextro- and levorotatory electrons, but it is the latter which „prefer“ to react with matter. This effect had already been predicted independently several years previously and at about the same time by Steven Weinberg and Abdus Salam [Close (1978)].

The theoretically postulated and empirically confirmed *CPT*-invariance indicates that electrical charge, spatial symmetry and time direction must have a common basis. This common principle appears to be chirality: The three dimensions, electrical charge *C* (*C* harge), space *P* (*P* arity) and time *T* (*T* ime), could contain chiral objects or contribute to chiral processes, i. e., those which differ from their mirror image. There are three types of mirror: the familiar, common space mirror *P* reverses parity and changes the left into a right hand, the charge mirror *C* transposes positive and negative charges, thus converting particles to antiparticles and vice versa, and the time mirror *T* reverses the flow of time, thus exchanging the past with the future. The laws of classical mechanics are mirror invariant with respect to all three mirrors, i. e., all processes should be equally possible, with left as much as right, with matter as with antimatter and with forwards as with backwards. Classical mechanics is *P*-, *C*- and *T*-invariant. Experiments with *K*-mesons have shown that this is not always the case, but that during certain processes, all three invariants can be violated, yet that the simultaneous mirroring in the *P*-, in the *C*- and in the *T*-mirror in all known cases do not change the validity of physical laws (Figure 13, p. 108).

Difficult to understand is that the parity violations portrayed occur almost exclusively in weak interactions. If only these weak forces are sensitive enough to be measurably influenced by the parity violation of space, then the much weaker forces of gravity would have to react to it even more so. Gravitation is *P*-invariant, however.

The smallest and most simple elementary particle known is the neutrino. Apart from spin $\frac{1}{2}$, it has a lepton number and perhaps a very small rest mass, it appears to have no intrinsic properties. The neutrino violates parity, since it has a clearly defined handedness or helicity: the neutrino is always left-handed, the antineutrino is always right-handed. Right-handed neutrinos and left-handed antineutrinos have not been observed. (For details, especially the distinction between Majorana and Dirac neutrinos see Section 8.2). It is unclear whether the neutrino always travels at

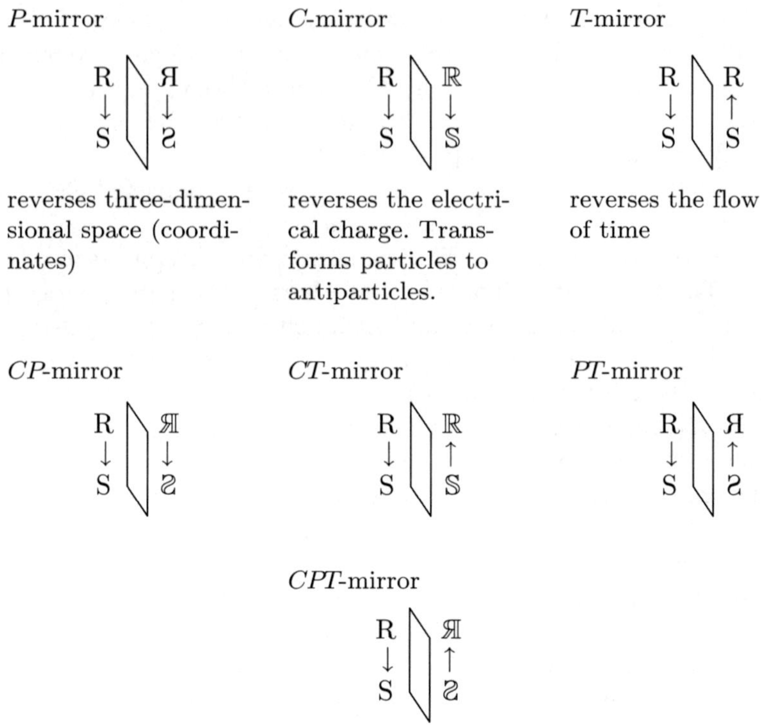

Figure 13. *CPT–**invariance:** Only during CPT-mirroring are all physical laws always conserved, including the weak interactions.*

light speed and in doing so, turns anticlockwise, or whether it has a mass and consequently has a chiral internal structure like a left hand. I suspect that the latter is the case, indeed on theoretical grounds which I will elucidate in Chapter 8. There are also increasing reports, according to which a finite neutrino mass of about $5 \cdot 10^{-7}$ electron masses, i. e., approximately 0.01 to 0.1 eV could be evidenced [Musser (1998)].

Parity violation in weak interactions applies not only to neutrinos, but also to electrons and quarks. The right-handed electrons and quarks, as well as the left-handed positrons and antiquarks, may exist and have exactly the same electromagnetic properties as their chiral counterparts, but they have no weak charge and cannot participate in weak interactions. The properties and the structure of a left-handed electron with spin $-\frac{1}{2}$ are consequently completely different from the properties of a right-handed electron with spin $+\frac{1}{2}$. The former has a weak charge and can be involved in a weak interaction, the latter cannot [Georgi (1989b) pp. 425–445].

4.8 Chirality is the universal duality of being

In summary, we can state that the phenomenon of chirality is found everywhere in the cosmos, from the smallest elementary particles to galaxies and in the simplest living matter as well as in thinking brains and applies universally. Even the transcendental realm, thinking and perception, has a chiral aspect. It is no great surprise, therefore, that for Kant (1768/1960), chirality constitutes the essential nature of space. For Kant, space is real, absolute, infinite, Euclidean and a condition for every perception. By „real“ however, Kant doesn't mean „belonging to the world and empirically observable“, but „really existing and thereby transcendent“. The numerous later critics of Kant have mostly misunderstood this point [Reidemeister (1957); Koch (2004)]. For Kant, chirality is a prerequisite for every sensory perception and every external sensation, even if he doesn't use the term chirality himself. There are few other phenomena with a universal validity of this kind; quite likely none at all. Not even time is so fundamentally and generally valid. Photons for example, which are so important for physicists, are timeless, but chiral. Thus, it is legitimate to place chirality, this duality of all being, right at the beginning of philosophy.

Chapter 5

The Dualism of Body and Soul

To be is to be perceived. „esse est percipi."
George Berkeley (1710/1948–1957)

It may be that the interface of mind and matter will turn out to be the most challenging legacy of the New Physics.
Paul Davies (1989) p. 6

5.1 The body/soul problem

„The title is already wrong! At the very least, the question mark is missing!" some specialists would exclaim at this point. Is the problem a problem at all? Fact is, philosophers have been pondering this core philosophical question for thousands of years and disagree more than ever. The oldest known pictorial representation of the body/soul problem is a drawing in the cave of Lascaux in the Dordogne, dating from about 15,000 BC. A dead hunter lies in front of a fatally wounded aurochs. Before it is a staff with a bird, which probably symbolizes the killer's free soul [Furger (1997) p. 27]. Wherein lies the problem?

Some philosophers regard body and soul as substances which are not reducible to each other. The soul is essentially characterised by thought, matter however by its extension [Descartes (1641/1960)]. Yet since the soul cannot be conceived of as localised in space, and matter not as thinking, it is impossible to explain the obvious interaction between body and soul, particularly not if one – as is usual – assumes the material sphere to be a causally closed system. This difficulty constitutes the body/soul problem. Approaches to solving the problem are called dualistic or monistic, according to the number of principles on which they are based. However, the question is actually based not on two, but on three principles, namely body, soul and the interaction between the two. All three are often not clearly defined or definable, which results in numerous misunderstandings. Before I explain my view of things, I would like to give a rather rough overview of the possible solution variants.

We differentiate four dualistic and four monistic solutions, which partly come in different sub-variants [Sayre (1980)].

The dualistic solutions

In texts of the 2nd millennium BC, humans are already seen as consisting of two parts, a kind of dualism of material and spirit, expressed as body and soul [Hasenfratz (1986)]. Also, Plato, as a dualist, occupied himself greatly with the structure of the soul [Böhme (2000) pp. 311–314].

1. The oldest dualistic view is occasionalism, according to which God is the cause of both physical and mental processes. An occasionalist is also of the view that the interaction between body and soul is indirect, with God as the mediative causal activity.
2. Spinoza's and Leibniz's solution is parallelism: God created the world in such a way that a perfect harmony exists between mental and physical events like between two watches, one as a measure for the mental, the other as a measure for physical activity, which were both set at the same time on the day of creation [Spinoza (1677/1990)]. It is also possible, in principle, to describe all psychological phenomena physically with a so-called „psychophysical parallelism" at least as long as one remains within the same universe and does not consider (quantum)gravitation [Mehlberg (1980) vol I, pp. 259–261)].
3. Popular among natural scientists is epiphenomenalism: here physical events are viewed as the cause of the mental, whilst a causation in the opposite direction is excluded. Mental events can thus be explained with scientific principles.
4. There are only a few modern dualists, for example Karl Popper and John Eccles (1977), who regard a subtle, quantum mechanical action of the soul on matter as possible. One could even call Popper (1979) a „Trialist". For him there are three worlds: First of all, the world of physical objects; secondly the world of states of consciousness; thirdly the world of objective thought. The second world is the link between the first world and the third world.

The monistic solutions

The monistic approaches reject every characterisation of body and soul which portrays them as fundamentally different substances. Body and soul are defined in expressions of the respective other category or with a common third term.

5. The purest monistic form is the mentalism of Berkeley (1710/1948–1957). He denied the existence of matter. Causal relationships between physical objects were portrayed by Berkeley as conditional statements about mental events.
6. Every attempt to abandon the category of the soul is referred to as materialism.

6.1 Behaviourism is the materialistic doctrine according to which the observation of the behaviour of objects produces the only reliable data for the investigation of mental phenomena [Dewey (1929/1958); Quine (1969)]. Language must be adapted accordingly. The notion of „shame“ for example must be empirically traced to something observable, e. g., red face, down-turned eyes.

6.2 Identity theory holds that the processes of the brain and functional roles of the nervous system are relevant to the causal explanation of physical as well as mental behaviour. Mental events are not necessarily identical, but at least consistent with events in the brain.

6.3 The physicalist asserts that all events (including mental) are ultimately explainable through natural science without the need for the brain to determine identity.

7. The approach of „artificial intelligence“ (also „mechanicism“) identifies mental activities as complex transformations between data which are presented on the sensory periphery of the organism and the intelligent answer of the organism to its environmental reality. The brain is ultimately a complex calculating machine. David Chalmers (1996) thinks robots could be capable of a consciousness if they just have the appropriate programming.
8. Neutral monism is based on neither matter nor the soul, but on a common third.

8.1 In Russell's selection theory, these are the „sensibilia“, (perceived and not perceived) sense-data, which are neither mental nor physical per se. When the sensibila are perceived, i. e., are selected, they change according to the relationship framework into mental or physical objects.

8.2 Information realism designates the intellect and matter as structures of information states which are organised and connected through information channels in such a way that the information at the mental and at the physical terminal can be the same.

8.3 Whitehead (1936) pp. 449 f., says this about his organistic philosophy: „The disastrous separation of body and mind is avoided in the philosophy of organism by the doctrines of hybrid physical feelings. A hybrid physical feeling originates for its subject a conceptual feeling with the same datum as that of the conceptual feeling of the anteceding subject. But the two conceptual feelings in the two subjects respectively may have different subjective forms."

8.4 McGinn (1996) calls for a new understanding of the concept of space, the concept of „extension", in order to resolve the contradiction between res cogitans and res extensa, however he is unable to make any concrete suggestions for it. My philosophy will develop in this direction.

For the sake of completeness, it should not go unmentioned that there are also philosophers, such as Wittgenstein (1969) for example, for whom the body/soul problem is not a problem at all, but rather, has arisen from philosophical misconceptions of discussions in everyday language about mental activities.

I have already stressed earlier, that there is no actual proof in philosophy. It cannot be objectively determined who is right. The theory is dependent on the language, the definitions and axioms, the logic used, the aim the theory pursues and on the subjective taste of the philosopher. A theory should be beautiful, comprehensible and useful. It is useful when it can explain the most possible in the simplest possible terms.

5.2 The aim of my philosophy

The goal of my philosophy is to find the ultimate source of the laws of nature. The natural scientist wants to explain the processes observed in nature and to make predictions about the future. To this end he formulates natural laws based on his experiences. Nothing proceeds in natural science without observations, perceptions, measurements and experiences. All these terms designate processes by which information is conveyed from a material object to a spiritual-mental subject. So, distinctions are made, rightly or wrongly, between object and subject. Should the division between subject and object be an artificial one, it will possibly influence the outcome of the information transfer. The type or state of the subject can also influence the perception [Gibson (1979)]. Moreover, the information transfer is directed: The object sends an information signal to the subject. So, the object „causes" a signal, which generates an „effect" in the subject. Perhaps the object is provoked into this by the subject, because the subject has decided when something is to be measured. In any case, the directed information also generates an order: The cause comes before the effect. The subject must be in a position to recognize this order, otherwise it is lost to the subject and the information received becomes chaos.

Thus, the impression of time is created. This ordering of time is a condition for every perception. By all of that, nothing is said about whether objects and a directed time, independent of perception, actually exist or that time is a continuum. The order, however, the chirality of time, is a philosophical condition for perception in the conventional sense.

There may be other forms of perception as well, such as where no subject has determined anything before the perception or where the transfer of information is not directed. Instead, it may be that for some reason, simultaneously or otherwise, the same information „surfaces" in the subject and the object, and if need be, consciously or unconsciously, is stored.

It could be rather worthwhile at some point to thoroughly analyse which basic philosophical conditions must be met for which form of perception, so that the perception can occur at all. The hope would be that the general conditions, found theoretically, turn out to be nothing more than our laws of nature. Is there reason for such a hope? I believe there is! The fact that chirality, at least the chirality of time, is not only a philosophical precondition for perception, but also a universally valid principle of nature, makes one suspect that a correlation exists between conditions of perception and natural laws, especially since according to the *CPT* -theorem, electrical charge and space parity also have something to do with the direction of time. In the event that our hope is fulfilled and the laws of nature can be traced back to a single ultimate principle, then nature becomes a unified whole [Weizsäcker (1999) pp. 7–22; Feynman (1965)]. Max Planck (1949) however, did not believe that the laws of physics could be derived from metaphysics. He was convinced that the exact opposite had to be the case: metaphysics is to be derived from the laws of nature.

Whether we can therefore also fully understand it, is considered doubtful by well-known philosophers and physicists however. Also, for the French philosopher and founder of the positivism, Auguste Comte (1880), the highest goal of the natural sciences existed in the unification of the scientific laws to a single law of nature. He considered this goal unattainable, however. Wigner (1950) believed: „In order to understand a growing body of phenomena, it will be necessary to introduce deeper and deeper concepts into physics. This development will not end by the discovery of the final and perfect concepts. I believe that this is true: we have no right to expect that our intellect can formulate perfect concepts for the full understanding of inanimate nature's phenomena." According to Turing, the human memory, and with it its capacity for understanding, is finite, and thereby limited as well. Gödel however, holds „that mind, in its use, is not static, but constantly developing … Therefore, although at each state of the mind's development the number of its possible states is finite there is no reason why this number should not converge to infinity in course of its development." Our capacity for greater understanding is thus potentially unlimited [Dawson (1997) p. 232]. Putnam (1998) wishes all philosophers and physicists who want to derive the laws of nature from an *a priori*-philosophy, „good luck", thinks nothing, however, of this approach.

In pondering all this, the subject, the soul, the intellect and consciousness all play a decisive role. It is important for a proper understanding of my ideas to know how I choose to interpret these terms (and how not).

5.3 The structure of the soul

In Chapter 3 I opted for the path of solipsism and declared that my consciousness exists. Before I consider the connection between consciousness and the world, I wish to elaborate on its relationship to the soul. The soul unifies the conscious with the unconscious. Jung (1967) pp. 163–183, distinguishes seven different functions of the soul, which can all occur in the conscious mind as well as the subconscious, to wit:

1. Sensations: sensory perceptions which have streamed into the consciousness from outside
2. Feelings as a comfortable or uncomfortable emotional outcome of evaluation processes
3. Memory
4. Thinking links different memories and leads to insight
5. Intuition, i. e., the perception of future possibilities held in a present situation
6. The will as a directed impulse, which is under the control of a so-called free discretion
7. Instincts come from the unconscious or directly from the body and are impulses with the character of bondage.

The transition from unconscious to conscious is a fluid one. Classical examples of unconscious psychic activity are provided by pathological conditions such as hysteria, compulsion neuroses, phobias or schizophrenia. Dreams can be understood as signals from unconscious processes surfacing in the conscious mind. Archetypes are a special kind of instinct which regulate unconscious psychic processes: they are „patterns of behaviour“. They effect excitations which diminish the clarity of consciousness. Thus, the excitations give the unconscious the opportunity to intrude into the space which has been vacated. It appears that the ego-connection advances from unconscious to conscious. Whereas the unconscious can certainly be connected with other souls, if necessary even in a world soul (*anima mundi*), perhaps via extrasensory perception or through archetypes, the conscious part of the soul is ego-connected. When I forget, I lose thereby something from my consciousness, from my ego. That happens very easily. But what is the ego? It concerns a composition of mental elements, the ego-complex, but in no way encompasses the entirety of the human essence: above all it has forgotten infinitely more than it knows. It has heard and seen an infinite amount without ever being aware of it. Thoughts sprout beyond its awareness, indeed they are often simply present, without the ego knowing too much about it. The inflow of unconscious contents of the soul enlivens and enriches the personality and, in a sense, excels the ego in scale and intensity.

The will is motivated by instincts on the one hand, and influenced and guided by thinking on the other. The more thinking dominates one's instincts, the freer the will becomes. I don't believe however, that a person can act completely free of every instinct, although many people strive towards this ideal. All humans are somehow dependent and subject to influences, otherwise they wouldn't be humans, but gods. This is the reason why even healthy, intelligent and perfectly astute people can sometimes behave completely irrationally. Often this occurs when they are not alone, but act in a group, mob or a crowd, where they get ensnared by their herd instinct and follow a charismatic leader, often without being aware of it. Such situations can lead to conflict or war, with all its horrors [Canetti (1960)]. Humans have a body and hormones which influence the instincts and desires, even if it is only hunger.

The free, and in a certain sense, highest component of the soul is called mind. The soul stands somehow between body and mind, constraint and freedom. Here also, the transitions are fluid [Feest (2017); Lohse and Greve (2017)].

So much for the soul from a psychological perspective. I am aware that the soul also has a theological aspect which is concerned with questions such as immortality, eternity, moral values, transmigration or karma. Within the scope of my work however, it is not necessary to go any further into these.

5.4 The nature of matter

The body in the body/soul problem is actually substantially more than just the body which houses the soul. Here, the term body usually refers to matter. What is that? Matter originally meant wood. The Greeks selected wood as the primary matter probably because by burning, wood is converted into gas („air"), water, ash (earth) and heat (fire), so that one can conclude that earth, water, air and fire together make up the different types of matter. Up until the 20th century, most scientists no longer regarded heat as matter. Only, in the context of the special theory of relativity are thermal energy and mass energy once again regarded as the same entity: The mass of the matter increases with its temperature. If one abstracted from the woodenness of the wood, there remains an empirically perceptible, more or less impenetrable object with a certain extension and form in a specific location. Empirical perceptibility, impenetrability, extension (*res extensa*), form and location are the features which distinguish matter from the soul. The experts are divided over what is real, matter or mind, but by and large, there is agreement about the difference. With the advance of physics, views about impenetrability and about form have changed, however. Impenetrability or materiality in modern physics have become merely names for points in a field.

For Aristotle (1984) 424a, form was what endured: concrete things come and go, in that matter assumes the form and then relinquishes it. According to Aristotle, matter itself is not perceivable, but only its form: „Perception is the receptor of the perceivable forms without matter, like the wax takes on the seal of the signet ring without the iron or the gold." The form itself is eternal in that it is continually assumed by new entities. Biological species, whose individuals continually procreate their own kind, are classic examples of this. We now know that similarly, molecules develop from atoms and then decay, and even elementary particles are perishable or can be reconstituted, whereby they assume the same form over and over again.

Today, that which endures is no longer called form, but initially substance, later energy and often information as well. The form has become in-form-ation, which as a pure number, i. e., without a physical dimension, can be measured. According to Weizsäcker (1986) pp. 170–173, the measure of the information is a dimensionless number, thus no longer a mass or an energy, as one would actually expect for matter. The number is calculated as follows: An experimental alternative with a factor of K is given, i. e., K possible alternative events x_k ($k = 1, 2 \ldots K$). We expect an occurrence of x_k in the case of a decision of the alternative with the probability p_k. The „piece of information" I_k is meant to measure the „news value" of the event x_k. An event which has taken place contains the least news value, the more probable it was before. If it was certain, then the news value is zero. The news value of an event resulting from two combined independent events is equal to the sum of their news values. The usual definition gives an event with the probability $\frac{1}{2}$ the news value of 1 (one bit). In addition, one must place $I_k = - \ln p_k$ (ln = logarithm to the basis 2). The expected value of of I_k, i.e., the average news value which can be expected of the unique determination of the alternative over many attempts, is then $H = \Sigma_k p_k I_k = -\Sigma_k p_k \ln p_k$. This measure of the information is called entropy, or roughly expressed – the measure of potential knowledge. Just like energy, the entire information in a closed system remains constant over time. As soon as a simple alternative is determined, a new open alternative emerges automatically; every answer to a question immediately generates a new question. Just like energy, the entire information in a closed system remains constant over time. Order can only be established by the simultaneous creation of disorder. Anyone who has ever done any cleaning could sing a song or two about this.

In quantum theory, information is described mathematically as a wave, and it is no longer clear whether matter is in reality composed of particles or waves. Particles with a finite rest mass, such as the electron for example, indeed have the characteristics of a wave, which is made use of in electron microscopy. In contrast, massless light waves have shown corpuscle characteristics in experiments, which Einstein demonstrated in 1905. That this dualism of particles and waves demands a radical break with classical physics, was probably first seen by Niels Bohr. He jokingly said on occasion: „If Einstein sends me a telegram saying he has finally proven the particle nature of light, then the telegram only arrives because light is a wave." One might ask what it means for the nature of matter, that in many cases, one can represent it better mathematically with the help of waves, rather than mass points.

The philosophical characteristic of a wave is that something about it changes, i.e., the phase, whereas something else, i.e., the wavelength, remains unchanged. We know that like in the atom, the position of the electron continually changes. Its average distance from the nucleus remains constant, however. These electronic states are described as stationary waves. The protons and neutrons within the nucleus and the quarks within the protons and neutrons are in motion as well, which is why it is not surprising that the states of nucleuses and nucleons can also be and quarks, are best described as waves, this could be an indication that they in turn are composed of even smaller moving particles, although no experiments to date directly suggest an internal structure of these particles. For sixty years physicists have known about the particle/wave dualism of matter, and despite great effort, to the present day no satisfactory solution to the contradictions has been found [Hawking et al. (1996a) pp. 121–135]. Most scientists have long since resigned themselves to this. That this somehow unsatisfactory situation has become a matter of course could also suggest, however, that we have not yet really understood the nature of matter. The sum of all existing matter is called, depending on the context, nature, world, universe or cosmos. Thus, consciousness, the ego or the subject stands face to face to the world. That need not necessarily preclude it from being simultaneously a part of the world [Hölling (1971a) pp. 212–218; Weizsäcker (1999) pp. 360 ff.].

5.5 Perception as the flow of information from matter to subject

Alongside body and soul, a third principle plays a part in the body/soul problem, namely perception. We have established that it is contentious, to what extent body and soul, matter and consciousness, concern two really separate, fundamentally different things. However, philosophers and natural scientists agree that there are structures, material and/or transcendent, which can exchange information amongst themselves. Information is conveyed continuously in nature, for example via heredity or when recording a concert on tape. Perception is more, however. It comprises a whole complex of processes, beginning with the reception of sensory data through our sense organs, then the assessment of these data in the form of sensations, followed by the development of a conception and a stored memory of the received information.

All these processes together form the information bridge from the object, the world, to the subject, the consciousness. Attempts have been made to divide the complex of perceptions into sensory, aesthetic and mental perception. The sensory is purely a transfer of information, the aesthetic effects a feeling in the subject, which is generated by the information received; the mental firstly provides the assessment criteria for the aesthetic perception and secondly, processes the information into a conception and a representation in the memory.

A distinction is made between external and internal perception. With the external, the subject and the world are two fundamentally different things. With the internal, the subject is on the one hand faced with the perceived world, and on the other, simultaneously a part of this world. It perceives itself as an object. The internal perception is a purely psychic and mental process. The subject can have various interests, moods, expectations and focusses of attention; accordingly, it will variously perceive the object sensorily, psychicly and mentally. So, it is that different persons perceive and register the same process completely differently. The same person will even perceive the same event, for example the film „Gone with the Wind", at different times quite differently. Perceptions can also take place unconsciously; indeed, they are probably more often unconscious than conscious. The soul thus holds many perceived conceptions, of which the conscious mind has no inkling.

The natural scientist concerns himself with observations which are repeatable, objective, and as rule, measurable. Due to these three conditions, the subject's influence on the result of the observation is restricted as much as possible and the objective validity is increased. Above all, this involves reducing the particularly subjective aesthetic perception to a minimum. Aesthetic criteria shouldn't play a part until later, during the development of an attractive theory. This differentiates modern natural science from the science of Goethe, which first and foremost sought the truth in the aesthetic and in feelings. Natural scientists forgo such concepts as love and hate, beauty and happiness, good and bad because they are subjective. But what is meant here by subjective? Perhaps not belonging to nature and mere appearances? If we dispense with the description of unique events and with all feelings, and limit ourselves to simple measurements, then that won't leave very many perceptions which objects of natural science can be. First of all, all unconscious perceptions are omitted. Then we have to consider what a measurement is, exactly. All of these restraints are so substantial, that one may wonder to what extent natural science can lead to true results when it simply ignores the greater part of all experience. This is the so-called „qualia objection" to the materialistically oriented theories where feeling-stressed qualities of experience and the free will of the observer have to take a back seat [Wechsler

(1999) pp. 30 f. and 51 ff.; Huber (1995)]. Even so, I regard the modern approach of natural scientists as legitimate, at least as long as they remain conscious of where, owing to the strict basic conditions, the boundaries of their science lie. The benefit of this method lies in the simplicity of its results, in their reproducibility and therewith in their significance for predictions about the future. It gives less cause for misunderstandings than a theory with subjective and aesthetic aspects and is therefore much simpler to communicate. This approach does not exclude the theory, at a later stage, from being supplemented, enriched or even being put on a completely new level by its extension to include emotional perceptions which are not easily objectifiable. A lovely attempt at such an extension of the theory, which leads to a new understanding of the mental quality of experience, has been undertaken by Dietmar Wechsler (1999). A program for an „understanding" or „romantic" scientific theory, which can take in aesthetic and ethical criteria has been formulated by Müller (2004). In addition to the criteria of the classical scientific theories (for example objectivity, value liberty, universality, falsifiability or predictive potential), criteria such as holism, reflection of subjectivity, respect, comprehensibility and ethics based on natural philosophy are included.

So, perception is a transfer of information from the object to the subject. Moreover, the content of the information yielded by the object must be at least as great as the content of the information received by the subject. Descartes (1619/2006) p. IXa32, formulated this principle as follows: „Now it is obvious by the natural light however, that at least just as much information content must be present in the entire working cause as in the effect of this cause."

Pieces of information are the answers to questions. All questions can be so formulated, that the following yes/no answers are the answers to a series of alternative statements. Instead of asking, „What colour is that?", one must ask, „Is that red? Yes or no." In natural science, the statements have to be free of all aesthetic aspects, so that they become measurable. Thus, such statements cannot take the form, „This tomato is red", since red is a partly aesthetic concept, but must run, „The colour of this tomato has a wavelength of *x* metres" or „The light reflected from this tomato activates molecule *y* in my eyes". Subject and object must both have a structure which permits the representation of such information. The representation of the information in the subject is completely different to the image in the original, the object, but the scientifically relevant part of the information, comprising the questions free of aesthetic aspects and the yes/no answers to them, is conserved in the representations. The mathematical theory which deals with representations of this kind is *group theory.* In this rather abstract and difficult theory, structured sets are modelled on themselves through clearly defined operations. The more such different operations are possible, the more symmetrical the structure of the set under consideration is and the smaller the information content of this structure is. With a few questions and answers, the structure can then be described completely.

1. A group consists of a set M and a linkage, which we indicate with „$\circ$". In so doing, the following 4 conditions must be fulfilled: If a and b are two arbitrary elements in M, then $a \circ b$ is also an element in M (closure).
2. If a, b and c are three arbitrary elements in M, then: $(a \circ b) \circ c = a \circ (b \circ c)$ (asociativity).
3. There is an element e in M, such that for all elements a in $M: a \circ e = e \circ a = a$ (existence of an identity element).
4. For every element a of M, there is an associated element a' of M, so that: $a \circ a' = a' \circ a = e$ (existence of an inverse).

An equilateral triangle might serve as an example. The following symmetry of operations can be made, i.e., the triangle can be brought to coincidence with itself, thus leaving it unchanged: One can rotate the triangle through 120°, 240° or 360° clockwise or anti-clockwise, or it can be reflected at any one of the perpendicular bisectors of the three sides. The set of these six operations is a group; the operations are the elements of the group. The three conditions required above are obviously met. This group is for its part, structured in turn: It consists of three sub-groups, the trivial identity element, i.e., the rotation through 360°, the rotation sub-group with the three turns through 120°, 240° and 360° respectively, and the reflection sub-group with the three reflections. Each congruent image, i.e., each element of the group, can also be expressed as the consequence of two or several other congruent images, for example as „rotational reflection", whereby it sometimes depends on the sequence of these images, and sometimes doesn't. The relationship between the different representation operations is often portrayed by so-called matrices, which for their part can have different symmetry characteristics and structures. In particular it can be shown that groups which are in essence completely different, for example those of geometrical and of arithmetic (often including complex numbers) operations, can have the same information content. Thus, the mother group of all physical operations, the special orthogonal group of SO(3) is algebraically identical to the complex group SU(2). Good explanations about group theory, its terms and applications can be found in textbooks, for example with [Wagner (1998) or Sternberg (1997)].

With internal perception, the information transfer is, similar to how it is with a representation in group theory, in fact an internal representation of structures within the conscious mind or the brain. With external perceptions by contrast, the structures of the object and the subject are completely different. We can never directly perceive an electron, for example. Rather, the electron's information is initially transferred to measuring instruments, from there via our sense organs to the brain and finally into our consciousness. The physicist hopes that the information is distorted as little as possible over this lengthy route, and he speaks of the electron as though he had perceived it and as though it actually existed. Ultimately, however, he only knows that in his consciousness there are certain information structures which he has called electrons. Apart from the information content, the structure in the conscious mind has nothing more in common with the structure of the electron.

Such structures can also be subjects of group theory where sometimes to the general surprise of mathematicians, it can be shown that completely different structures, for example geometric and arithmetic, can contain the same information, so that a proof which is valid in one mathematical theory is also correct in the other structure, although there, an analogous line of reasoning is not at all possible.

Three different cases of information transfer can be differentiated:

1. A copy is created in the object and transferred to the subject. The information is thereupon available in the object as well as in the subject. The entire amount of information has increased, for which in physics, energy must be expended. This means that the world as a whole must have changed during the perception, since energy, or information, has been transferred from the rest of the world to the

object. Moreover, the subject must take into account that during the release of energy from the world to the object, the latter has also been changed, so the information transferred no longer describes the original object exactly.

2. The information is transferred from the object to the subject and is thereby lost to the object. Although no external energy inflow is needed for this, the object, without this information, is still no longer the same as before the information transfer. The subject does not know the object's current state.
3. The information is transferred from the object to a subject, which in turn is part of the object. In this case, the total volume of information remains unchanged, yet the object as a whole has changed, since the information is now stored in another location in the object.

The first two cases concern an external perception, whereas the third concerns an internal one. In all three cases, an artificial division was made twice during the information transfer, i. e., the first between the object and the information sent, and the second between the information and the subject. Should one assume that the information has no separate existence independent of the object and subject, the two divisions collapse into a single one between the object and the subject. However, one could also argue the opposite and contend that the only thing that exists with certainty is the transferred information, i. e., that which is perceived. The subject and the object could be described, according to this theory, as constantly changing quantities of information, whereby only the change exists for certain, since only this change is perceived. From the natural scientist's point of view, this appears to me to be the most honest approach. The subject (thinking, consciousness) thus becomes in solipsism a stream of information, a stream which perpetually changes and yet is always present as the same thing.

In quantum theory, information is usually described mathematically as wave function ψ [Wechsler (1999) pp. 79–156]. The wave functions form a language which describes all knowledge that we have received by objective observation, i. e., by measurement, and which is relevant for describing the future behaviour of the system. This description cannot take the form of an exact prediction; it only makes statements about probabilities of future events [Wigner (1970)]. Every observation changes

ψ in a way that first becomes apparent when the perception is taken up in the consciousness of the subject. At this point, consciousness inevitably becomes a part of quantum theory. Wechsler assumes that in the brain there are neural interconnections that force self-referential situations whereby consciousness becomes a representative system of infinite complexity. For this self-reference, quantum theory requires an eigenfrequency of ψ with overlapping and cophasal processes, which in principle are not measurable, because each measurement dissolves the system [Wechsler (1999) pp. 180–190]. In the words of Niels Bohr (1960): „The word consciousness, applied to oneself as well as to others, is indispensable when describing the human situation."

Consciousness plays a different role from an inanimate measuring instrument which records data. Ultimately, a consciousness is required to perceive these data and in so doing, changes ψ. As a consequence, ψ is subject to constant change and with it, also our consciousness and our predictions with respect to the probable future. The laws of quantum theory therefore, no longer deal with elementary particles themselves, but only with the knowledge that our conscious mind can have about the particles. With every perception, an artificial division must be made somewhere between the observed system and the observer himself. The propositions of quantum mechanics do not, however, depend on which place this division occurs. The conclusion suggests making the cut as close as possible to the consciousness defined as extra-physical and not in the middle of the physical system [Neumann (1932)]. The information contained in a wave function is communicable. If another observer somehow determines the wave function of a system and informs me of the result, probabilities of future measurement results prognosticated by ψ will be exactly the same, irrespective of whether I myself or the observer carries out these measurements. In this sense it can be said that the wave function ψ really exists [Penrose (2004) pp. 507–519].

5.6 The conditions for perception

To summarise, the following conditions must be met for perception to be possible:

1. The subject and the object must be distinguishable, i. e., the unity which is no longer exclusively an object and in which observer and observed system are conflated, must be destroyed. This is equally valid for realists and positivists. (The realist imagines knowing *a priori* what „reality" means; the positivist imagines knowing *a priori* what „experience" means.)
2. The direction of the information stream is from object to subject, i. e., the perception process is chiral.
3. Subject and object must have a structure which enables them to store information. For this they must be chiral.
4. The standard measure of the information is the alternative, i. e., the answer to a yes/no question.
5. Perception is the transfer of information. In the process, something changes, namely the information carrier, whereas something else, namely the information content, remains unchanged.

5.7 Reality

So, what really exists? The body, the soul or the perception? Before I answer this question, it must be stressed once more that there is no such thing as a definition of the concept of absolute reality [Wigner (1973)]. As long as we are not practising natural science, however, we may freely assume that body and soul exist and that they are even perhaps different aspects of the same thing. Stent (2002) pp. 237–261, suggested a new understanding of the duality of mind and body, that is, the body/soul problem. Accordingly, psychology and neurobiology would be two complementary sciences describing the same object, the human mind. In principle they are incompatible, however, both are necessary since they embrace different aspects and only thus make an integrated picture possible. The concept of complementarity was introduced by Niels Bohr. Complementarity means that different possibilities go together to experience the same object in different ways. Complementary properties belong together, insofar as they are properties of the same object; they exclude each other, however, insofar as they cannot be measured together at the same time. Perception is real. It is function and Be-ing at the same time. Metaphysics ought to form the connection between these two aspects of perception [Geiger (1966)]. The body or the world appear as a reality provided for the soul to contend with. This contention is not only possible, but vital. According to Jung there is only the *one* world. Humans have to artificially split this into consciousness and object in order to be able to recognize it [Primas (1995)]. Hence the world is a form of appropriation and *Be-ing*-in-the-world is the comprehension of reality [Heidegger (1977) p. 206]. The Be-ing of the existent is in one way or another always a subject's imagined state of Be-ing [Heidegger (1961) pp. 455 ff.]. The world and the soul are both real, but not independent of each other. The conscious mind, as a subject, knows things about matter as an object. Matter is explained as information by the subject. The information content of matter in turn cannot now be drawn upon to explain knowledge, since it is itself the knowledge. Every explanation of this sort about the soul, consciousness, knowledge, the world and matter would be circular.

For the natural scientist, such definitions of reality are unsatisfactory, since they do not lead to objectifiable perceptions. Weizsäcker (1971a) pp. 288 f., turns this sentence around and assumes: „The laws of physics are nothing other than the laws which formulate the conditions of the possibility of the objectification of the event." Thus, real is that which is objectifiable. Consciousness serves as an anchor for the solipsist. Russell (1921/1991) had expectations that a yet to be found fundamental science of the causal laws of mental events in the subject would lead to a standardized and simplified science which also contained the laws of nature. Physics would thus become a derivative science. The information stream of perception does not go round in circles, but flows from outside into the conscious mind. When the conscious mind registers the inflow of information, it becomes reality. To what extent a real world exists and what characteristics such a world has had, apart from the information which has flowed into the conscious mind, the natural scientist cannot know. That is to say, for him it is not the world itself, but the information received which is real. Plato (1981), advocated a similar view: „When we designate something as „being", we have to say it is for something or in relation to something", namely in relation to us as the perceiving subject or to the self. Other philosophers however, are convinced that an external world also exists independently of our consciousness and define reality as „existence or ongoing existence or simply being as it is, independent of a relationship to mental activities in form of the use of concepts and the making of judgements“ [Wingert (2003)]. No phenomenon is a phenomenon unless it is an observed phenomenon. With the help of such information, the natural scientist makes predictions for the future, which he can also communicate. In so doing, the observation and the knowledge is objectified.

The solipsist is a monist in the following sense: the ego and the world are one and they are a comprehensible reality only in as much as they are connected. This monistic view corresponds to the so-called Copenhagen interpretation of quantum theory, according to which the perceiving subject is part of the object perceived [Weizsäcker (1971a) p. 235]. And Karl Jaspers opines: „For one never knows the whole, since one stands in the whole“ [von Salis (1979)]. And Henri Berr adds: „Instead of wanting to be one's self, one must strive to be all [von Salis (1975)]“. The Self is a part of the world which it perceives.

At the same time, however, the solipsist is also a dualist in the following sense: the ego and the world can be differentiated, in that the conscious mind takes up and distinguishes new information flowing in from the outside [Wiehl (2000) pp. 70f. and 134f.]. Einstein (1977) put it differently: „The belief in an external world independent of the perceiving subject is the basis of every natural science. Since the sensory perceptions only give indirect information about the external or the „physical-real“ world, then we can only grasp this in speculative ways. It follows from this that our views of the physical-material can never be definitive. We must always be prepared to change these views, i. e., to change the axiomatic foundation of physics in order to do justice to the facts of observations in the most logically perfect way possible. Indeed, a look at the development of physics shows that this axiomatic foundation has experienced profound changes in the course of time.“ Now whether mind and matter are two realities or one, is a question answered by Goethe (1920) in his „Divan“[43]:

[43] Goethe, Johann W.: Gingo Biloba. In *West-Östlicher Divan*: „Ist es ein lebendig Wesen das sich in sich selbst getrennt? Sind es zwei, die sich erlesen, Und man sie

Is it a living creature
which divides itself into itself?
Is it two which choose each other
and are known as one?

Such a question to answer,
I probably found the right sense;
Do you not feel in my songs
that I am twofold and one?

This duality of Be-ing is beautifully expressed in the Chinese yin-yang symbol (Figure 14) which also appears on the South Korean national flag. The light and the dark areas in the asymmetrically divided circle are called Yin and Yang, respectively. They are symbols of all the dualities of life: good and evil, beautiful and ugly, true and false, masculine and feminine, odd and even, left and right, body and soul – the list is endless. The two small round spots were added later to express the idea that on each side of a duality, there is a little bit of the other side. Every good deed has a moment of evil, every evil, something good; every ugliness comprises something beautiful, every beauty a little ugliness, and so on. The spots remind the natural scientist that every true theory harbours an element of ignorance. The philosopher recognises that the body and soul may well form a unity, but that they can also be seen separately, whereby in every body there remains some soul and in every soul there remains some body [Gardner (1964) pp. 249 f.].

Figure 14. The yin-yang

als eines kennt? Solche Frage zu erwidern fand ich wohl den rechten Sinn; Fühlst du nicht an meinen Liedern, dass ich eins und doppelt bin?“

Chapter 6

To Measure is to Count

God doesn't do calculus, he only counts.
Richard P. Feynman

6.1 Measurement

The physicist must register his perceptions, i. e., the information which has flowed in from outside in such a way that it is unequivocally communicable. Unequivocal means that the information, as far as possible, is to be stored in the form of numbers, since there are no misunderstandings about numbers, at least as long as it only involves finite numbers. This compels the physicist to measure. What does the physicist do when he measures? What does he know when he has measured? Metaphysical foundations deal with the conditions under which such a physical interpretation of the numbers is possible. In the description of measurement theory below, I essentially follow Patrick Suppes (1980).

The measurement links the perceived properties of objects with numbers. The numbers say something about the structure of the objects, or rather, about the perceptions of them. For this purpose, the object is conceived as a set M, with a finite series of relations R_i between the elements of the set. Such an object $S = [M, R_1, \ldots, R_n]$ is called a relation structure. So, the systematic features of the measurement process are to be characterized in the form of finite relations R_i, which must be shown to permit a numerical representation. A simple example may illustrate this.

An object S_1 comprises a set M and a single binary relation $R = \geq$, thus $S_1 = [M, \geq]$. So, for the structure of the object S_1 to be homomorphous to a numerical structure $S_2 = [\mathbb{R}, \geq]$ (whereby $\mathbb{R}$ is the set of real numbers and $\geq$ the usual numerical greater-or-equal-to relation), the relation $\geq$ must be reflexive, transitive and connex in M. The relation $\geq$

is reflexive when $A \geq A$ is always true. The relation $\geq$ is transitive when it follows from $A \geq B$ and $B \geq C$ that $A \geq C$. The relation $\geq$ is connex when it is true for all A and for all B that either AB or BA. In this case there is a numerical correlation, that is a representation φ of M in $\mathbb{R}$ so that for all x and y from M, $x \geq y$ applies precisely when $\varphi(x) \geq \varphi(y)$. The perceived structure of the object can thus be described in the form of real numbers.

The information is only unequivocal when the representation φ is unequivocal. By no means does this go without saying, since $\mathbb{R}$ is chiral, and it is not inconsequential, whether one reads the numerical series from left to right or from right to left, from the negative numbers to the positive, or from the positive to the negative. The value of negative number is of equal magnitude to the value of the same number with a positive sign. Therefore, there is no objective reason for assuming that $+1 \geq -1$. One could define the relation equally well, so that -1 becomes $\geq +1$. The convention is that a real number is positive when its root is also real. In this way, the orientation of chirality of $\mathbb{R}$ can be unequivocally determined.

In the one-dimensional space of the set M, the one-dimensional relation is also chiral. Thus, an agreement is required as to how the chiral orientation of the object's structure should be modelled on the chiral orientation of $\mathbb{R}$, and this convention is arbitrary. As I explained previously in Section 4.1, also the definition of left and right in the physical world is ultimately a random one, and that two observers can only agree on the definition of left and right when they are able to discuss this face to face. Without such a convention, there is no clear communication and consequently, no objectivity. In contrast to number space $\mathbb{R}$, in M there is no non-arbitrary definition of the orientation.

The measuring process itself is always a counting process [Kitcher (1983)]. For this, the physicist must know three things, namely which elements of an object he wants to consider, which relations R he wants to measure and in which units he is counting. A speed camera's radar only captures the vehicle's tailgate and the assumption is made that the whole car is travelling at the same speed as the back of it, although the individual atoms are moving quite differently and the front of the vehicle is not measured at all. The radar only measures the speed; the temperature relations and other parameters are of no interest. Measuring, or rather, counting is done in kilometres per hour. When the swimming pool attendant measures the temperature of the pool, he analyses only a small sample of water and assumes that this is in thermodynamic equilibrium with the rest of the water. Water currents or contamination are of no interest when measuring the temperature. Measured, or rather, counted, are the degrees Celsius, whereby these can be only indirectly determined, for example by measuring the length of a column of mercury in the thermometer. When the market stallholder weighs a bag of cherries, she measures weight indirectly via the deviation of the scale's pointer needle, i.e., via an angle measurement. Digital scales can immediately convert this angle to a number, for example the number of grams. A Geiger counter, which measures the amount of radioactive decay, counts irregular events. The result is an absolute number. Clocks also ultimately do no more than measure angles or count periodic events. So, in the case of most measurements, nowhere near all the numbers are recorded which would fully describe the state of an object. For the sake of simplicity, one is content with a few key figures and hopes that the results of the measurement will not be excessively distorted thereby.

But it could also be that the physicist is interested in a single, quite simple object and wants comprehensive information about its state, for example a neutrino. This can be (presumably) fully described by three location figures, three velocity figures and a figure for each of the spin, the mass and the point in time, yet it is impossible on quantum mechanical grounds to measure all nine figures at the same time. Every measurement of one of these numbers is a crude interference with the state of the neutrino, which due to its minuteness, is highly sensitive and can alter the values of the other eight figures. So, it is unfortunately impossible to ever completely know the state of a neutrino. One may naturally wonder whether one can say that a state, which one can basically never quite know, actually exists. This is a question about the meaning of „Be-ing". Most quantum physicists, unlike Einstein are of the opinion that the concept of Be-ing is not appropriate for such states. They say, then, the different values of the neutrino are complementary and thus not defined. In principle this quantum theoretical constraint applies to all measurements on any object. For states which require a great number of figures for their complete description, but where only one or a few of these numbers are of interest, this constraint is of little consequence, however. The state of the vehicle is scarcely altered by the radar, and during the measurement, the temperature of the swimming pool's water also remains practically the same.

All physical measurements can be attributed to three types of counting: units of length, periodical or non-periodical events are counted. There are no other measurements. Physics can thus be described entirely by these three types of numbers. According to the pertinent branch of physics, such numbers and also combinations of numbers are grouped in classes and given names such as mass, acceleration, charge, heat, entropy and so on, but only lengths and the quantity of periodical or of non-periodical events can really be measured. The axiomatic theory of the different measuring structures and number combinations was extensibly described by Krantz et al. (1971). If physicists were to limit themselves in their theories to these three single measurable categories of numbers, the theories should actually become simpler with a more general validity. Everything which goes beyond that is basically no longer physics at all, since it is no longer verifiable by measurement.

Just briefly, I would like to touch upon the problem of errors. During measurement, four kinds of errors occur: instrument error, for example as a result of an imprecise calibration, personal error, as a result of the observer's personal idiosyncrasies, the systematic error, for example due to an incorrect calculation of a constant, and the random error, the cause of which is not understood, but which can arise from random natural phenomena.

Still unanswered is the question posed at the opening of the chapter: What does the physicist know, when he has measured? In classical physics, it is assumed that reality is observed and measured and that it is not influenced by the measurement. Since, however, every measurement is based upon an interaction between the object and the measuring instrument, and since every information transfer from one object to another alters both objects, the assumption of classical physics is not quite accurate, especially not when it concerns very small, and therefore easily influenced, objects. Every act of measurement must be irreversible; otherwise there would be no physical documents of the past, no facts of observation. Judging by quantum theory, the description of a measuring procedure as irreversible, however, is only an approximation which offsets the interference of the probabilities between the measured object and the measuring instrument. The validity of classical physics is actually in this very approach. To this day it remains unclear and controversial, what a measurement means exactly for the physicist in terms of quantum theory [Weizsäcker (1999) p. 202; Görnitz (1999) pp. 171–180)]. The physicist can only know something when the information from the measuring instrument has arrived in his consciousness. In this sense, the observer is a measuring instrument which can gather and document information, but which in addition, possesses a consciousness. In terms of quantum theory, that is a new, additional quality, which is vital for a perception [Wigner (1970)]. For this, it is not necessary to have several consciousnesses from different observers which are intersubjectively connected. A single consciousness is sufficient, just as is claimed in solipsism.

6.2 Geo-chronometric conventionalism

6.2.1 Counting non-periodic events

What is the difference between counting units of length and counting events? It is simplest to count events whereby it can remain provisionally open, what is to be understood as an event. We will return to this in Chapter 7. In any case, the events counted must share a certain similarity for the counting to make any kind of sense. One can, for example, count apples, ballot papers, shots, galaxies or atoms. However, it adds little to the information when I know that I have perceived six things, two apples, a ballot paper, a galaxy, an atom and a bang. To be able to count, in principle I only need perceptions. As the subject, I don't necessarily have to be active to be able to perceive anything but can simply wait and see. As soon as a perception follows an event which belongs to the class of events I want to count, I must somehow record this perception, on a slip of paper, in a measurement device, in the brain or in the conscious mind. This can occur as a picture, a check mark or a number. Whatever the case, at the end of the counting process I must be in a position to add up the events counted or to read them off as final totals. This, in principle, is how a Geiger counter works. So, it is always perceptible things which are counted, and every other observer at the same time at the same place with the same measuring devices within the parameters of measurement accuracy would arrive at the same result.

Counting these kinds of perception events is only ever possible if there is an intrinsic measure for it, i.e., when the quantity to be counted is divisible into single elements which I can perceive separately and count that way. When I have counted electron charges, I can thus multiply the number by the charge of the single electron and thereby obtain the aggregate charge. However, measuring in physics is usually not so simple. To determine the mass of an object, for example, it is not sufficient to know the number of atoms and the atomic weight. Due to the binding energy between the atoms, the mass of the object is somewhat smaller than the sum of the atoms' masses. Masses can, on the other hand, be measured by a comparison with other masses by means of a suitable set of scales. When two objects balance each other out on the scales, then they are said to have the same mass. If an object pushes its side down more, then its mass is greater. This is certainly correct when the two masses being compared are close to each other on their scales. Widely separated masses cannot be compared with each other like this, since how can it be proven, that a mass doesn't change during its transportation to the location of the other mass? It is theoretically quite conceivable, that a mass, subject to the spatial geometry or time, changes during such a transportation, i.e., that it is dependent on the time and place of the measurement. In order to be able to compare masses which are separated from one another, we need an axiom, or rather, a convention. It states:

> The mass of an object is independent of the time and place of the measurement.

This is called a geo-chronometric convention.

It makes sense because one can count the number of atoms in the different places and at different times, and this number cannot change during transportation. If something changes therefore, this must be the binding energy. Under extreme conditions, for example in a very strong gravitational field, such changes are actually conceivable and the mass of an object can then change, whereby the atoms possibly even change into neutrons, radiate energy and the object consequently becomes lighter. In this extreme case, however, we are no longer dealing with the same object with the same number of atoms, but with part of a neutron star.

6.2.2 Counting periodic events

To count, one requires time. If the events proceed periodically, then their quantity forms a measure for the time [Aristotle (1987) book IV (219b and 221a)]. Periodic events are, for example: resting pulse, sunrise, the solstice, a pendulum swinging, the movement of a watch hand, the vibration of a crystal or the frequency of a radiating atom. The more exactly the single periods are attuned and the shorter these periods are, the more precisely time can be measured by counting the periods. In the course of this, time itself is never perceived, but rather the quantity of events. Without events, periodic or non-periodic, it cannot be determined whether time passes or stands still. When the different periods are always perceived the same and no external influence which could alter anything about the length of the period is apparent, it is assumed that the periods are of equal length. This cannot be proven, however, since it is not the length of time between beginning and the end of a period which is perceived, but only the event of the period change, i. e., the pulse beat or the return of the pendulum, for instance. A more precise time measuring device might reveal that the periods are irregular, or that over time, they become increasingly shorter or longer.

Two new problems in the measurement of time arise in the comparison to the counting of non-periodic events. First of all, time doesn't have an intrinsic measure of its own, unlike electrical charge, for example. Time is not granular. There are theories today, such as for example „loop quantum gravity", which postulate a granular time, but past experimental cosmological examinations speak against such a theory [Lieu (2003)]. Time is therefore neither perceivable nor countable. Should an absolute time in itself exist, then it could not be determined how quickly it passes, indeed not even whether it passes regularly. For if time were to slow down for example, then it would not just do that for the events and the objects, but also for the measuring instruments, the event counters and the observers. The velocity of absolute time would have to be measured by an observer with the help of a clock. Since such a measurement is always a comparison of the frequency of the clock with the frequency of the observed events, and since both frequencies must change in equal measure with a possible change in the absolute velocity of time, it is not possible to perceive „time in itself" or even to measure it at all.

A further result of time's lack of an intrinsic measure is that the quantity of periods counted is not an absolute one but must be compared with the inner clock of the observer or of the measuring device. It is thus dependent on the state or the pulse of the observer, or on the difference between the state of the observer and that of the object measured. The special theory of relativity states that the greater the relative velocity between the clock and the observer, the slower the time measured passes. If the clock travels at the speed of light, then from the observer's perspective, its time appears to stand still. The number of the clock's inner events, measured by the number of the observer's inner events („the pulse"), is then nil. Conversely, the observer's pulse, from the clock's perspective, is also nil [Møller (1969) p. 48; Sexl et al. (1987) pp. 31 ff.]. The formula for the time dilatation of moved clocks reads

$$t_A = t_B \sqrt{\left(1 - \tfrac{v^2}{c^2}\right)} ,$$

whereby t_A is the time of the moved clock, t_B the time of a resting clock, v the velocity of the moved clock and c is the speed of light.

The findings of the general theory of relativity are quite similar: the stronger the gravitational field in which the clock is located, the slower it appears to run. If the clock is located at a distance of a Schwarzschild radius from the centre of a black hole, then time also appears to stand still [Einstein (1922) pp. 24 f.; Møller (1969) p. 247]. The Schwarzschid radius $\mathscr{R}$ is

$$\mathscr{R} = \tfrac{2GM}{c^2}$$

whereby G is the gravitation constant, M is the mass which causes the gravitation and c is the speed of light. The time dilatation caused by the gravitation is

$$t_A = t_B \sqrt{\left(1 - \tfrac{\mathscr{R}}{r}\right)} ,$$

whereby t_B is the time of a clock resting in infinity and unaffected by the gravitational field and t_A is the time of a clock at a distance of r from the centre of a mass with the Schwarschild radius $\mathscr{R}$. If $r \gg \mathscr{R}$, then the relation presented becomes

$$t_A = t_B \left(1 - \tfrac{\mathscr{R}}{2r}\right) .$$

If $r \simeq \mathscr{R}$, then $t_A = t_B \left(1 - \frac{\mathscr{R}}{r}\right)$, that is, all clocks stand still at the Schwarschild radius [Fritzsch (1996) pp. 213–222]. Even when the length and the quantity of the periods perceived are relative, something also remains the same during the measurement of time for both the clock and the observer: the sequence of events and this is chiral! Consequently, *time* is never perceived objectively, but rather a chronologically ordered, more or less regular sequence of events. Expressed succinctly, one could say: The events exist and they have a sequence; time does not exist [Zwart (1973)]. This also means that it cannot be objectively determined whether two different events are simultaneous [Poincaré (1898/2003)]. Different observers could very well have different opinions about this, since they are in different locations and in other states of motion. The concept of simultaneity is not objective. Gödel (1983) formulated it as follows: „The existence of an objective course of time, however, means (or is at least equivalent to it) that reality consists of an infinite number of layers of the „now existing", which successively come into existence. If, however, the simultaneity in the sense described above is somewhat relative, reality cannot be split up into such layers in an objectively determined way. Each observer has his own series of such layers of the „now existing", and none of these different layer systems can exclusively claim to represent the objective course of time."

So, a clock and an observer only register time the same if they are in the same state of motion and in the same location. This condition is usually pretty much the case in practice, but never 100 %, since it is impossible for two things to be simultaneously in exactly the same place and because their states of motion often differ. In different places, however, the gravitation field, and therefore also the rate at which time passes, is not exactly the same. Using the face of the clock, an observer is able to determine the time by measuring the angle of the hand. In order to best meet the condition of the location and state of motion of observer and clock being the same, the observer positions himself in the centre of the clock. From here, he looks towards twelve o'clock when the hand is also on twelve. Unlike time, angles have an intrinsic measure: when the hand returns to the same position, it will have detailed an angle of 360°. If one assumes that the hand's angular velocity is constant, such a circumvolution is a measure of time. It is an objective measure, since any observer at all in the centre of the clock must also make a 360° turn to once again face the twelve. The duration of time the hand needs for a circumvolution can be given a name by the observer, for instance, an hour. If he wants to know when half an hour is up, then he first turns until the hand is in front of him. From then on, time begins to run. The observer now turns 180° and waits for the hand to be right in front of him. At this point in time, half an hour has passed. This detailed description may come across as a little bit fussy, but it shows two things: both the observer and the clock must be in the same place during the time measurement so that the result will be unequivocal, and although the observer must turn around during the measurement, he must not move away. By turning, the observer also performs a periodic movement, which he then compares with the clock's periodic movement. The measurement of time is always based on comparing two periodic movements – the observer's with that of a clock. As in the case of the digital scale, the counting of periodic events can be automatised so that the observer can read the hour count directly from the clock's digital display. However, in order to be certain that the digital clock performs periodic movements, an observer must at some point personally perform a periodic movement as well and thus have indirectly checked whether the clock's movement is really periodic. If the observer is blind, then he can, for example, measure the time by counting the clock's ticking. Whether this ticking is also regular, however, can only be determined by the observer if he compares it with another periodic event, such as his pulse, for example.

Therefore, time measurement requires also an axiom, or a convention. It states:

> The periodic time between equal consecutive events remains unchanged in the same place and in the same state of motion.

This is the chronometric convention.

A period of time is a chronological characteristic. Since time per se is not perceivable, it is unsatisfactory to the physicist to have to make use of such a non-physical characteristic as the period of time for the axiom of time. However, as long as one formulates physical theories in which a time continuum features, the chronometric convention is unavoidable. Evidently, we still haven't quite yet grasped the nature of time. „What then is time? If no one asks me, I know; if I wish to explain it to one that asketh, I know not“ [Augustinus (396–400 AD/1946)].

6.2.3 Counting units of length

Even more complicated is the measurement of distance, the counting of units of length. As boy scouts, we always had to carry a so-called measuring cord with us; this was a two-metre cord which had a knot every ten centimetres. To make such a cord, it is sufficient to take a single measure in order to place the second knot exactly ten centimetres from the first knot. For all further knots, measurements can be made based on this initial standard length. To measure a distance using the measuring cord, one end of the cord is laid at the beginning of the section to be measured, the cord is then pulled taut so that it touches the other end of the section, and then the number of knots, including the first, between both ends of the section being measured is counted. This figure is a measure of the distance between the two ends of the section, whereby the accuracy of measurement is ten centimetres. With this measurement, the following questions arise: Why must the cord be pulled taut before measuring? Can an inaccuracy occur because the measured section and the measuring cord can never lie in exactly the same place at exactly the same time, but at best, make contact? Does the length of the measuring cord remain unchanged during transportation from the observer to the section to be measured? And does it also remain the same length during the time the observer needs for counting? The answers are not all trivial. As with the measurement of time, problems arise because space itself is not granular. There are theories today, such as for example „loop quantum gravity“, which postulate a granular space, but past experimental cosmological examinations speak against such a theory [Lieu et al. (2003)]. So, one cannot simply count the minimum number of „grains“ between the two ends of the section to obtain the distance, rather an arbitrarily knotted standard length is required, which – in contrast to the clock used to measure time – must be transported to and from between the observer and the object, whereby yet another physical variable, namely time, changes.

The cord must be pulled taut because distance is defined as the shortest path between two points. That does not mean, however, that the path is a straight one, nor that there is only one shortest path. If the space in which the object is located is curved, for example the surface of a sphere, then there can be several shortest distances and they are all curved. Thus, on the earth's surface there are an infinite number of shortest paths from the North to the South Pole.

The next two questions concern how a change of location affects our measuring cord. It can never be proven whether the standard-length changes during transportation. To double check this, the observer must move to the new location, together with the cord and the standard measure, and make the comparison there to see whether the cord, measured against the standard measure, is still the same length. Should the distances change during transportation, for example because of the spatial geometry, then the length of the standard and that of the measuring cord change equally. Therefore, one cannot verify the change by comparing these two lengths.

The congruence of lengths is not necessarily dependent on the metric used. One speaks of a metric if the following three conditions are fulfilled: 1. The distance from *A* to *A* is zero. 2. The distance from *A* to *B* is the same length as the distance from *B* to *A*. 3. The sum of the distances from *A* to *B* and from *B* to *C* is the distance from *A* to *C*. Also, number spaces, for example a set of real numbers, can be metric; one then no longer speaks of the distance between two points, but of the difference between two numbers. In the metrics of the Euclidean plane, distance *ds* according to Pythagoras' theorem is

$$ds = \sqrt{dx^2 + dy^2}\ .$$

Poincaré showed that two congruent objects measured with this metric would also remain congruent if they were measured with quite a different metric, in which

$$ds = \sqrt{\frac{(dx^2 + dy^2)}{y^2}}\ .$$

It can be proven mathematically that also in curved spaces, there is not only several, but indeed an infinite number of metrics where the congruence of like objects, independently of the metric, remains the same [Grünbaum (1973)].

How the term „spatial distance" is defined is also a matter of convention, and the definition is normally composed such that the resulting theory, which is meant to describe the physical perception, is as simple as possible. With most physical theories, this geometric convention states:

The length of the distance is independent of the location.

(That does not mean, however, that it is also independent of the state of motion of the distance.)

In the general theory of relativity, the consequence of the geometric convention is that space is bent by gravitational fields. Still, one could easily also formulate the theory of gravity in Euclidean space as well; then, however, the distances become dependent on the location in the sense that a standard length becomes all the shorter, the stronger the gravitational field is (Figure 15), [Sexl et al. (1990) p. 29]. The length contraction of a standard-length is described by the equation

$$l_A = l_B \sqrt{\left(1 - \frac{\mathscr{R}}{r}\right)} \ ,$$

whereby l_B is the length of a standard-length resting in infinity and unaffected by the gravitational field and l_A is the length of a standard-length at a distance of r from the centre of a mass with the Schwarschild radius $\mathscr{R}$. If $r \gg \mathscr{R}$, then the relationship becomes

$$l_A = l_B \left(1 - \frac{\mathscr{R}}{2r}\right).$$

If, however, $r \simeq \mathscr{R}$, then the relationship remains $l_A = l_B \sqrt{1 - \mathscr{R}/r}$. That is, when $r = \mathscr{R}$, all lengths l_A shrink to zero [Fritzsch (2000) pp. 213 ff.; Einstein (1922) pp. 24 f.; Møller (1969) p. 28].

Both general relativity theories, the one with bent spaces and the other one with length contraction in gravitational fields make the same, correct predictions (Figure 1, p. 144, Figure 16, p. 145), so one's choice therefore, remains a matter of taste [Brandes (1955); Mittelstaedt (1972); Mittelstrass (2001) pp. 65 f.]. Physicists are accustomed to working with curved spaces. The calculations become much simpler than those employed by theories which concern Euclidean spaces, in which the lengths of the standard-length are constantly changing. On the other hand, curved spaces are unsatisfactory in two respects. First of all, a physical space can only be curved if it exists. As we have seen, space itself can never be perceived, however, and accordingly, does not exist in a physical

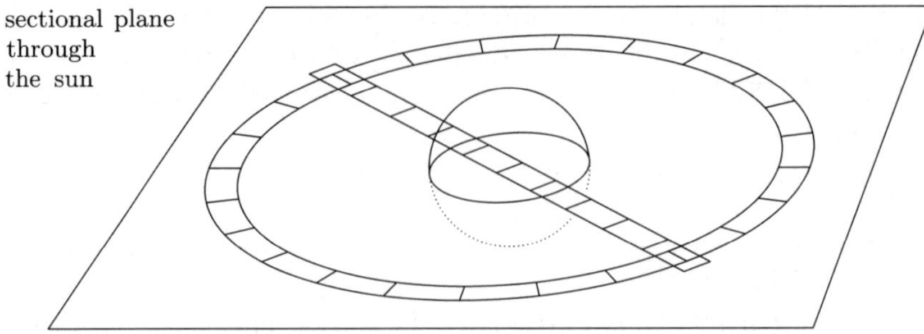

Figure 15. Geometry in the solar environment: *Standard-lengths shrink all the more in Euclidean space, the closer one brings them to a heavy mass. A series of standard-lengths is laid along a sectional plane, serving to measure the circumference and radius of a circle. The picture clearly shows that the measurement of the radius requires more standard-lengths than would normally correspond to the circumference. If one makes note in the usual way of the radius* a *and the circumference* u *by the number of standard-lengths laid down, this results in a ratio* $\frac{u}{a} < 2\pi$ *[Sexl et al. (1990) p. 29].*

sense [Einstein et al. (1950)]. Secondly, the assertion that space exists a priori as a condition for every perception is questionable also. Kant himself was actually convinced that space is real in this sense; however, it was equally clear to him that this could only be a Euclidean space. For experienced physicists as well, it is not easy to really imagine a curved, three-dimensional space. On the other hand, the calculations become very complicated when one formulates gravitation with a Euclidean space. Even so, Brandes (1995) p. 270, surmised that Euclidean space ultimately described nature better than a curved space. The area of conflict between the theories of space and our sense of reality would seem to indicate that we have not yet really understood the nature of space.

The question still remains concerning how the result of length measurement might be influenced by the time the observer requires for counting the units of length. This influence is a subject of the special theory of relativity. The observer has essentially two possible measuring procedures: either he must move along the measuring cord at a fixed velocity while counting knot by knot – this takes time – or having nicely laid out the measuring cord along the section to be measured, he occupies a location outside the measured distance and counts from there the

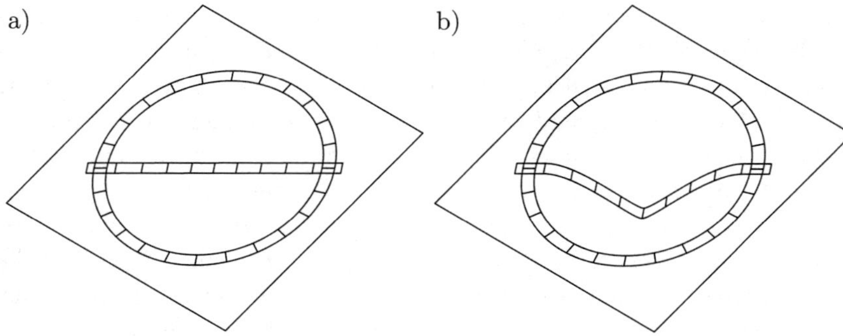

Figure 16. The behavior of standard-lengths: *The pictures a) and b) of the sectional plane through the sun are two different interpretations of the empirical fact that the ratio of the circumference* u *to the radius* a *is smaller than* 2π. *The two models are mathematically equivalent. In the theory of general relativity textbooks, the viewpoint of 16 b) is usually taken. Here the length of the standard lengths remains unchanged, but instead, space becomes curved.*

light signals which are emitted by the individual knots and by the object to be measured. He can count these signals indeed, he can even photograph practically all of them simultaneously and count them afterwards. The special theory of relativity asserts that as a consequence of the constant, non-additive speed of light, these two methods of measurement lead to different results. Common to both procedures is that events are counted, be it the passing of knots or the reception of light signals. The difference between the procedures is that in one case the observer moves relative to the section measured, whereas in the other, he doesn't.

When the observer remains outside the section measured and counts the measuring cord's knots between the two ends of the section, then all he needs to do is to multiply the resulting figure by the length of his standard-length and he will obtain the aggregate length of the whole section measured. This is so simple because we have postulated that according to the geometric convention lengths are independent of location. Counting light signals emitted by the knots is a consequence of regular perception events, which is periodic in the sense that the observer shifts his attention from knot to knot, i. e., that he must turn through a certain number of degrees. Similar to reading a clock face, the observer performs a turn, which in a certain sense is also periodic. Through this kind of counting, knots ordered periodically in space are represented as knots with a chronological periodic arrangement which the observer compares using his own, internal measure of time. The process can, as with the digital clock, be delegated to the measuring instrument so that the observer can read the result directly as a number. It is fundamentally impossible, however, to conduct length measurement outside time since there is no counting outside time. Turning the head is only exactly periodic for the observer if he is such a long way from the object that the angles, when turning from knot to knot, remain practically the same. Naturally, this condition can only be met approximately at best.

A further finding is not as trivial as it first appears: when the observer turns his head to count the knots, he must always know the direction in which he has to turn. He must not turn his head to the right from the first knot to the second, count the second knot and then turn the head back to the left and count the first knot a second time by mistake. He also cannot mark off the knots when counting in order to avoid such mistakes, since he cannot move from his position. Accordingly, the observer must be able to distinguish between left and right so that he is able to correctly portray the number of knots in the chiral time of the counting process.

Of course, there are other possible ways of measuring length in which observer and measuring instrument are stationary relative to the section measured during the actual counting process and at most, turn on their own axis, e. g., triangulation. However, that makes no fundamental difference to the end result.

So, what happens then when the observer or measuring instrument move along the section measured counting the knots? Naturally, the number of knots cannot change, since this is based on the intrinsic measure of the measuring cord and there is no-one there who could tie additional knots or undo any between the two end points. The question is just whether the length between the knots, i. e., the standard-measure remains unchanged, although it moves relative to the observer. Such a change would not violate the geometric convention in any case. According to the special theory of relativity, the standard length becomes all the shorter, the greater the relative velocity between the section measured and the observer is. This reduction is called the Lorentz contraction. It is the result of experiments showing that the speed of light c is constant – independent of the realative velocity v between observer and luminous object. The length of a moved body is measured smaller in its direction of motion by a factor of $\sqrt{1-\frac{v^2}{c^2}}$ than the proper length measured in the rest frame, whereby v is the speed of the body relative to the observer and c is the speed of light [Sexl et al. (1987) pp. 77–85].

As we have seen, clocks on bodies in motion run slower than at rest, i. e., time t measured by them is shortened by a certain factor. For the speed of light ascertained by the observer to remain constant, the intervals l of the body, measured in the direction of motion, must accordingly also become shorter by the same factor, since only thus will $c = \frac{l}{t}$ remain constant. So, the observer can say exactly how many knots there are between the ends of the section measured. But since he doesn't know how big the intervals between knots on the moving measuring cord are in comparison to the length of the standard at rest, he cannot know how long the section measured is in reality. If the observer were familiar with the relative velocity, then he would even be able to calculate from this the distance between the knots. In order to determine the velocity, however, he would, conversely, have to know the distance: the calculation goes round in circles. The length of a moving section is not absolute; there is only a maximum length, which according to velocity is shortened and even approaches zero when the velocity nears the speed of light. There is a limiting condition with this method of measurement as well, which can be only approximately met at best: the observer must move precisely on the section measured. However, this is not at all possible, since this section is actually occupied by the object to be measured. Thus, the observer always views the section measured from the outside.

Since the limiting conditions for both methods of measurement cannot be completely fulfilled, it is worth looking briefly at what happens when the observer is situated neither at an infinite distance nor exactly on the section measured. Is he able to perceive the Lorentz contraction and even to photograph it? Oddly enough, that is not the case. The crucial fact here is that with visual observation, the transit time of light from the object to observer must be taken into account: the light which is simultaneously emitted from different, variously spaced points of an object does not all reach the eye of an observer at the same time (Figure 17).

When a railway carriage passes an observer at high velocity, the light which is emitted by the far frontal corner of the side facing the observer is absorbed and covered by the rest of the side in motion. The observer only sees the corner closest to him. The view of the other side is quite different: all beams of light from this side are able to reach the observer unhindered, so he is able to see this side quite clearly. As a result, the moving carriage is not contracted when observed visually, but appears rotated. If the carriage were to travel at the speed of light, then the rotation would appear from the observer's perspective to be 90°, which means that he would only see the back of the carriage.

Thus, different observers can obtain quite different results in their measurements of the same section with the same measuring cord. However, there is one constant for all observers: the number of knots counted. Only this is objective. Could one disregard space and distances and limit the theories to such objective numbers?

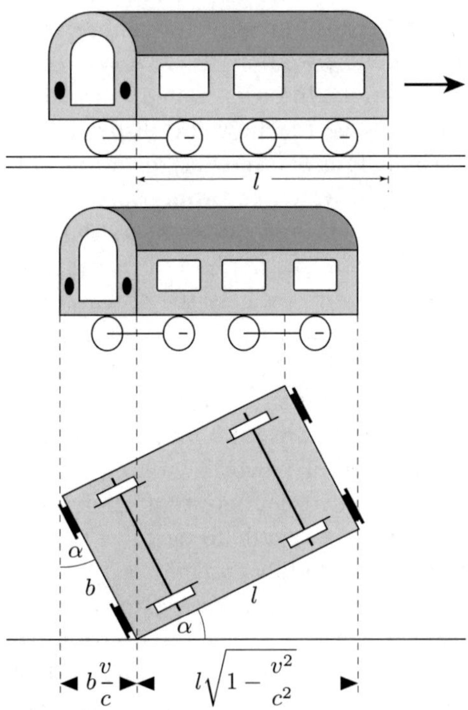

Figure 17. Rotation by Lorentz contraction: *The fast-moving carriage, viewed by an observer standing a long way from the tracks, will not appear contracted, but will instead appear to be rotated.*

6.3 Analogy between the special and general theories of relativity

The shortening of rigid standard-lengths and the slowing down of clocks appear somewhat unreal to us. Even so, these phenomena occur both in the general and special theory of relativity, at least as long as we remain in Euclidean space. This raises the question as to whether or not the phenomenon common to both theories also has a deeper common basis. To analyse this question, both theories have to be made comparable. The general theory of relativity must therefore be formulated in Euclidean space as well, which means we must bid farewell to our beloved curved spaces. According to the general theory of relativity, the fastest time and the maximum length are measured with objects which are uninfluenced by the gravitational fields of other bodies, i. e., the distance between these objects and all other objects must be infinite. According to the special theory of relativity, such objects with fast time and maximum length must be at rest, relative to the observer.

In contrast, the distances reduce to zero and time stands still with objects which according to the general theory of relativity are situated at a distance of one Schwarzschild radius from the centre of a black hole, and – according to the special theory of relativity – with objects which travel at the speed of light. These four extreme cases appear to share commonalities in pairs: an object at rest corresponds to a body in a space removed from a gravitational field, i. e., a body without external interactions. An object which flies at light speed corresponds to a body which passes the boundary of a black hole (Table 1). Perhaps a fundamental common theory can be found which describes all four extreme cases. The relationship between observer and object always plays an important role in the phenomenon of distance reduction and the slowing of time, and this role differs in the four extreme cases described above. The role of the observer in the special theory of relativity is clear: relative to the object, he is in motion or at rest. Less clear is his role in the general theory of relativity. Here it may well have to be newly defined. I will come back to this in Chapter 10.

Table 1. Analogy between the general and special theories of relativity

	General theory of relativity (described in Euclidean space)	Special theory of relativity
Greatest possible lengths and fastest possible time	Gravitation is zero, i. e., no interaction between object and other bodies	Object is at rest, relative to the observer
Smallest possible lengths and slowest possible time, i. e., all lengths are zero and time stands still	Object is situated at a distance of 1 Schwarzschild radius from the centre of a black hole	Object moves at the speed of light, relative to the observer
Time dialation	$t_A = t_B \sqrt{1-\frac{\mathscr{R}}{r}}$	$t_A = t_B \sqrt{1-\frac{v^2{}_A}{c^2}}$
length concentration	$l_A = l_B \sqrt{1-\frac{\mathscr{R}}{\mathrm{r}}}$	$l_A = l_\mathrm{B} \sqrt{1-\frac{v^2{}_\mathrm{A}}{\mathrm{c}^2}}$

t_A, t_A = object time
t_B, t_B = time of observer *B*
$\mathscr{R}$ = Schwarzschild radius
r = distance of the object from the black hole
l_A, l_A = length of the object moving with velocity v_A at location *A*
l_B, l_B = length of the object at rest at location *B* of the observer *B*
v_A = velocity of the object relative to the observer *B*
c = speed of light

6.4 $A \equiv A$?

In every metric, the distance from *A* to *B* must by definition be equal in length to the distance from *B* to *A*. The direction of measurement, from right to left or left to right, is irrelevant. On the other hand, we have ascertained that when measuring, an observer must clearly know which is left and which is right, otherwise he could miscount. The requirement that the distance in both directions must be the same length is therefore always a postulate which has been artificially introduced into the theory.

It may be that the distance itself is independent of direction, but perception of it never is. In physics, therefore, there is very much a distinction to be made between the distances AB and BA. The fact that both distances in a given theory are the same is expressed mathematically by the equation $AB = BA$. One can also give the distance a name, for example $a = AB = BA = a$, or abbreviated, $a = a$. If it concerns one and the same distance a, one says that a is identical with itself, or $a \equiv a$. However, the reader is only in a position to read the equation because he is able to distinguish between the a to the right and the a to the left of the equals sign. The equation $a \equiv a$ must thus be read precisely as follows: „We wish to disregard the fact that we are able to differentiate between the a to the left and the a to the right of the identity sign." Although the equation is entirely symmetrical in itself, the perception of the equation is chiral, i. e., the reader always knows which a stands on the left of the equals sign and which a is on the right. Readers of the symmetrical equation $a \equiv a$ are chiral themselves, and by their perception, they induce in the equation their intrinsic, chiral coordinate system, whereby the total system reader + equation becomes chiral [Domotor (1973)].

Mathematicians, when they write equals signs (and they write them very often), are rarely aware of what the equals sign means exactly. As a result, chirality is at most a marginal subject in mathematics. Without chirality, however, mathematics is not even possible, since without chirality there is no language. When the equation is not read, but spoken and heard, the spatial orientation of the chirality is represented by an orientation in time and it is also easily understood because both speaker and hearer can distinguish quite easily between the past and the future, just as writer and reader can distinguish between left and right. The intrinsic structure of both *a*'s is indeed the same: they have the same form, the same size and the same colour. Their mutual relationship to each other, to the equals sign and to the observer varies completely, however. Only such relationships are ever perceivable, never the objects themselves. When we write equations in physics therefore, we always bring an albeit practical, but yet somehow artificial element into our theory: we are saying that two things are the same, although they obviously cannot be the same, since we can always differentiate them. They are at most „equal in part". The relation sign „=" is to be understood such that $x = y$ means: The elementary experiences (thus elements of the constructional system)

x and y are equal in part [Carnap (1928/1998) p. 108]. Small children – and frequently their adult teachers as well – often struggle therefore to understand the meaning of the equals sign.

A perception of real things is only possible when the reality, i. e., that which exists, has an identity, when there is a certain constancy in Be-ing. Such a constancy in the things we draw from our experiences is requirement for any knowledge, knowledge which would not be possible amidst completely random, chaotic and continual change. It must also be possible to recognize and to describe what remains the same and what changes. The internal structure of the thing can change, as can the moment of observation and the place, i. e., the relation to the other things or to the observer. The identity sign „≡" means that the inner structure, as well as all elements which go to make up this structure, are the same for the things which are signified by the symbols to the left and right of the identity sign. There must be no exchange between the objects and the environment; otherwise they will lose their identity. On the other hand, the time and place of the observation may vary. This is not to say that such variations of time and place without any change of the inner structure are at all physically possible. (In most cases they are not.) Applied to itself, the equation $A \equiv A$ is always false, since this equation does not concern a single, identical A, but two of them. The equals sign „=" means that a quantity (to be defined case by case) of the structures left and right of the sign is the same. In physics, this means that a method of measurement must be defined which permits the measurement of the quantity concerned. The equation $A = A$ in isolation may not be false, yet it is incomplete, since it doesn't indicate which quantitative aspect of the two A's are supposed to be the same. The fundamental difference between identity and equality is that identity concerns just one thing, whereas equality concerns multiple things.

6.5 Physics without distances?[44]

If it is never really clear whether two distances or lengths are the same, one may ask whether the concept of distance is necessary in physics at all. To be sure, perceived objects are somehow arranged amongst themselves, they can be counted and the resulting figures can be presented as a number space, but whether therefore a physical space must also exist, be it as a physically independent entity or as a condition *a priori* for every observation, remains rather doubtful in light of the previous considerations. An essentially distanceless theory of spatial structure would be topology. For physics, a simple form of topology, whose elements are points, would have to suffice. But this topology must work without any concepts of continuum and infinity, since a continuum and infinity are not perceivable. For this, physics requires a concept of chirality which should be defined not by means of distances, but on the basis of rotational direction. The distances between adjacent points would all be the same; all 1, for example. The result would be a finite, chiral order topology, a mathematical theory, which to my knowledge has not been completely formulated, but which is substantially closer to physical reality than conventional theories of space. As an illustration, we can look at two objects, each comprising three points, in one-dimensional space (Figure 18).

First of all, two such objects can be topologically equivalent, but not identical; in this case they are distinguished by the orientation of their chirality. Secondly, they can be topologically identical; they then have the same structure, but they are still two separate objects. The third possibility is that the two objects are neither topologically identical nor equivalent. The only thing they then have in common is that they each have three points. These distinctions, however, are only possible if the individual points have been named so that they can be distinguished from one

[44] The idea of starting from chirality and arriving at differences in distance via chirons was a brainchild of André Dreiding in the sixties. He developed the idea further in a working group with Peter Hohler, Dimitri Pazis and Karl Wirth at the beginning of the seventies in Zurich. In 1973–1975, the idea was discussed further under the direction of André Dreiding with input from Alex Häussler, Martin Huber and the author. It concerned particularly the question of how the idea could not only be applicable in chemistry, but also in physics and mathematics. The text of Section 6.5 is not expressly authorized by these persons, however.

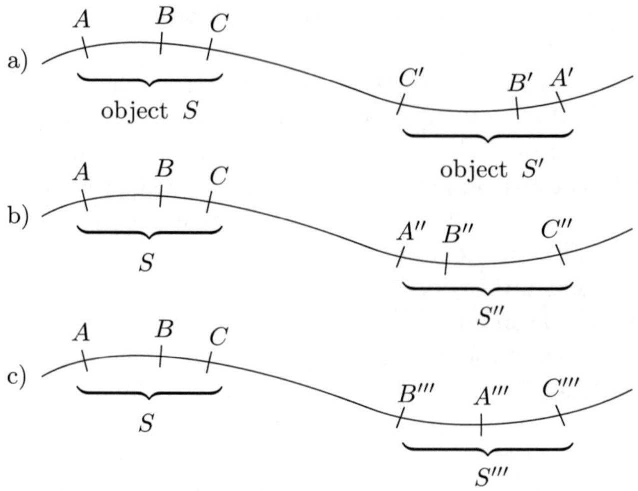

Figure 18. One-dimensional discrete topology:
a) The objects *S* and *S'* are topologically equivalent, but not identical.
b) The objects *S* and *S"* are topologically identical.
c) The objects *S* and *S'"* are neither topologically identical nor topologically equivalent.

another. Of course, physical points are not labelled, which is why other ways must be sought to make them distinguishable. It cannot be the task of this book to develop such a theory, particularly since a different approach is taken in Chapter 7, which is a further step away from the usual theories of space.

Chirality is usually defined by means of distances: an object is chiral when it is isometric to its mirror image but is different [Kelvin (1893/1904)]. This means that the mirror image of a chiral object cannot be brought to coincidence with the original by an actual movement. In order to determine whether an image and its mirror image are isometric, we must measure distances [Reidemeister (1957)].

Dreiding had the idea of reversing the definition of chirality and the concept of distance, so that distance is defined by means of chiral units, so-called chirons. The term „chiron“ as „the smallest chiral unit“ was coined by André Dreiding in the middle of the 1960s and described in numerous public lectures. Together with Karl Wirth, Dreiding et al.

a) Mirroring of an object composed of two chirons (triangles with rotational direction) in two-dimensional space.

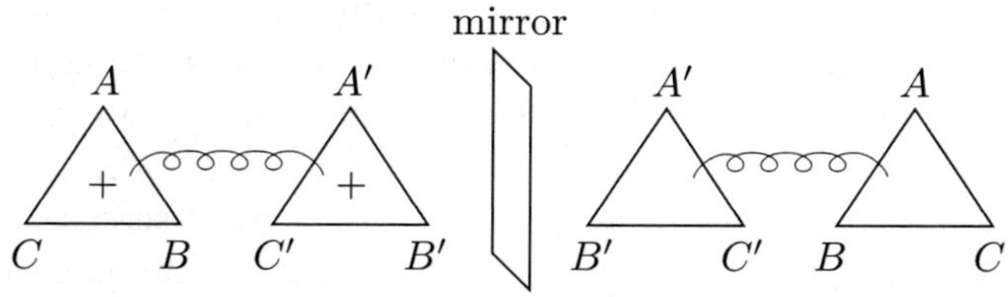

All distances remain the same; only the orientation of the chirons changes.

b) Mirroring of only one chiron.

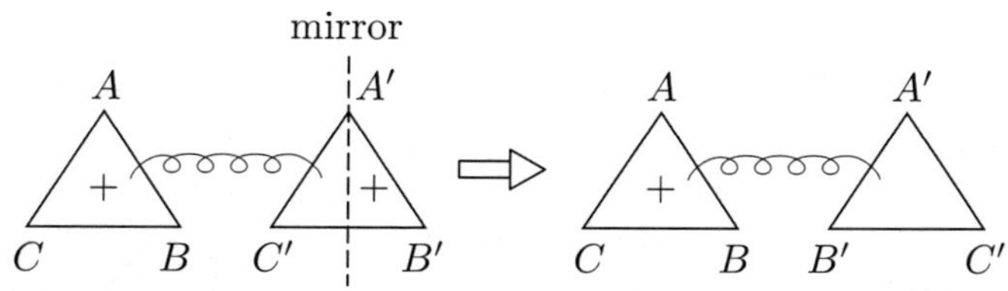

The relative orientation between the two chirons changes by a partial mirroring. In the process, the distances between the corners of both chirons change as well.

Figure 19. Chirality and distance[45]

(1980) later defined the term somewhat more closely: A chiron is an n-dimensional simplex, whose edges are all different in pairs. In three-dimensional space, the chiron is thus a scalene (lopsided) tetrahedron. An object composed of two chirons is converted by a process of reflection into its mirror-image. Applying this process to the mirror image reproduces the original once more. That which changes during this operation is called orientation. If the process is only applied to one chiron, then something changes between the two chirons which previously remained unchanged. This something is called distance (Figure 19). This theory can only function, however, if the points which make up the chiron can

[45] Figure 19. is essentially the reproduction of a drawing by Dreiding from the 1960s.

be distinguished from one another. That generally presents no problem in chemistry where one deals with different atoms or ligands. In physics, however, where corners don't have names, this theory of distance is not practicable. Therefore, I will opt for another approach in Chapter 7.

Humans are chiral; they primarily see chirons and judge distances by them. Children very probably develop their perception of distance in this way: first, they realise that the order and the topology of perceived objects differ or can change, and only learn much later on the basis of such positioning to distinguish distances as well.

6.6 The probabilistic outcome of measurements

Probabilistic outcomes of experiments happen for four reasons:

1. Depending on the definition of reality and existence something might exist but not be real (see definitions in Section 1). In chirality theory, this is the case for single points whose only intrinsic property is their existence although they are not observable. This means that the presumed observer with his mathematical model describes an existing, finite, natural world whose structure is in principle not perceivable. Einstein proceeded this way: „God does not play dice". The real observer with his mathematical model then describes only his observations of real nature and makes corresponding probabilistic propositions. Bohr proceeded that way.
2. Every perceived object is changed by every perception. Therefore, it is impossible to know the precise present state of any object.
3. In chirality theory, there is one and only one universe. Every point is related to all the other points of the universe. A separation of the universe into systems or objects is always an artificial one and falsifies the outcome of the measurements. Nonetheless such separations are inevitable because the universe is too big and too complex to be observed as a whole and to be precisely described mathematically.
4. Measurements never are perfect.

All this is in harmony with the philosophy of quantum theory. The first two causes of probabilistic measurements are indeed related to quantum theory where the states have no definitive values for the measurement outcome. There is in fact zero-probability in nature itself, but not in the observation of nature. This is another example for the importance of strict distinction between nature, observation of nature and the mathematical model for the description of the observation.

Chapter 7

The Event as a Mathematical Unit

> Thus, I ponder whether all of nature and heavenly delicateness are not symbolised in geometry.
>
> Johann Kepler (1595)[46]

7.1 Metaphysical presuppositions for a new physical theory: chirality theory

In this chapter I will develop a strategy towards a new physical theory. I will call it chirality theory. It is closer to reality as we perceive it than the conventional theories. Before I begin with this therefore, it is worth recapitulating once more what metaphysics requires of such a theory and to what extent the familiar theories depart from this. These questions are the subject matter of epistemology, which points out the conditions necessary and sufficient for being able to explain perception as empirical knowledge of the external world [Baumgarten (1999)].

In order to present the highest possible degree of truth, the theory is to have a direct ego-connection, in line with Sections 2.4 and 3.2 and it should be simple, immediately evident and logical. The things which the theory describes must be perceivable, so they must be linked coherently and causally with each other. In Chapter 4, it was argued that chirality, the duality of all Be-ing, should be an important basis of every physical theory.

As stated in Section 5.5, information quantities must be the subject of the theory which makes propositions about how such quantities change. In the course of such a change, however, something else must

[46] „*Also dass es einer auß meinen Gedanken ist, ob nicht die gantze Natur und alle himmlische Zierligkeit in der Geometrie symbolisirt sey.*" Letter by Johann Kepler (1595/1619).

remain constant. The structures of subject and object must be such that the information can be copied and stored by them and that it can flow from object to subject. For this the perceiving subject and the perceived object do need to be linked to each other, but nevertheless clearly distinguishable. The standard measure of the information is the alternative, i. e., the answer to a yes/no question. According to Chapter 6, the units are finite and countable.

These conditions are, as we have already seen in Section 2.3, largely violated by the conventional physical theories. Space and time as a continuum are neither perceivable nor countable. Photons and neutrinos are not perceivable either; only their destruction is perceivable. Interactions, gravitational forces and fields may be convenient, but they are certainly not immediately evident and logical. The same applies to the shortening of standard lengths, the slowing down of clocks and to curved spaces. In some theories chirality plays no part at all, or only a minor role. In Section 5.4 I also commented in the discussion about the particle-wave dualism that we are yet to understand the nature not only of space and time, but of matter as well.

From earliest times natural scientists have described the world by means of four entities: time, space, matter and interaction. Since Aristotle there have been repeated attempts to reduce the number of entities to three, the latest example being string theory [Greene (1999); Johnson (2006)]. The attempts have all failed up until now. Even so, there appear to be similarities between the four entities. Space and time are both chiral to the observer and appear never independently of each other. The curvature of space influences matter and vice versa. Are space and matter really two completely different things? The photon, with its particle-wave dualism, is half particle, half interaction. So, are space, time, matter and interaction really four different entities, or are they rather four different aspects of one and the same entity? And what, then, is this entity? I will call it an „event“ and I assert the following: Viewed in terms of physics, there is nothing apart from events.[47] Only events can be perceived. Only

[47] Most philosophizing mathematicians use the term event in the same sense. In probability calculus, event means something else, however, for example throwing a six with a die.

they are real. Indeed, the act of perception is itself an event. To measure means to count events.

Methodically I will now proceed as follows with the development of a new physical theory. I begin with the composition of a mathematical language, logic and an axiomatic theory, which for the time being, only have the condition of having to be as simple as possible. Axioms must not contradict thereby the metaphysical principles which must be fulfilled for perceptions in accordance with the remarks in the past six chapters. The axiom of infinity is therefore omitted and the axiom $A \equiv A$ must be replaced by a new „axiom of chirality". With these few first steps, only the simplicity of mathematics and the metaphysical basic conditions are thus considered. Physical experiences, i. e., concrete perceptions, play no role at all for the time being. Very soon it will be shown that mathematics, developed in such a way, not only fulfils all metaphysical conditions but that surprisingly, it also can be interpreted as the laws of nature. Applied skilfully to physical perceptions, the new mathematics is able to describe the theories of relativity, quantum theory, the natural constants, elementary particles with their interactions and cosmology with amazing beauty and uniformity. And all of this, without the need for physical experiences as the basis for the development of the theory. At most, these experiences still serve as the inspiration.

7.2 Axiomatics for space, time and events

It cannot be a task of this book to construct an axiomatic system for chirality theory. A mathematician once told me that for this goal 200 mathematicians and physicists would have to work for 20 years. But in a rather philosophical book like this it is necessary to list the contradictions between the usual axioms and physical perception. This concerns mainly the axioms of infinity and identity. The first is the main cause of the incompatibility of relativity theory with quantum theory, the latter for the neglecting of chirality as the basic principle of order in physics. However, there is no need to build up a complete axiomatic for chirality theory to understand the main ideas of its strategy.

Since a physical theory, as stated in Section 2.3, must be formulated mathematically, i. e., if possible in terms of set theory, a mathematical definition needs to be found for the concept „event“ that satisfies all metaphysical demands and which, applied in a simple theory, makes the correct predictions about the future. I cannot formulate a complete axiomatic of this theory, however, I would like to give an indication of the extent to which the conventional axiomatic systems need to be changed. Such systems can be found in the writings of Alfred Tarski (1959) pp. 6–29, Patrick Suppes (1973) and Stewart Shapiro (1997) for example, with many further references. Most axiomatics are concerned with a continuum, yet there are also finite axiomatic systems. The event is usually simply a point in space-time. However, in some systems there is also the requirement that the event should be a space-time-*region* rather than a single point [Earman (1973)]. Instead of chirality, axiomaticians generally speak of causality, which then transforms an intrinsically symmetrical one-dimensional time into a chiral time [Domotor (1973)]. Chirality is introduced as the relation between the events, whereby one event can be the cause of the other. Naturally, however, pairs of events which are not causally linked are also possible. They are usually classified as simultaneous. The causal relationship is often represented by an arrow. $A \Rightarrow B$ means that A is the cause of B. The primacy of either the event or time is founded on different presuppositions. Just what the elements of set theory should be is also contentious. It could be simple points, or it could be the relationships between the points. Spaces come about as a special abstraction of sets comprised of point configurations. Some, but not all, of these axiomatic systems give rise to a metric.

Walker (1959) has a clear conception of what an observable should be, namely a mapping from a differentiated particle, called the observer, on this very observer, resulting from a chain of signal-mappings and inverse signal-mappings.

An overview of a possible new type of mathematics and axiomatics for such theories can be found in Section 14.7. It needs no longer necessarily be based on set theory.

7.3 A point is a point

As we have seen in Section 2.4, a theory's degree of truth is greater the simpler the theory is. Let's begin therefore with the simplest thing mathematics has to offer, with a point. Since Leibniz, there have continually been philosophers who have held the opinion that it is logically impossible to compose a space or a time period solely of points. Russell (1903/1956) pp. 445–455, demonstrated the incorrectness of this view. A single, lonely point: not a lot can be done with it. A point has neither dimensions, nor location, nor time, since location and time must be described as a relation to other points, which according to our assumption, don't even exist. So, it's neither a matter of a point in time, nor a point in space, but simply a point. The point is effectively suspended in nothingness. Since it has no dimensions, neither can it have properties, not even a colour. It is a completely structureless, boring point. Nevertheless, we can assert that it exists. However, for this we require an observer who perceives the point or at least imagines it, so that he can say it exists in his imagination, in his ego, in his Self. If, on the other hand, the point is completely alone, then nobody knows of its existence, not even the point itself, since it has no structure and therefore no consciousness.

Most mathematicians probably would prefer to begin with an n-point space or even a ∞-point space which would better correspond to our universe than a few single points. But compared with a 4-point space the n-point space is much more complicated. It allows so many relations that the model would have to be simplified for calculations possible and for describing physical observations, e.g. by artificial perturbation and renormalization tools which have nothing to do with nature. This would contradict the metaphysical requirement of simplicity and compatibility with nature. The n-point system will be discussed later in Section 8.13.

Over 2,500 years ago, Anaximander (610–546 BC) sought to explain the world as composed of a single primary material, devoid of properties. There were various suggestions for such a primary matter: water, fire, air. Yet the point is probably the only thing in existence devoid of properties, indeed this very lack of properties characterises a point and suffices to define it.

The concept of the point has essentially virtually the same characteristics featured by Plato's „One" [Plato (1953); Böhme (2000) p. 129]. Yet, Plato never really explained exactly what he meant by „One", and Plato's One would probably not be definable at all. The „One" is a self-identical something which has no intrinsic relationships within itself, perhaps the negation of nothingness. Certainly, Plato's One is not simply a point, yet a point could very well be an example of the One. There is nothing to be gained by describing the concept of the point any closer, linguistically, philosophically or mathematically. Nonetheless, I cite Whitehead's (1936) p. 456, definition: „A geometrical element is called a ‚point', when there is no geometrical element in it. This definition of a ‚point' is to be compared with Euclid's definition: ‚A point is without parts'." Children, when they are still very small, have a pretty good grasp of what a point is. I expect, as Plato also did, that the reader knows what a point is and that his conception of a point is not all that different from mine.

7.4 Two points

I will now add one point after another and investigate what can generally be said about the growing quantity of points. Surprisingly enough, this will bring us very quickly to a rather simple mathematical structure which could prove interesting for the construction of a physical theory. Initially it involves neither points in space nor time and not even objects, but solely mathematical points. Two points can appear in two different states: they can both coincide, or not. If they coincide, then the outcome is not, as such, a point with a Siamese twins structure, since points can't have any structure at all; on the contrary, the outcome is a single point, of which an observer possibly knows that it originated from two points. Such an observer would also be in a position to distinguish both states of the two points, e.g., by counting the perceivable points or by saying that one of the states precedes the other temporally. However, as long as the two points are present, completely alone and without an observer, there is no distinction between the two states.

The set of the two points I call space. This involves a purely mathematical space. There is no time. For the time being, a relationship between the two points is not required either. The state in which the two points coincide is zero-dimensional. If the points are separate, then the space is one-dimensional. There is no distance between the two points in the conventional sense, since the distance would have to consist of a (possibly infinite) number of further points, and there is an absence of such in the two-point space. Nevertheless, a distance can be defined mathematically as follows: if the two points coincide, then the distance is zero; if they don't coincide, then the distance is not zero. Perhaps it is useful to also assign the distance „not zero“ a number, most simply, the number one. So, one cannot construct from two points alone longer or shorter distances, but only ever the distances zero and one. Apart from that the two points cannot be closer, nor further apart from each other. I have tacitly presupposed that the distance is the same, irrespective of the direction in which it is measured. That would have to be explicitly required in an axiom of reflexivity, because it would actually also be possible to say that in the one direction the distance was +1, and in the other one –1 [Russell (1903/1956) pp. 171 ff.].

Were we to employ these mathematical propositions to describe empirical perceptions, the mathematical two-point model would be reduced further and thereby simplified: infinity is fundamentally not perceivable, so neither is the infinitely short distance zero. Two points which coincide are mathematically, but not physically possible, since the distance between the two coinciding points is infinitely short and thus not perceivable. A point is, from a physical perspective really only one and never a pair of points. That would have to be explicitly required in an axiom of individuality. It follows from this that two points also are only ever two points, without any internal structure of the two-point space.

Most remarkable, and taking some getting used to, is the use of the term space. The two points are not *in* the space, rather they *constitute* their space. They cannot move before a spatial background. Each change of state is actually a change of the space. There is no distinction between the points as things in the space and the space as a framework for the things.

7.5 Three points: between

Thanks to the third point there are possibilities for relations between the points, since the third point can lie between the other two. Gauss, in a letter to W. Bolyai in 1832, criticised the absence of ordering axioms with Euclid, i. e., essentially rules for the use of the word „between“ [Janich (1997)]. Mathematically, I wish to follow Tarski (1959) and Suppes (1973) p. 387, and express the relation „between“ thus: $\beta(xyz)$ is to signify that point y lies between points x and z. A segment between each two points need not exist, and should there indeed be a segment, then it's not necessarily straight. There is nothing to distinguish one curved segment between end-points from another curved segment between those end-points. One is no straighter than another [Whitehead (1936) p. 591]. With the help of the relation „between" the important term „straight line" can also be defined: The straight line is complete, it contains points (these are things which do not have parts), it is clearly defined by each pair of points and the straight lines can cross each other at one point. Thus, all points which lie between a pair of points must be on the same straight line [ibid. p. 594]. It is stressed once more however, that all axioms which mean or use a continuum, i.e., an infinity, do not apply in this form for chirality theory. In nature there is no infinity, because infinity is never perceivable [ibid. p. 590].

Moreover, I wish to use the following symbols, which are usual in set theory: logical implication (if …, then …) $\Rightarrow$, the universal quantifier („for all …, … applies") $\forall$ and the equals sign =. I would also like to adopt a distance relation: $\delta(xyzu)$ should be read as „the distance between x and y is equal in length to the distance between z and u." With these symbols, the following axioms can be formulated in the three-point space:

A1 Identity axiom for betweenness

$$\forall xy\,[\beta(xyx) \;\Rightarrow\; (x \;=\; y)]$$

The axiom is read as follows: When there are two points x and y which have the characteristic that y lies between x and x, it follows from this that points x and y coincide. Incidentally, three points are not necessary for this; the axiom already applies in two-point space.

A2 Reflexivity axiom for equidistance

$$\forall xy\,[\delta(xyyx)]$$

This axiom can also be formulated in two-point space.

A3 Identity axiom for equidistance

$$\forall xyz\,[\delta(xyzz) \;\Rightarrow\; (x \;=\; y)]$$

In axioms A1 and A3, points x and y coincide in each case. Although such cases are, as we have seen, mathematically possible, physically, however, they are invalid. For physics, therefore, axioms A1 and A3 cannot be used; all that remains is axiom A2.

If none of the three points lie between the other two, the points form a triangle, i. e., a two-dimensional space. The three distances between the corners are all 1. If the three corners have names, e. g., x, y and z, then they can be arranged clockwise or anti-clockwise. In two-dimensional space, these two states cannot be brought to coincidence and can therefore be distinguished. Since points don't have names in nature, the distinction is physically impossible, however. On the other hand, there is a third state, which is clearly distinct from a triangle, namely the case where a point lies between the other two, e. g., $\beta(xyz)$. Here the three points only form a one-dimensional space, and the clockwise positioning is no longer distinguishable from anti-clockwise. If the three points do not coincide, $\delta(xyyz)$, i. e., y divides the line xz in two parts, each will have the distance 1. Thus, distance $xz = zx = 1 + 1 = 2$, and in this way, it is clearly distinguishable from distance 1 in the triangle. In this way, all three conditions for a metric are fulfilled: First, the distance between points x and x is zero and therefore quite invalid in physics. Second, the reflexivity axiom applies for equidistance, and third, the distance xz is $xy + yz \geq xy$. The meaning of the term distance in a space with four or more points is explained in Sections 8.12.2, 8.13.2 and 8.13.5. The reader may wonder why I have defined the term „between" in such an unusual way. Normally one would simply say y lies between x and z if $xy + yz = xz$. That presupposes, however, that it is clear what the term „distance xz" means. I have already explained that with respect to a three-point space; concerning a space with more dimensions and any desired number of points, however, the concept of distance has not yet been precisely defined, although I present possible solutions of the distance problem in Sections 8.12.2 and 8.13.5. It will be seen that this definition remains the main problem of chirality theory for the time being. I proceed from the term „between", use it to define the term „event" and then call for the distance to be defined by an event count. This makes sense because a physicist can only count events; there is no other method of measurement.

Now I permit the three points to move freely and to adopt different states in turn, both singly and as point sets. The individual points have four different states of movement relative to the two other points; they are each characterised by an arrow symbol:

1. The point can move away from the two other points: (↦) = „move away".
2. The point can move towards the two other points: (⇤) = „move towards".
3. The point lies between the two other points: (↔) = „between".
4. The point can turn back: (↩) = „turn back".

How the terms „away" and „towards" are to be understood in the context of a space where the distance between any two points equals one will be described in Section 7.6 on the basis of the term „event". With the between-point (↔) it is intentionally left open as to which of the points is between the other two (compare Figure 21, p. 171). (↔) designates the transition from (⇤) to (↦), whereas (↩) indicates the transition from (↦) to (⇤). So, the individual point moves in a defined sequence through the four states: it moves towards the other two points, passes between them, moves away from them and then turns back. What causes it to turn back? It is important to always keep in mind that the points do not move in absolute space, but only relative to the other two points, which form the point space for the third point. This distinguishes chirality theory from most conventional theories of space which require a background space as part of the model. It is not points moving in space, but rather – since the points themselves form the space – space moving through itself. Each time a point moves through between the others, the space turns itself inside out like an inverted glove. If a point is situated at the tip of the glove's index finger and it moves away from the glove in the direction the finger is pointing, then as soon as the glove is turned inside out, this point will abruptly turn back to the glove and fly back into it. Exterior and interior have been transposed by the inversion of the glove. From the point's perspective, the direction of motion has not really changed; through the inversion of the space, the direction of motion, however, has reversed relative to the other two points, which form the surrounding space for it. The cause of the inversion of the space is always the point in the (↔)-state, i.e. the one which migrates through the other two. A further remark on the (↔)-state: it occurs in two variants, because the symbol means that one point lies between the two others, without specifying which point it concerns thereby. In the following, I will assume that the (↔)-point itself lies between the other two. The variant, whereby one of the two other points lies in between, I will discuss in Section 7.6 which will cover the four-point space, where not only two, but even three different variants of the (↔)-state are possible (cf. Figure 21).

The inversion of a point space I define as an **event**. The physical interpretation of the model will be given in Chapter 8. Events will be interpreted as the basic ontological real objects in physics. Should the physicist find these ideas plausible, then he can formulate the following *axiom of chirality*:

A4 (↤) ⇒ (↔) ⇒ (↦) ⇒ (↩) ⇒ (↤) ⇒ and so on

The axiom of chirality corresponds neither in the form nor the content to the usual mathematical or physical axioms. It is quite possible that it could be divided into sub-axioms or that a somewhat different axiomatic system leads to a similar theory. „We must always be prepared to change the axiomatic foundation of physics in order to do justice to the facts of observations in the most logically perfect way possible“ [Einstein (1991b)]. Since the axiom of chirality portrays a process and consists solely of arrows, it belongs most appropriately to category theory, which can thereby probably play an important role in theoretical physics [Dümont (1999)]. The four states of motion of the individual points are thus ordered and have a direction. The symbol ⇒ („if …, then …“) denotes in this context the phase change from one (motion)-state to the next. This process was precisely analysed by Aristotle (1987) vol. VIII, § 11. The „in between“ is referred to there as the „middle“. The states (↦) and (↤) are moved, the states (↩) and (↔) have „stopping“ characteristics.

The axiom of chirality is intended to make propositions about many points. In the definitions of the five different axiomatic arrows there were only 3 points. To become applicable to an n-point space the axiom of chirality probably must be somehow mathematically modified (see Section 8.13).

The symbol ⇒ denotes in the context of the axiom of chirality the phase change from one (motion) state to the next. According to Russell (1903/1956) p. 471, in mathematics a „state of change“ is impossible in principle. For him, a change is not a state itself, but the difference between two points (in time), i.e., between two states. Precisely such changes however, constitute the nature of the axiom of chirality. A completely new type of mathematics should result from this. A branch of mathematics which is not based on set theory, but rather on morphisms expressed by arrows is category theory [Mac Lane (1969/1998)].

In order to be able to show further characteristics of point motion, let's now suppose that the process was not stopped here, but continued indefinitely. Since the three-point space is two-dimensional, the consequences can be illustrated and understood more easily with a two-dimensional figure as this will be possible later with the three-dimensional four-point space (Figure 20, p. 170). The motion of the points according to the rules of axiom A4 results in a rotation of the triangle which is phase dependent. The triangle rotates within two phase transitions in each case by 120°, either clockwise or anti-clockwise. So, two different, phase-dependent, chiral states of the triangle can be differentiated from each other. The triangle's „clockwise" state (↻) can be differentiated from its „anti-clockwise" state (↺). By shifting by a phase, one state passes into the next.

A presumed observer would note that certain directions in the observational space are distinguished. The space is not isotropic, since the corners of the triangle repeatedly point in the same six directions. Following every six phase transitions, both the triangle as a whole and the individual points are back in their original state: the triangle remains itself. Similar to the case of a standing wave, something changed through the six phases, whereas something else remained unchanged.

Thus, in our simple model of the three-point space several of our conditions for a physical theory are already fulfilled: the theory is simple, albeit a little unusual, but yet logical and clear (at least for me). Chirality plays a fundamental role. It provides information which is countable and which can be partly assessed with simple alternatives, for example with the question: Does the triangle rotate clockwise or anti-clockwise? After all, the theory dispenses with the concept of infinity and a continuum.

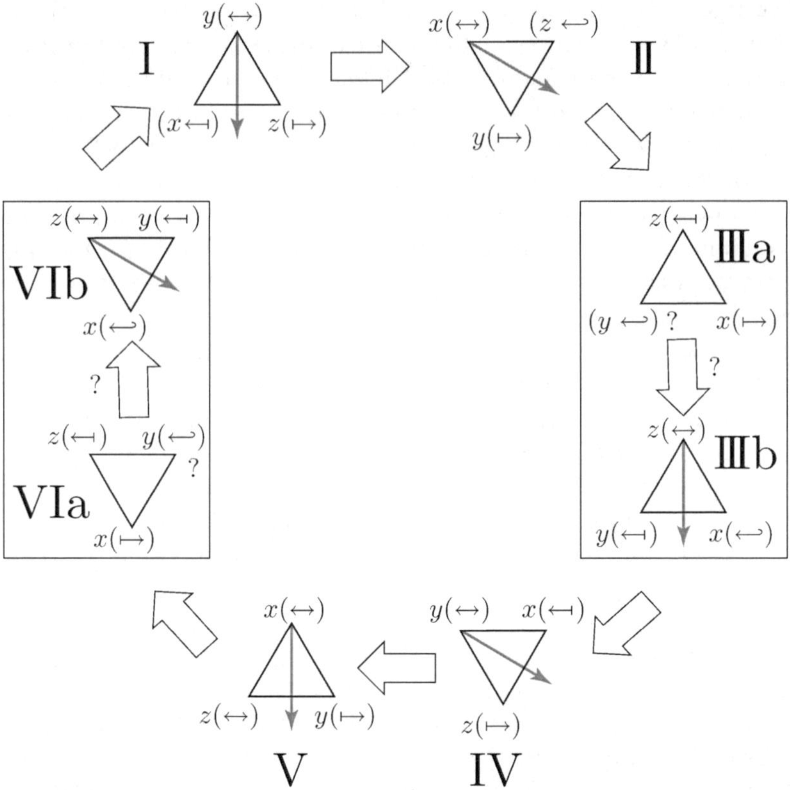

Figure 20. Motion state of the points in the three-point space: *Following the 6 phases* I *to* VI, *the space returns to its original state* I. *The phase transitions* (IIIa ⇒ IIIb) *and* (VIa ⇒ VIb) *are actually forbidden, since no point can turn back if no other point is in the „between" (↔) state at the same time. In both of the forbidden transitions, the orientation and the position of the triangle remain unchanged. Throughout the odd phases* I, III *and* V *the triangle (x, y, z) rotates in an anti-clockwise direction; throughout the even phases, however, in a clockwise direction.*

7.6 Four points: definition of the term event

When not all of the four points lie on the same plane or indeed on the same line, then they form the corners of a tetrahedron, i.e., a three-dimensional space composed of four points. A tetrahedron is the flat locus defined by four points which are not coplanar. The four points are called the corners of the tetrahedron [Whitehead (1936) p. 467]. In terms of group theory, the characteristics of the tetrahedron are described for example by Sternberg (1997) pp. 27–35. In the tetrahedron the corners can also adopt the four different states of motion (⇤), (↔), (↦) and (↩). The three fundamentally different locations where the (↔)-point can break through the surface, which is determined by the triangle of the other three points (Figure 21), can be characterized with the help of our simple relation „between": The sides of the triangle consist of the sets of all points which lie between the corners. The inside of the triangle consists of the points which lie between the three sides. The extension of the side beyond a corner consists of the set of all points with the characteristic that the corner lies between them and the side concerned. The points over a corner lie between the two sides extended beyond this corner. The points over a side lie between this side and the extensions of the two other sides. The plane, which is determined by the three points, consists of the set of all points specified above. Concerning the axiomatic of point theory, see Tarski (1959) pp. 16–29.

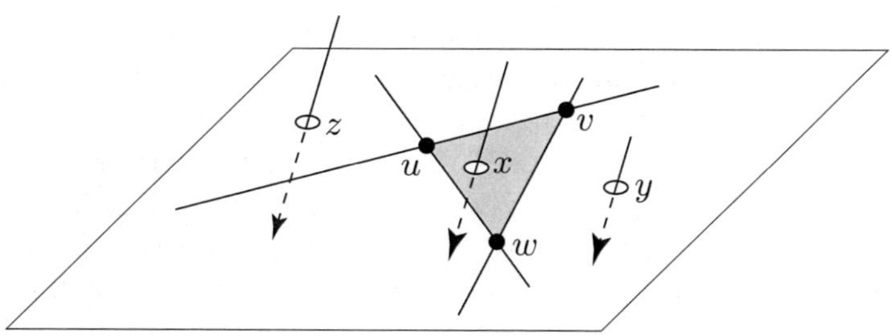

Figure 21. The three possibilities for the point „between": *The plane which is defined by the triangle* (u, v, w) *can be pierced in essentially three different places by a fourth point: inside the triangle* x (↔)*, above a side* y (↔) *or above a corner* z (↔).

One can ask whether it is legitimate to speak of a tetrahedron although the (↔)-point is between the three other points, thus actually lies on the same plane as these. Even so, I believe that the tetrahedron model is appropriate because the (↔)-point also has, relative to the other three, a direction perpendicular to the two-dimensional three-point space, a characteristic which is quite well represented by the three-dimensional tetrahedron. For the (↔)-point, there are three different possibilities. It can pierce the surface which is defined by the other three points in three fundamentally different places: in the interior of the triangle, above a side of the triangle or above a corner of the triangle. (Figure 21). Each time a point pierces the surface, which is marked out by the other points, the three-dimensional space which is defined by all four points inverts.

Definition: This inversion of the three-dimensional space I call an event.

The four possible states of motion of a point can now, with the help of the term event, also be characterised in the following way: (⇤) is the state which leads to an event; (↦) does not result in an event; (↔) and (↩) mean that in this state, an event occurs whereby (↔) is the cause and (↩) is the consequence of it. This view has the advantage that the term event can be defined without requiring the term distance: „to move towards the other points“ no longer means that there is a distance between the points which becomes shorter, but simply that the motion leads to an event. So, distances only ever change as a result of events and thereby become countable.

This perspective recalls Zeno's arrow paradox, ca. 450 BC. How can an arrow actually fly at all? After all, it is always in a specific location. In order to fly, however, it must change its location. In so doing, it is neither in its old position, nor in its new position, so it doesn't have any position [Aristotle (1987) vol. VI, § 8 and 9]. The contradiction has been discussed over and over to the present day and appears unsolvable. In the four-point space, two points (↔) and (↩) are each in a specific location, whereas the points (↦) and (⇤) change their locations. Still, the individual points are never empirically perceivable, but at most the event as a whole, the inversion of the space is and this is countable.

For counting, numbers and their axioms play an important role. With chirality theory, only finite rational numbers are required because all measurements are described by numbers of counted events. Thus, only few axioms are required, namely the simple Peano axioms. If surds, imaginary or infinite numbers were introduced, e. g. to describe waves or information, this would be part of the mathematical theory and not an aspect of real nature.

For the mathematical description of measurements, a metric is required. Since infinity and continuum should be excluded in physical models the metric must define distances as rational numbers and it must be defined what entity is counted by these numbers.

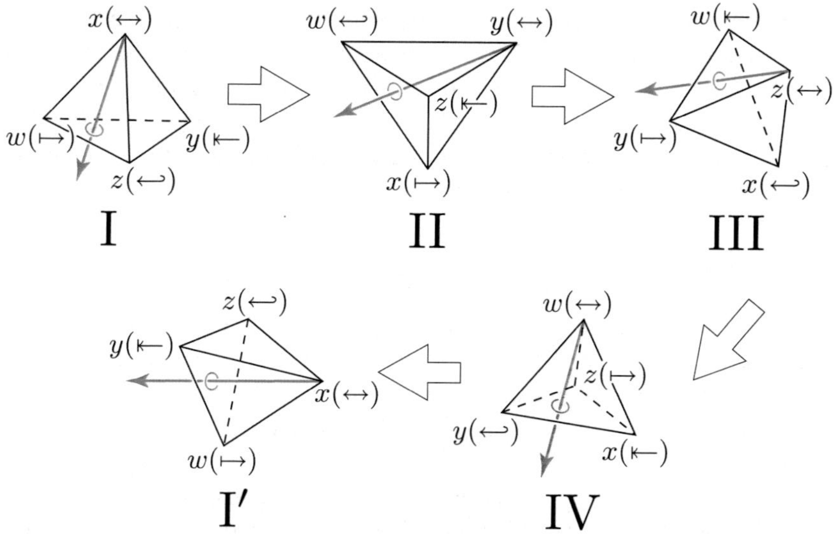

Figure 22. Four-point space: *Following phases* I *to* IV, *the four-point space returns to the original state* I, *although in another position from the perspective of the external, presumed observer. If one disregards the designations* x, y, z, *and* w *of the four corners, then the state of the four-point space is the same following each phase change, only its position has changed, whereas the chiral orientation has remained the same: viewed from the* (↔)*-point, the triangle* [(↦), (↩), (⇤)] *always rotates in an anti-clockwise direction.*

In the four-point space all four possible states of motion can be assigned to one of the four points. Since one point is thereby in the (↔)-state following each phase change, the motion never stops. The chain of events cannot break down and after every event the result is once again a tetrahedron with the four corners in the four different states of motion. Only a disruption from outside could stop this process. An outside does not exist in the four-point space, however. In Figure 22, such an unrestricted sequence of periodic events is illustrated. As in the case of the three-point space, the motion of the four-point space is chiral, whereby – in contrast to the three-point space – the orientation of the tetrahedron's motion as a whole remains unchanged during the phase change. When compared with the three-point space, the four-point space reveals two further interesting characteristics: First, the process of periodic events does not cease. The tetrahedron with its four different corners retains its identity throughout every event, similar to a physical particle. According to Smolin (2001) p. 53, the universe consists simply of events. By an event he understands a smallest possible change. The changes can be calculated. Second, the tetrahedron's motion from the perspective of an external observer is probably isotropic: No particular direction in space is distinguished in the three-dimensional space, similar to the case with physical space. However, this assumption is yet to be proven.

Thus, the six different states of the four-point space can be differentiated: three are different in that the (↔)-point either pierces the triangle's surface in the triangle's interior, or above a side or above a corner. Each of these three states can occur in two variations, (↺) or (↻), which differ through the orientation of the chiral three-dimensional point motion.

What is the philosophical significance of the event? Heidegger (1977) says: „Time itself is the Be-ing" and he wanted the „is" to be read as a transitive verb, in the sense of „Only time yields Be-ing". To the German sentence „*Es gibt Sein*", literally „It gives Be-ing" (i. e., „There is Be-ing" in natural English), he attached the question, what „it" is it that „gives"?[48] And he answered, „The event gives", since the event yields time [Han (1999) pp. 86 f. and 103].

In the same way that a point is an example of the One of Parmenides and Plato, the four-point space is an example of Heraclitus' „everything flows": it is in perpetual internal motion, yet it remains itself. The statement that „everything flows" was made independently at about the same time by Heraclitus (550–480 BC) and Confucius (551–479 BC/1972). The philosophy of Confucius goes back to the *I Ching*, the age-old Chinese book of wisdom from the Shang dynasty (16th–11th century BC), commonly known as The Book of Change. It concerns systematic, though esoteric teachings, yet is quite comparable with modern group theory [Wilhelm (1912/1923)].

[48] Compare the colloquial expression, „What gives?" One might also ask what the „it" is that rains when „it" is raining.

Plato (1961–1963c) in Philebos, said: „… those things which are from time to time said to be existent are made up of one and many, with a determinant and indeterminacy inherent in them“ [Böhme (2000) p. 149]. The One in chirality theory of the four-point space is the point; the many are the multitude of points, the determinant is the space composed of four points which remain together without an external effect in perpetuity and the indeterminacy is this perpetuity. Aristotle (1987) described most characteristics of the four-point space in his Book Physics, vol. VIII, § 3–8.

In order to avoid having to repeatedly mention the relativity of the concepts of space and time, physics has now the advantage of being based on the absolute concept of an event. An event, according to Minkowski is something that takes place at a certain time in a certain place. Of course, a localisation of the events is only possible relative to a frame of reference, but the event itself is absolute, i. e., it is one and the same in all frames of reference [Hölling (1971a) pp. 29 and 103 ff.].

I have now reversed Minkowski's procedure: from the absolute, namely the event which is the same for all observers, I will derive space and time. It is important to always bear in mind that the space, as I have just described it, is not a three-dimensional space with four points in it, but that the four points themselves form the space, which is comprised exclusively of these four points. There are no lines, neither straight nor otherwise, no planes and no volume in this space, but only the four points which are arranged according to definite rules and this arrangement continually changes. In the process, a connection exists between the arrangement before and the arrangement after the change. In contrast to most other theorists, Mehlberg (1980) vol. I, p. 205, defines the event as an object, that is connected with other objects by a causal relation. Thus, his event is not an infinitesimal point in space-time as in most other physical theories. More concrete details as to how conventional, set theory-based infinite mathematics should be changed will be found by the reader in Section 14.7.

7.7 Dimensionality

The dimension of a mathematical space or object is informally defined as the minimum number of coordinates needed to specify any point within it: The dimension of x is the set of physical quantities that one obtains by multiplication of x with a real number [Suppes (1973) p. 400]. For metric vector spaces this is the number of countable basis ($\mathbb{R}^n$) [Reinhardt (1994) p. 87]. Topological spaces can only be homeomorphic if they have same dimensionality [Reinhardt (1994) p. 233].

The problem of the meaning of dimensionality with the 4-point space of chirality theory is threefold:

1. Because there is no manifold as background space there is no countable basis ($\mathbb{R}^n$).
2. The 4-point space inverts with each event.
3. Since there are only events to be counted it is not evident why the dimensionality can be other than 1.

It follows that in chirality theory other criteria must be found for determining the dimensionality of a space. This is not unusual in mathematics where various definitions for the notion of dimensionality are used. This situation resulted in rational, irrational, imaginary or negative numbers for the dimensionality in specific theories.

Since the 4-point space consists of 4 points in 4 different physical states relative to the states of the other 3 points the 4 points can be characterized or illustrated by a tetrahedron. Each of the 4 points requires a dimension of its own to distinguish it from the other points. The single first point is 0-dimensional; together with the second point a 1-dimensional space with a distance 1 between the points is formed; the third point gives rise to a 2nd dimension and with the fourth point the space becomes 3-dimensional. A set of n points in n different states relative to each other results in an (n-1)-dimensional space. The 4 points in 4 different physical states are the 4 basic *existing* ontological objects. The basic *real* ontological object is the 4-point space as a whole. Because 4 points are required for an unrestricted sequence of periodic events to happen, only such events are perceivable and one wants to know how many dimensions are required to describe this observation mathematically.

According to the axiom of chirality, in the 4-point space each point changes its state with each event whereas the space as a whole remains unchanged. How this change is accomplished, by an inversion or by a 3-dimensional rotation of the space, is left open and cannot be observed. By the inversions of the space the events produce a chiral rotation of the space that can be described by two polar coordinates ϕ and ψ as shown in Figure 23. To describe such a rotation a 3-dimensional space is required and the orientation or rotational direction of both angles must be defined. The events can be counted. Such a chiral process with four different point states is only possible in a 3-dimensional space. In a 2-dimensional space no tetrahedron could be constructed and in 4-dimensional space the process would not be chiral any more.

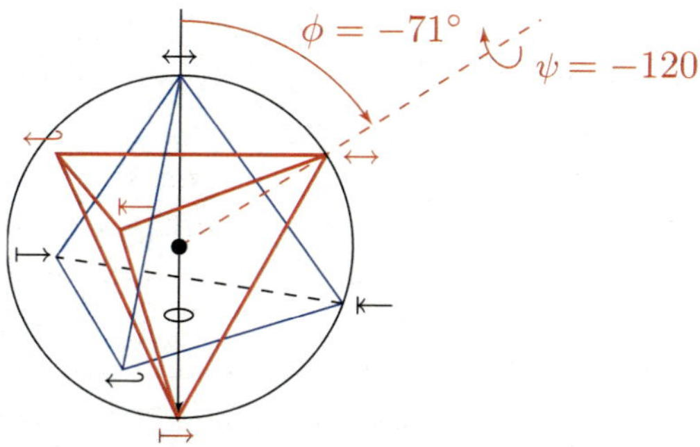

Fig. 23. Chiral rotation of the tetrahedron in 3-dimensional space.
Φ = clockwise rotation about the axis through the center of gravity (parallel to the blue side of $\hookleftarrow$ to $\mapsto$) of –71°.
ψ = clockwise rotation about the axis through the center of gravity and $\leftrightarrow$ of –120°.
The inversion of the blue tetrahedron relative to the red one corresponds to a rotation through the angles Φ and ψ.

In four- and higher-dimensional space the 4-point model would be unable to function, because there the right- and left-handed spaces are no longer chiral and so they would not be distinguishable; they could be turned into one another by simple rotation, i. e., they could be brought to coincidence with each other. Since a point can only take up one of four different states of motion, it is only possible in three-dimensional space to construct a chiral tetrahedron with four different corners in such a way that the internal chain of events never stops of its own accord. This is probably the underlying reason why physical space as we believe we perceive it is three-dimensional. In four-dimensional space perception would be impossible, since there can be no distinguishable chiral particles there. Every empirical perception is probably based on precisely this possibility of distinction. Only three-dimensional space permits perception.

In quantum theory, the three-dimensionality of the space can be derived from the quantisation of the Hilbert space of an individual *Ur*-object (i. e., a simple alternative). It is mathematically represented as SU_2-symmetry, or as a two-dimensional complex vector space and this corresponds to a three-dimensional space with SO_3-symmetry. This derivation of three-dimensionality is ultimately possibly equivalent to mine. [Weizsäcker (1971a) pp. 222 and 271; Finkelstein (1969) pp. 1261–1271].

For our further considerations regarding nature as it is perceived we can confidently confine ourselves to three-dimensional spaces, for without a three-dimensional space there are no chiral particles; without chiral particles there is no perception, and without perception there is no physics.

With the between-state as it is drawn in Figure 21. the drawing itself is 2-dimensional. But the drawn space is 3-dimensional because the between-point is not only at a location of that space, it also has a direction (arrow) perpendicular to the 2-dimensional space formed by the other 3 points. For that a third dimension is required.

It must always be made clear the dimensionality of what is considered, that of nature *per se*, that of the perception of nature ot that of the mathematical model which describes the perception. The dimensionality of these three aspects of nature may vary from 3 to ∞ (see Section 11.1). In this section it was the dimensionality of the model of a 4-point space obeying the axiom of chirality.

7.8 Pauli's symbol of sublime harmony

Wolfgang Pauli (1961), who postulated the neutrino in his famous 1930 letter, „Dear radioactive ladies and gentlemen", regularly dreamed about certain symbols [van Erkelens, (1995)]. As a Jew, he originally often dreamed about hexagonal Stars of David, yet he realised – influenced by C. G. Jung and trying to become at one with himself – in an archetypal dream on the 9th of November 1953, that there must be something quite correct, yet also incorrect about this Star of David. It is correct that the number of lines is six, however, it is incorrect that the number of points is also 6. A Chinese lady, who often appeared in his dreams und who likely symbolised the duality of Be-ing, showed him instead a square with clearly drawn out diagonals and said: „See now finally the 4 and the 6, that is, the 4 points and the 6 lines – or 6 pairs from 4 points. They are the same 6 lines which are to be found in the I-Ching. The Chinese book of wisdom, the *I Ching*, originated during the Shang dynasty (16th–11th century BC). I Ching is based on the coaction of the principles of Yin and Yang and of an internal world and an external world. It is a guide to the Tao (way, sense, law of all changes) [Wilhelm (1912/1923)]. There the 6 is correct, which also latently contains the factor 3. Now look at the square some more: 4 of the lines are of equal length, the two others are longer – in an „irrational ratio", as you know from mathematics. There is no figure composed of 4 points and 6 lines of equal length. Therefore, symmetry cannot be produced statically, and *a dance ensues*. This change of location in this dance is called coniunctio; one can also speak of a game or of rhythms with rotations. Consequently, the 3 must be expressed dynamically, which is already latently present in the square. Therefore, Jung's formula of 4 squares is perfect in its way, because indeed, the dynamics are expressly discussed there." (Figure 24).

Pauli's psychoanalyst Marie Louise von Franz (1988) commented on the dream as follows: „This dream and the fantasy which follows it clearly alludes to a numerical underlying rhythm of life ...". And: „First of all, particularly important is the emphasis on the three or six, as a sequence of figures which enable the symbol of harmony to manifest itself in a spatio-temporal succession in all its latent possibilities and thereby not to rigidify into a static symmetry and harmony" [von Franz (1970)]. Pauli found such psychoanalytical interpretations of little use and complained that psychologists understood nothing of physics. On the other hand, he opined that it wasn't his task to give psychologists free lessons in physics [von Franz (1995)].

To my knowledge, Pauli did not attempt to directly use such symbols as models for his neutrino, yet it seems reasonable to assume an unconsciously close connection, a mutual inspiration, between the neutrino's structure and his dreams. Pauli's square with the dancing corners and the two diagonals is very close indeed to my neutrino model. Pauli only had to allow his four points three-dimensions instead of just two and to have made his points dance, as his Chinese lady demanded, and he would have arrived at my four-point space. The tetrahedron is precisely the figure with four points and six lines of equal length, which in the opinion of Pauli's Chinese lady, does not exist. I invented my neutrino model in about 1975, also in dreams, whereby I initially dreamed about shapes resembling Stars of David as well, although three-dimensional ones. I first read about Pauli's dreams in 2000. The „irrational ratio", which the Chinese lady demanded, arises in the tetrahedron not through the irrational ratio of the line lengths, but through the „dance" of the four points, which results in an irrational chiral rotation of the tetrahedron in the three-dimensional space.

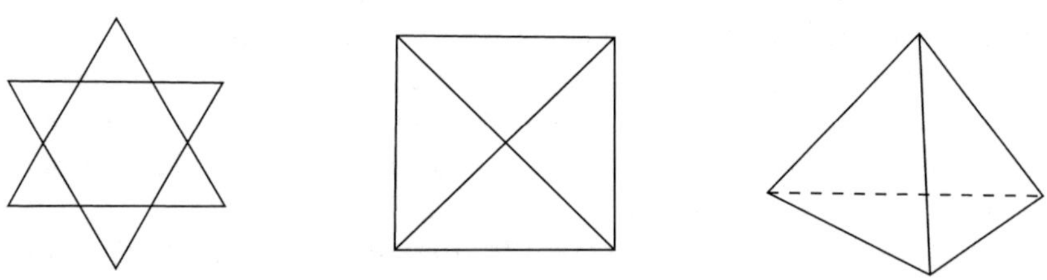

Star of David	Square	Tetrahedron
6 corners + 6 lines	4 corners + 6 lines	4 corners + 6 lines
2-dimensional	2-dimensional	3-dimensional

Figure 24. Dreamed symbols of sublime harmony[49]

[49] The square with the two diagonals as a basic element of the existent occurs as far back as Plato: It is one of the two-dimensional, static figures, of which the five Platonic solids tetrahedron, octahedron, icosahedron, cube and dodecahedron are composed, which are the elements of fire, air, water, earth and the universe, respectively [Böhme (2000) pp. 294–310]. However, Plato (1961–1963b) Timaios 53c–55b, obviously did not succeed in also convincingly justifying this. He therefore essentially placed its acceptance open to scrutiny: „… but another man, considering other facts, will hold a different opinion. Him, however, we must let pass."

Chapter 8

Physical Interpretation of the Chirality Theory Model

> Following a lecture by Pauli in 1958 about his unorthodox particle theory which was intended to explain the parity violation of weak interactions, younger researchers directed sharp criticism at Pauli. Then Bohr spoke up and said to Pauli: „We are all agreed that your theory is crazy. The question which divides us is whether it is crazy enough to have a chance of being correct. My own feeling is that it's not crazy enough."
>
> Freeman Dyson (1958)

8.1 The presumed observer

A real observer is himself subject to the laws of nature. By means of his real measuring apparatuses he describes his empirical observations. He counts the perceived events. As shown in Chapter 5 a presumed observer who describes the system of both the observer and the observed events together is inevitable for metaphysical reasons. For the moment, it is assumed that one and only one such presumed observer is present. He has an inner clock and can count, but the counting is possible without any interaction. The real observer is physical and real, whereas the presumed observer is as transcendental as his mathematical model. The presumed observer can view the 4-point space from two different locations: either he perceives the tetrahedron from outside as a whole how it inverts or rotates chirally in leaps from event to event (Figure 22), or as an internal observer he accompanies one of the 4 points, and perceives how it repeatedly pierces the plane defined by the other 3 points and proceeds through a translation. Thus, in the 4-point space, the observer is located either outside or inside the black mini-hole (see Section 8.3).

In order to count events, the presumed observer must have a transcendental structure that enables him to order information. To be able to do that he must be chiral. This ordering structure is most simply defined as an internal clock with a subjective time direction. But if the theoretical physicist prefers some other chiral structure as e.g. the transcendental ordered letters of a book he is free to do so. There is no need of an internal clock or a background time in chirality theory, at least not more than there is a time in every mathematics because the mathematician must have time for his thinking. For the description of a sequence or for a causality an order or an arrow is required. The metaphysical source of the arrow is the flow direction of information from object to subject. This flow must be ordered, i.e. chiral. This does not have to be an arrow of time. It might be any chiral duality.

The relation between the chiral order of the presumed observer, i.e. the arrow of time of his internal clock, and the order of the counting process must be defined by a convention. If the counted events are periodic, they are the measure of time (not time itself). If they are not periodic, they are the measure of space (see Section 8.4). The direction of the time arrow and the orientation in space are subjective. They are not an inherent aspect of nature. So, the time arrow of a perceived antiparticle might be opposite to the time arrow of the observer. Mesons consist of a quark and an antiquark. This means that within the same meson time might flow in two opposite directions as postulated by Feynman (1985). So, the presumed observer „lives" in his own time with his subjective time arrow. But to order his perceptions he uses a time model that allows both directions of the arrow. Time per se is not an inherent aspect of nature (Einstein 1979). The inherent aspect is chiral order.

8.2 A neutrino model?

The mathematical model which has been presented in Chapter 7 is not an end in itself. Rather, we are actually looking for models which on the one hand, come as close as possible to our basic metaphysical conditions, and on the other hand, describe physical objects, i.e., empirical observations, without unnecessarily distorting these observations. Are there physical objects which already correspond to the four-point space model? The simplest known particle today is the neutrino. It occurs in six different states, namely as an electron-, muon- or tauon-neutrino, which can all be left or right-handed in three-dimensional space, without a space direction being distinguished thereby. So, the space of the neutrino is isotropic. The neutrinos have a spin corresponding to their spatial orientation, but no poles and no axis. The three neutrinos with a right-handed spin are also called antineutrinos. By reversal of the orientation, i. e., of the spin, the neutrino would turn into its antiparticle. The three neutrino types are probably able to turn into one another spontaneously, and the three antineutrinos like-wise. However, a neutrino probably never turns into an antineutrino [Gibbs (1998)]. Thus, all the intrinsic characteristics of the neutrino known today are described quite well by the four-point space model.

However, there are still current theories which hold that a neutrino can turn into its antiparticle, according to its direction of motion. Whether a neutrino is perceived as a particle or antiparticle is therefore relative in these theories. One then no longer speaks of the chirality, but of the helicity of the neutrino, since in contrast to chirality, helicity depends on the direction of motion. Such neutrinos, which are identical to their antiparticles, are called Majorana neutrinos in contrast to the Dirac neutrinos, which due to their intrinsic properties – as in chirality theory – are fundamentally different from their antiparticles. Today it is almost certain that neutrinos have a mass and accordingly cannot travel at light speed. That leads to some virtually unsolvable difficulties with the Majorana neutrinos since all experiments indicate that neutrinos and antineutrinos behave fundamentally differently in their weak interactions, which is unexplainable if this behaviour only depends on the relative direction of motion [Schmitz (1997) pp. 14 f. and 27–32].

There is a certain fascination in comparing my neutrino model with Plato's model for the element fire, which is composed of – admittedly static – equilateral tetrahedrons [Plato (1961–1963b) Timaios 53c–58c]. Timaios is Plato's representation of the school and philosophy of the Pythagoreans.

8.3 Black holes

No point can leave the four-point space in which the four points are in the four different states of motion. Whenever a point, which is moving away from the others looks to break out, the space inverts and the point moves towards the other three again.

A three-dimensional space from which no point can escape, because the space inverts too quickly, I call a **black hole**.

The neutrino is the smallest possible black hole. The black hole, so defined, has the same characteristics as the conventionally defined black holes, as will be demonstrated in the following chapters. The same applies to my new definitions of space, time, mass, spin, energy and so on.

8.4 Space and time

Every time a new point acquires the ($\leftrightarrow$)-state, a new event takes place. Seen from the perspective of the presumed external observer, the events of the four-point space are periodic and can, insofar as they are observable, be counted. The observation and counting of the event sequence are one-dimensional and chiral, that is, they have a direction.

The dimension in which the periodic events are counted I call **time *t***. Weyl (1923) defines time as the „archetype of the stream of consciousness", i. e., as a simple category of consciousness in the consciousness of the external observer.

The number of periodic events is the measure of time. So, time is a dimensionless number. Periodic means in this context that the object, i. e., the four-point space, is in precisely the same internal state following the event as before the event. Whitehead (1919), likewise for philosophical reasons, came to the conclusion that the ultimate (i. e., smallest) objects cannot be uniform, but must consist of parts which for their part generate events. As a consequence of these events there must be time minima, so-called time quanta. Geometrically, this means that periodic events must involve a rotation [Aristotle (1987) vol. VIII, § 265a]. Since before, during and after each event, the external presumed observer sees a tetrahedron with four corners in the four different possible states, he perceives exclusively periodic events in the defined sense. Each event, without any external input, automatically generates the next event. Kritias (2000), an uncle of Plato, formulated this very elegantly: „Time, tireless and brimming with an eternally flowing stream, rolls by, giving birth to itself." A duration of time only ever concerns a number of counted events; it is consequently a dimensionless number. If one were to ask which of the two concepts, event or time, the most fundamental is or has primacy, then that is without doubt the event. The definition of time uses the term „event", whereas the entity event can be explained without using the word „time". This was already clear to Aristotle (1987) vol. IV chapter 11, 219b–221a, although he usually spoke not about events, but about motion.

Every point of the neutrino model regularly moves towards the other three, between them and thereafter away from them. Since the space then inverts, the same point then approaches the other three once more and so on. From the point's perspective (and that of the presumed internal observer who accompanies and watches the individual point), it moves constantly straight ahead between the three others, a translation which never ceases. The individual point does not turn back, rather the space turns inside out. The presumed internal observer, who follows the individual point, sees it successively from phase to phase in the four states; (⇤), (↔), (↦) and (↩). He doesn't count the events as time, since the state of the point before, during and after the event is always another one and thereby not periodic in the sense defined above. For the individual point, an event only ever occurs when it is situated between the other three points. Incidentally, in each instance there is an analogy between the two states (⇤) and (↦) on the one hand, and the two states (↔) and (↩) on the other: whilst the points in the state (⇤) and (↦) change places, they are in the states (↔) and (↩) in one location. In Section 8.9 about the boson I will come back to this state of affairs.

Whilst the presumed, external observer in counting events measures time, the sequence of events, from the perspective of an observer who moves with the individual point comprises a change of location over a certain time: whenever it pierces the plane defined by the other three points, it reaches „a location further on". Each of these four points can be thought of as having its own internal observer who accompanies it. The events, which are counted by these four internal presumed observers as changes in location are the same ones which the individual external observer counts as time events.

The dimension in which the changes of location are counted I call **space**. Since, from the external observer's perspective, the four-point space rotates during this process without a direction in space being especially distinguished, we receive, viewed from the outside, the impression of an isotropic, three-dimensional space. The number of location changes is a measure of the distance covered. Since this measure is nothing other than a quantity of counted events, distances involve dimensionless numbers.

So, I have derived space and time from the entity event and mathematically defined how I choose to understand what an event is. Usually, physicists proceed conversely and define events as aspects of space and time, generally as a simple space-time point. In so doing, they usually don't explain what they understand by a point. In the general theory of relativity, the coordinates of space and time then lose every physical meaning; they merely portray a certain arbitrary, but unambiguous counting of physical events [Møller (1969) p. 226].

Thus, according to the standpoint and state of the observer, he counts an event as being a periodic rotation of the tetrahedron or as a change of location of an individual point. Put succinctly: whether an event is perceived as a change in space or time is relative and depends not on the internal state of the object, but on the standpoint and state of the observer. What the observer outside of a black hole perceives as a length of time is for the observer inside the black hole a distance. For the object itself, there is no difference between time and space, there are only events.

A **clock** is a periodic event counter. It runs most quickly for an observer who perceives and counts all events as rotations. The **standard-length**, however, counts changes of location. Every event which it perceives and counts not as a periodic rotation, but as a change of location, is lost to the clock, for the observer cannot count any event twice, partly as time and partly as a change of location.

Consequently, there is a fastest possible time and a longest possible distance. The fastest time is measured when all events are perceived and counted as periodic rotations. The observer in this case does not perceive a change of location for the object: the object is at rest and can be described as a standing wave. The longest distance by contrast is measured when all events are perceived as changes of location. The presumed observer perceives in this case – excluding his internal sense of time – no events as rotations; time stands still for the object observed. Such objects are, for example, the graviton or the photon. This state of affairs corresponds precisely to the conclusions of the special theory of relativity.

8.5 Velocity

The ratio of the number of location changes to the number of periodic time events is called **velocity *v*.** The greatest velocity is measured when all events are counted as location changes and this number is compared with the event count of the presumed internal observer's inner clock. The corresponding velocity is called **light speed *c*.** It is the natural and unambiguous measure for all velocities. The observer's inner clock can be calibrated such that the ratio c of both numbers becomes 1. In this case every event which is perceived as a change of location corresponds to a time event of the observer's inner clock:

$$c = 1 : 1 = 1.$$

In four-point space this means that the inner clock of the presumed internal observer, who moves with one of the four points between the other three, always counts a time event of his inner clock whenever it is between the other three points. Consequently, the points inside the tetrahedron move at light speed. It is important to realise that the observer is a presumed one, since according to the special theory of relativity, a real observer would not be able to move at the speed of light. This is easy to appreciate: a real observer must also possess a real inner clock. So, he cannot only count the location changes as events, but he must also count the time events of his inner clock, which changes location with them. So, a fraction of the events in the counting of location changes are always lost with respect to the counting of time events.

Since velocity does yet have a direction according to this definition, one would have to speak probably more precisely of speed. The term distance is defined more precisely in Section 8.13.2, so velocity in the conventional sense can't really be spoken of until then. In physics, speed is defined as the rate of increase (or decrease) of distance travelled, with time. Linear velocity is the rate of increase (or decrease) of distance traversed by a body in a particular direction. Since velocity is nothing more than a ratio of counted events, namely in each case an event count as the numerator and one as the denominator, velocity is a dimensionless number. Thus, Einstein's (1949/1983) p. 23, vision is fulfilled that natural constants (here the speed of light) would have to be dimensionless quantities.

8.6 Frequency and mass

Frequency ν is what I call the ratio of the number of observed events (periodic events and location changes collectively!) to the event count of the presumed observer's inner clock. The concept of frequency corresponds to Kant's (1781) A 182, substance, which is quantized and thus calculable: „The substance persists through all change of the features, and the quantity in nature neither increases nor decreases." The minimum frequency during the observation of an object is measured when no change of location is observed, i. e., when an object is at rest. This minimum ratio I call **rest mass m_0**. The frequency or rest mass of such a stationary particle is the result of the particle's internal events. It is thus independent of its state of motion. Since mass and frequency are nothing more than ratios of counted events, namely in each case an event count in the numerator and one in the denominator, mass is a dimensionless number. The external presumed observer's inner clock can be calibrated such that it always counts an internal time event whenever one of the neutrino's four points is located between the three others. This calibration results in the neutrino's rest mass becoming 1; the event count of the external presumed observer's inner clock concerning four-point space is exactly the same as the number of events in this space itself. Further aspects of mass will be discussed in connection with gravitation in Section 9.5–9.11.

Plato in his *Timaios* regards matter and space as being the same [Aristotle (1987) vol. IV, chapter 2, 209b,]. For Descartes, space and matter were *res extensa*, the extended thing [Weizsäcker (1999) p. 132]. Space is matter and matter is space. In chirality theory as well, distance (in space) and matter are the same, namely frequencies of events. Space and matter are a consequence of observed events. A distinction is made between space and matter only based on the state of the observer: what the internal presumed observer perceives as distance, is matter for the external observer, a standing wave which oscillates with time. According to Russell, the world is made up of elements which are limited in time as well as in space. We call such elements events. An event is neither permanent, nor does it move like the conventional piece of matter. It exists only for a short moment and is then gone. A piece of matter thus disintegrates in a series of events [Russell (1925/1958) pp. 134 f.].

8.7 Spin and angular momentum

The chirality of a particle is the consequence of its internal point motions in the three-dimensional space. These chiral motions can be described mathematically as the changes of two polar coordinates, the angles φ and ψ, whereby the rotational direction of both angles must be defined so that the orientation in the four-point space is unequivocal. The zero point of the coordinate system in the four-point space is the central point or centre of gravity of the tetrahedron. All four corners are on the surface of the same sphere with the radius 1 around the zero point of the coordinate system (Figure 25a, p. 191). This coordinate system corresponds to the presumed observer's sphere of consciousness. Mathematically, it is naturally not the same whether the tetrahedron inverts or whether it rotates. Physically, however, the two mathematical models are indistinguishable, since the position of the four points after the event is the same in both models.

Now let us consider the positional change of the (↔) point on the spherical plane during a single event. One must bear in mind thereby that the (↔) point during the phase change always becomes the (↦) point, whereas the former (⇤) point turns into the new (↔) point. The angle φ is the tetrahedron's angle of rotation about the axis, which passes through the zero point and which lies parallel to the straight line through the tetrahedron's corner points (↩) and (↦) before the event. The new (↔) point then lies at the angle φ = ca. 71° turned clockwise away from the former point (↔), whereby the external observer orientates his view parallel to the rotation axis of (↩) facing in the (↦) direction.

It is probably demonstrable that the angle φ = ca. 71° is an irrational fraction of 360°, whereby the tetrahedron can never return to exactly the same position, irrespective of how many inversions the four-point space makes. This is yet to be proven, however. The corners of all tetrahedrons, which can be formed by an individual tetrahedron's successive inversions in space, describe a more or less continuous spherical surface around the tetrahedron's centre of gravity as the central point. This creates the impression of a spatial continuum, although the events counted are always intrinsically finite.

In order to bring not only the (↔) point, but also the whole tetrahedron **before** the phase change to coincidence with the tetrahedron **after** the phase change, it must still be rotated about the axis through the angle $\psi = 120°$, which runs through the zero point and the new point (↔). This second rotation also has to proceed clockwise. If the points (↩) and (↦) are alternated before the event, then one begins with the tetrahedron's mirror-image and new anti-clockwise rotations through the angle φ as well as through the angle ψ will take place (Figure 25b, p. 191). It becomes evident from this change of rotational direction that the rotational motion of the chiral tetrahedron in three-dimensional space is also chiral. Incidentally, the two angles φ and ψ are not independent of each other. Rather, they change periodically in mutual dependence. The greater φ, the smaller ψ becomes, and vice versa.

This chiral rotational motion in three-dimensional space I call **Spin**. The spin is usually described mathematically by means of spinors as „three-dimensional rotation" [Penrose (2004) pp. 549–566]. As we have seen, spin comes about as a result of four points moving between each other according to certain rules. I use the term three-dimensional rotation to express that it doesn't require a fixed axis, but that the points rotate simultaneously about two constantly changing axes, which is distinct from the conventional understanding of the terms rotation and spin. The rotational motion must then be mathematically described in such a way that it is described correctly by event counts, as it is measured by the presumed internal and external observer respectively. Presumably, the spin will be perceived thereby quite differently by the two observers.

It is important to realise that the spin does not have an axis as such. There is no North and South Pole and the spin is definable without the need to assign the particle a direction of motion. This view of the spin fundamentally differentiates the theory suggested from the conventional understanding of the concept of spin. Nevertheless, it must be borne in mind that for Pauli as well, the spin does not primarily mean a rotation or

a)

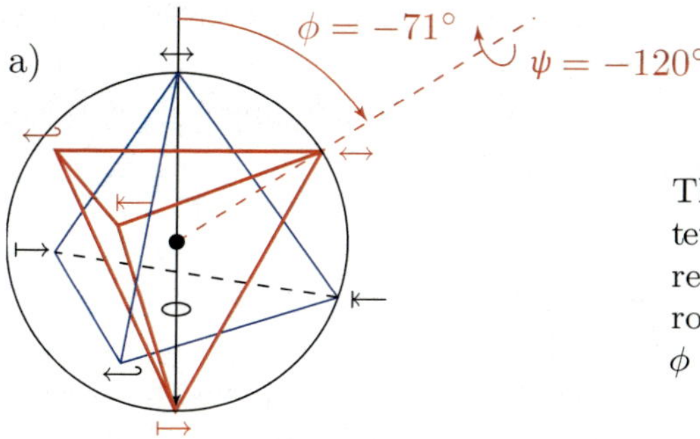

The inversion of the blue tetrahedron relative to the red one corresponds to a rotation through the angles ϕ and ψ.

$\phi =$ clockwise rotation about the axis through the centre of gravity (parallel to the blue side of $\hookleftarrow$ to $\mapsto$) of $-71°$.
$\psi =$ clockwise rotation about the axis through the centre of gravity and $\leftrightarrow$ of $-120°$.

b)

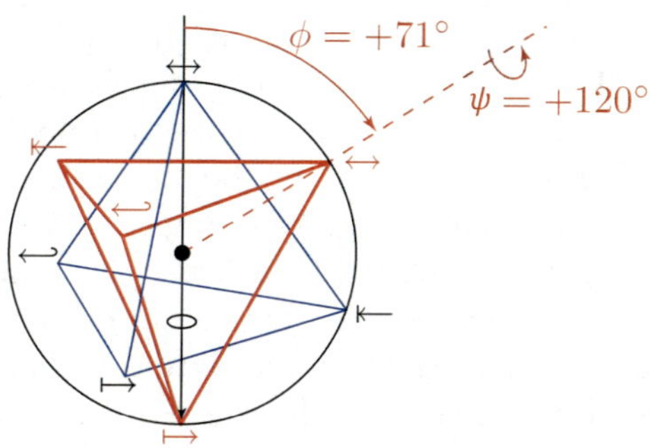

$\phi =$ counter clockwise rotation about the axis through the centre of gravity (parallel to the blue side of $\hookleftarrow$ to $\mapsto$) of $+71°$.

$\psi =$ counter clockwise rotation about the axis through the centre of gravity and $\leftrightarrow$ of $+120°$.

Figure 25. Chiral rotation of the tetrahedron in 3-dimensional space: *The rotations in a) and b) are chiral in 3-dimensional space and are in opposite positions. b) is the mirror image of a).*

an **angular momentum**, but rather the abstract form of the freedom which a rotation has to opt for one direction or the other [Fischer (2003)]. Chirality theory provides new guiding principles for what is to be understood by the term direction in this context.

Later we will see that in most cases, the tetrahedrons are lopsided and their radius is not 1, but greater, for example *r*. Thus, the rotation, or the event frequency with a rotation, becomes correspondingly greater. Such cases will be described in Section 9.4 and 9.9.

8.8 Fermion

For the presumed external observer, all perceived events with respect to the four-point space are the same. From the perspective of this observer, the rotational motion of the tetrahedron has three parameters, namely the two angles φ and ψ and the radius of the spherical surface which circumscribes the tetrahedron. Although the rotation for every event in the four-point space is the same, the two angles φ and ψ change constantly, according to the position of the tetrahedron in the polar coordinates system. The greater the two angles and the radii of these angles are, the greater the motion is. When a presumed observer measures angle changes, he does this by comparing the observed angle with the hand of his inner clock. The smallest possible angle change is that owing to one individual time event. The radius in the four-point space is 1 for both angles, since no planes between the centre of gravity and the tetrahedron's corners are defined which could cause a change of location if one measures the distance between the centre of gravity and the corners. It follows from this that a smallest, directly perceivable rotational motion must exist, and that is the rotation of the four-point space during a single event. It comes to

$$(1 : 1) \cdot (1 : 1) = 1.$$

The 1s in the numerators of this equation stand for the minimum event count which is required for measuring the length of the radius with the two angles; the 1s in the denominators stand for the event count which the presumed observer needs to perceive both angle changes.

The previous consideration of the rotation from the external observer's perspective has not yet taken into account that the motion of the four-point space is chiral. Mathematically, chirality can be introduced into the equation by adding the + or – sign to the angular momentum. With the aid of a chiral, three-dimensional object, such as a corkscrew or a right hand, it must then be defined which orientation of the movement is positive and which is negative.

So, the rotational motion of the four-point space, measured by the presumed external observer has been perceived during an individual event. It is the difference between the state of the four-point space before this event and its state after the event. These two states are essentially the same; only the states of the four individual points have changed during the rotation, not, however, the internal state of the rotating object as a whole. In Chapter 10 we will replace the presumed observer with a real one which interacts with the object. In the process, we will find that such an observer cannot directly observe the rotational motion itself, but only possible changes in it, i. e., events in which the rotational direction of the observed object reverses. This can be expressed mathematically such that the empirically observed minimum change of the rotation, namely quantity 1, is the difference between $-\frac{1}{2}$ and $+\frac{1}{2}$, whereby the quantity is defined as the **angular momentum** of the four-point space. A particle with spin $\pm\frac{1}{2}$ I call a **fermion.**

8.9 Boson

The internal presumed observer does not „see“ the four-point space as an object in its entirety but perceives the individual point he accompanies in a space which is formed by the three other points. There are four different such internal observers who perceive the events the same, but in different phases. For an internal observer, the successive events of the four-point space are different, since the state of the observed point changes during each event, as we established in Section 8.4: every fourth event the point reaches the ($\leftrightarrow$) state. Only the ($\leftrightarrow$) state is a countable event for the internal observer, i. e., an event where he can perceive the inversion of the space. He perceives all events as a translation at light speed and does not himself rotate. What rotates is not the observed object, i. e., the single point he accompanies, but the space in which this point, together with the observer, moves. The space is the other three points which in their respective ($\leftrightarrow$) states (during the three successive events that are not observable for the internal observer) fly past the observer, one after the other, on different sides at light speed at a distance of 1. Every fourth event, the observed point is itself in the (↔) state once again. All four points participate in every event.

So how great is the rotational motion of the space which encompasses the observed point? It must be just as great as the angular momentum perceived by the external observer, since the internal observer always flies straight ahead with the point he observes and turns just as little as the external observer does. However, since the internal observer only perceives as such every fourth event within the four-point space, from his perspective the rotational motion from event to event is four times greater than it is for the external observer: it is

$$4 \cdot \tfrac{1}{2} = 2.$$

The same result ensues when one presupposes that the points (↦) and (⇤), which change locations each event, fly past each other each time at light speed, thus bringing about the rotation of the space:

$$2 \cdot c \cdot c = 2c^2 = 2 \cdot (1 : 1) \cdot (1 : 1) = 2.$$

The 2 in the above equation is the result of there being 2 point-pairs which alternate between being in motion and being in one location, so that in each case their internal presumed observer only perceives the motion at every second event. The two fractions (1 : 1) are the light speeds of the two points which change places at each event. The greater this velocity, the greater the rotation per event. This description of the three-dimensional rotation by means of 2 points which fly past each other in different directions at the speed of light, is intuitive. For a more precise description, one would need an axiomatic system for this finite three-dimensional topology, which is yet to be formulated.

Since the three-dimensional rotation of the four-point space is also chiral from the perspective of the internal observer and is consequently able to assume two opposing directions, the spin can be positive or negative. With the four-point space, the spin from the internal observer's perspective is ± 2. I call the particles with integral spin **bosons**. Thus, the same four-point space has from the perspective of the external observer, a spin of $\frac{1}{2}$ and is a fermion, whereas the spin from internal observer's perspective is 2, thus the object is a boson. With the four-point space, that which the internal observer perceives as a boson probably concerns the graviton. The internal observer measures a four times greater rotation per time event than the external observer. Whether this particle involves a fermion or a boson is relative and depends on whether the observer is inside or outside a black hole.

8.10 Planck's constant *h*

In the descriptions of both observers, the expression $(1 : 1) \cdot (1 : 1)$ appears, in which all four number 1s stand for an event counted by the observer. I call this constant

$$c \cdot c = c^2 = (1 : 1) \cdot (1 : 1) = 1 = h$$

Planck's constant *h*. It equals one. Since h is nothing other than a ratio of counted events comprising two event numbers in the numerator and two in the denominator, h is dimensionless. h is the natural and unambiguous standard measure, or quantum, for the angular momentum. h is the smallest possible angular momentum a real observer can perceive, since at least **one** event must take place for a rotation to come about. There are no half events. With c and h, one and the same physical constant is involved, which is construed by the observer outside a black hole as the minimum angular momentum; by the internal observer on the other hand as maximum velocity.

8.11 Energy

Dirac recognized that the electron had to have a spin if one wanted to reconcile the claims of the special theory of relativity with those of quantum theory in a logical and proper mathematical fashion [Weizsäcker (1999) p. 284]. This state of affairs matches my derivation of the spin from the concept of an event. If according to Section 8.6 the mass m_0 is synonymous with the frequency ν and according to 8.10 $c^2 = h$, then

$$m_0 \cdot c^2 = h \cdot \nu = E.$$

This equation unifies the special theory of relativity with quantum theory. The quantity $(h \cdot \nu)$ I call **energy *E***. Energy is also a dimensionless ratio of counted events, comprising in each case three event counts in the numerator and three in the denominator.

8.12 The 5-point space

8.12.1 Location

A fifth point *P* can, in principle, be situated in 16 different locations in the space which is defined by the tetrahedron, namely within the tetrahedron, over each of the four faces, over each of the four corners or over each of the six edges and, if curved spaces are permitted, within a second tetrahedron which lies „opposite" the first one. The latter case is easy to visualise by means of a triangle which is situated on a spherical surface, i.e., is in a curved space: if the three sides are lengthened to extend all the way around the sphere, then these extensions will form a second triangle on the opposite side. An analogous situation occurs with the elongation of the four faces of the tetrahedron in a curved three-dimensional space. Thus, a new tetrahedron is created „opposite". All 16 locations can be defined by the relation „between". If we disregard for a moment the location within the tetrahedron and the place „opposite" then there are still 14 locations outside the tetrahedron where a fifth point *P* can be situated. Since the four-point space, seen from outside, is constantly turning without a direction being distinguished in particular the point *P*, over time, reaches all 14 possible locations. So, a location in the sense discussed is not a point as such, but a space defined by the relation „between", in which there could be many different points.

The fifth point *P* forms with each of the three points of the original tetrahedron (*ABCD*) a new, lopsided and chiral tetrahedron with its own rotational direction. Thus, four different tetrahedrons in all are created, namely (*ABCP*), (*ABDP*), (*ACDP*) and (*BCDP*). What happens then with these four lopsided tetrahedrons when the tetrahedron (*ABCD*) inverts to tetrahedron (*A'B'C'D'*), as is depicted in Figure 26? Whether or not the four lopsided tetrahedrons invert depends on the position of *P*. An inversion always occurs whenever, during the inversion of the tetrahedron (*ABCD*), one of the four points of the observed lopsided tetrahedron gets between the other three, whereby the lopsided tetrahedron also inverts.

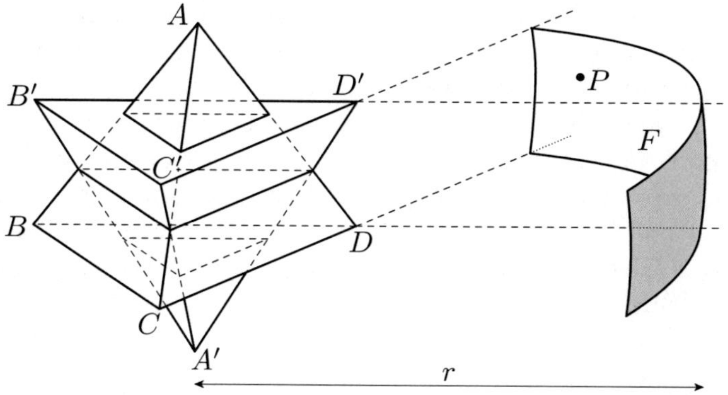

***Figure 26. The distance* r *between the four-point space and the fifth point* P:** *The tetrahedron ABCD inverts into the tetrahedron A'B'C'D'. Thus, outside the two tetrahedrons, 32 different spaces are formed which are defined by the relation „between" and where a fifth point P can be located outside the four-point space. The spherical plane F, which delimitates the 6 spaces between the parallel planes BCD and B'C'D' in the radius r from the tetrahedron's center of gravity, increases proportionally to r. The rest of the spherical plane, which delimitates the remaining 26 spaces, increases proportionally to r^2.*

If point *P* is a certain distance away from the original tetrahedron, then it cannot possibly get into the ($\leftrightarrow$)-state itself when the tetrahedron *ABCD* inverts; so, in its four lopsided tetrahedrons it remains always in one of the states ($\mapsto$), ($\hookleftarrow$), or ($\leftarrow\!\!|$). Since according to the axiom of chirality ($\mapsto$) $\Rightarrow$ ($\hookleftarrow$) $\Rightarrow$ ($\leftarrow\!\!|$), the fifth point will arrive sooner or later in the ($\leftarrow\!\!|$)-state, in which it will remain until it has approached the original four-point space to the extent that it gets into the ($\leftrightarrow$)-state, thereby falling into a black hole. (However, a black hole with five points probably cannot be stable.) The gravitation between the original tetrahedron and a fifth point which travels away from the four-point space results in the point sooner or later reversing course and travelling back towards the tetrahedron. The four-point space attracts the fifth point.

The circumstances surrounding the fifth point are distinct from those in the four-point space insofar as the fifth point does not immediately change its state with every event. Although the fifth point shares the sequence of the four possible states required by the axiom of chirality, it needs more events, i. e., it takes longer for it to change its state. Seen from the perspective of the external presumed observer, the four lopsided tetrahedrons do not rotate unceasingly, as was the case in the four-point space. The rotational direction of these tetrahedrons can change from event to event, and there can also be two different points of the same lopsided tetrahedron in the same (↦)-, (↩)- or (⇤)-state. Point *P* is separate from the four-point space in that it is always situated outside the black hole; it is connected to the four-point space in that it is situated in a clearly defined location relative to it and this location is dependent on the events of the four-point space.

8.12.2 Distance

Naturally, one would like to quantify these statements somewhat. Can the distance between the fifth point and the tetrahedron somehow be quantified, if possible, purely by event counts? The lopsided tetrahedrons experience such events whenever during the inversion of the tetrahedron (*ABCD*) they themselves invert as well. Every such inversion is a countable event. Now if it can be shown that the number, or the frequency of such events is dependent on the distance *r*, then we can define the distance as a function of event counts.

It can be seen from Figure 26 that the environment of the four-point space can be divided into three sections, namely into a planar disc between the two parallel planes (*BCD*) and (*B'C'D'*) and into two (approximate) hemispheres above and below this disc. The area of the disc outside the tetrahedron (*ABCD*) can be divided into 6 different spaces or locations, namely three above the „edges" (*BB'*), (*CC'*) and (*DD'*) and three above the surfaces (*BCC'B'*), (*CDD'C'*) and (*DBB'D'*). The strip of the spherical plane *F*, which delimitates these 6 spaces in the radius *r* from the centre of gravity of the two tetrahedrons, increases almost proportionally to *r*. By contrast, the surfaces of the two hemispheres which delimitate the remaining 26 spaces, increase almost proportionally to r^2. Whether one of the four lopsided tetrahedrons inverts or not depends on which of the 32 possible locations the point *P* happens to be in at that moment; the inversion itself, however, is for every individual location, independent of the distance.

The probability of the point *P* being situated at one of the six locations within the planar disc decreases with the distance r, since the ratio of the strip-like surface F to the surface of the spherical plane also decreases with the distance. At the six locations within the planar disc, the proportion of events with an inversion of the lopsided tetrahedron compared with events without such inversions is greater than at the 26 locations within the two hemispheres. This assumption is yet to be proven exactly, however. As a result, the inversions of the lopsided tetrahedrons, i. e., the events, are more frequent the more often P lies in the planar disc, i. e., the shorter r is. Thus, we have found a **measure for the distance r**: The length of the distance r between a point and a four-point space is a **function of the frequency n of the inversions of all tetrahedrons** which contain this point.

$$r = f(n)$$

I am not in a position to formulate the relationship between the number of events n and the distance r more exactly. I have shown, however, that such a relationship can exist and indeed, be found already in the four-point space and its surroundings. If the universe is composed of more than four points, then the relationship $r = f(n)$ becomes complicated and would probably have to be mathematically simplified, so the distances can be only calculated as an approximation. But the calculation is in principle also possible here with the help of a function $f(n)$. As a consequence of this, one can proceed with the development of the theory from the perceived events and then derive space and time from them. The theory is thus based on the perception and the conditions for the perception and not on a theoretical concept of space, unlike the situation with the conventional theories. The concept of time as well is then a direct consequence of this procedure, because space and time are connected by the natural constant c.

It is appropriate to calibrate the standard length for measuring r such that

$$r = c \cdot t,$$

whereby t is the number of events which the inner clock of the presumed observer counts when he shifts at light speed c from the beginning to the end of the distance to be measured.

The concept of distance allows locations to be specified more precisely. The more events n are counted, the more precisely the location of a point can be specified. To count, however, one needs time, i. e., a precise localisation means that it can no longer be said precisely when the point was at this location. This is the first indication of an analogy to the Heisenberg uncertainty principle, which will be examined more closely in Section 10.18.

8.13 The n-point universe: particles, distance, action and information transfer

The model of chirality theory described up to now primarily considers the simple 4-point space. This model illustrates the philosophical idea of the theory quite well but of course it cannot describe the universe as a whole. As explained in Section 7.3 the description of an n-point space is mathematically difficult, so that artificial simplifications have to be made that contradict nature. Nevertheless, in this section some rules regarding the mathematics and the interpretation of an n-point space are sought. Today only a qualitative description of terms such as distance, action and information transfer in a space with more than 4 points can be presented. For a mathematician, such „poetic" models might be unsatisfactory but I hope that some scientists will be motivated to think in the same direction and develop useful axioms and a complete mathematical theory for the description of physical observation. Even in the present form chirality theory can already metaphysically explain many physical phenomena despite the mathematics of the theory being rudimentary. This is in contrast to other modern theories with their beautiful mathematics which are difficult or impossible to be applied to physics and which cannot explain any physical phenomena at all.

8.13.1 Universe

The universe consists of an unlimited number of points. For forming an ontological existing object these must all be related to one another as well as to the physicist as the observer. An infinite number of points would contradict the metaphysical condition of being perceivable. All points are related to each other by their relative position and changes of position.

Every separation of points, particles, objects and subject is an artificial one and might falsify the description of a perception. Nevertheless, such separations are inevitable in physics because the universe as a whole is much too complex to be precisely described. Therefore, mathematical technics must be developed for simplifications which do not unduly falsify the result. Such technics are e. g. used in renormalization and perturbation theory. These technics are never aspects of nature but only mathematical tools to simplify the description of real nature. It is important for the physicist to remain conscious of the ontology of his observations and not to believe that the simplified description of nature is truth. It cannot be a task of this book to discuss such mathematical tools and they have still to be further developed.

8.13.2 Space, location, distance, event

Physical space is formed by any subset of $n \geq 4$ points. In a system of n points there are $\binom{n}{4}$ 4-point spaces. They can overlap. Sets of only 3 or less points cannot describe a real system obeying the axiom of chirality and therefore cannot be a model of reality. Space is not an aspect of nature, it is only one of the model. In that model the 4-point space is 3-dimensional. As long as we describe our physical measurements of space and time by the chirality theory model we describe 3-dimensional objects. This does not mean that the universe as a whole is 3-dimensional because there might be objects within objects called black holes. With every new level of such black holes 3 more dimensions arise (see Section 11.1).

Locations are defined by an *axiom of between*. An event occurs when a point gets between any 3 other points thereby changing its location. Thereby the 4-point space is inverted. Events are countable by natural numbers. *Distance* is the relation between any 2 points of the universe (see Section 8.4).

8.13.3 Measurement. Periodic and not-periodic events

Events might occur periodically or not. A physical measurement consists either of the counting of periodic temporal events by a clock or of not-periodical counting of events by marks on a measure stick. The relation between the number of counted events and the physical unit of the distance or time has to be based on a convention.

8.13.4 Single points and particles

Single points might travel chaotically through our universe. Possibly they are the source of the mysterious dark matter. No single point can leave the universe. If the universe is interpreted as the inside of a black hole the point cannot spontaneously leave this black hole. It remains related to all the other points of the universe. Regarding dark matter and the possible role of single points for the flavor oscillation of neutrinos see Section 14.3.

If a set of points forms a particle at rest, i. e. one with exclusively internal periodic events, then the internal dynamics and properties of the particle can be described as explained in Sections 8.1–8.11. All dynamical laws must be derived from the axiom of chirality. The axiom postulates that the inversion of a space consisting of 4 points in the 4 different possible states relative to one another prevents the point in the ($\mapsto$)-state from „moving away".

8.13.5 Distance. Modification of the axiom of chirality

All points of the universe are related to one another by a distance.

In an n-point universe there are many 4-point spaces. Every point belongs to many of them at the same time and the spaces may overlap. To account for this complex situation the axiom of chirality has to be appropriately modified. Such an extension of the axiom of chirality still has to be sought.

According to the axiom of chirality for the 4-point space the distance is the number of counted events during a translation. This should also apply to the n-point space. But there the weight of the inversions of the numerous different lopsided 4-point spaces depends on the distances between the points within the involved 4-point spaces and also these distances are countable. The larger they are the smaller the weight or action of the involved event.

With this model two types of distances have to be considered, namely the counted events of the presumed observer who travels by the shortest way along the measured distance and the distances between this observer and all the other points of the universe. Since in chirality theory all distances are counted numbers of events, the distance between two observed objects S_m and S_n must be a counted ratio $\mathbb{Q}_{m,n}$ of the 2 types of involved non-periodic events. This description is very vague and intuitive and the algorithm to calculate the distance is not yet known but in the end every distance must be a finite rational number of counted events.

This way the measured distance depends on the density of the observed system analogous to general relativity theory in which at the Schwarzschild radius of a black hole all distances become zero. But – in contrast to general relativity theory – in chirality theory there are space inversions and no space continuum.

8.13.6 Non-local action and information transfer

According to the axiom of chirality every inversion of a 4-point space with the 4 points in the 4 possible different states relative to each other exerts an attractive action onto all other points of the universe. The action is the result of a change of the configuration of all involved 4-point spaces. Such changes of configuration always happen instantaneously with every event. No point must thereby travel from S_m to S_n. Therefore, all actions are instantaneous. No time is required. But since the system S_n has no information about the changes in the system S_m it does not „feel“ the action (Figure 27).

For an action to be perceived the total system (S_m + S_n) must be considered and at least one point must travel from system S_m to system S_n. Only this allows a change of the internal configuration of the system S_n that it can „feel“. Since according to the axiom of chirality all points travel with light speed from event to event, the information cannot be transferred faster than light (Figure 28). This does not mean that the same

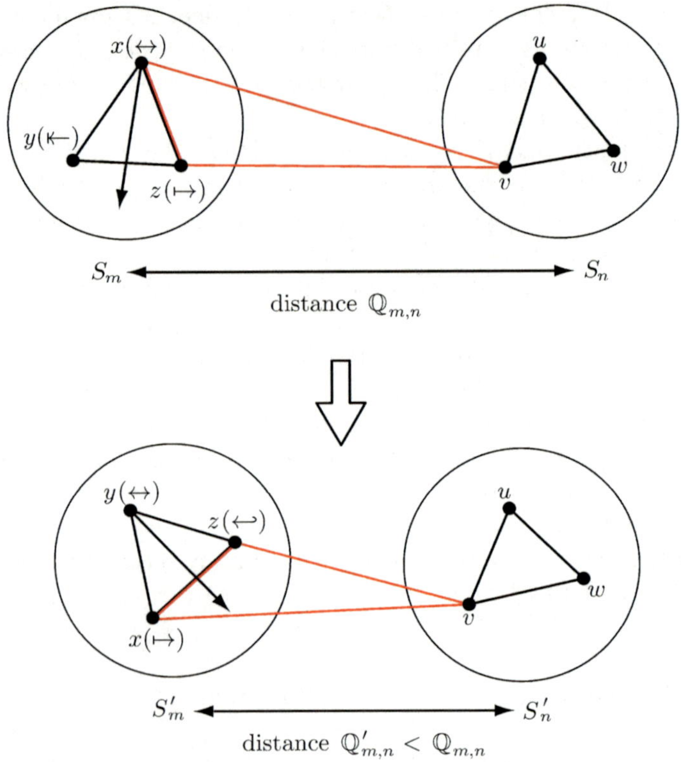

Figure 27. Instantaneous, non-local (inter)action between S_m and S_n.
The orientation of space (x, v, z) inverts during the event to (x, z, v), but all points stay within the original systems S_m or S_n. Before $v_{(\mapsto)}$ in the red space is transformed to $v_{(\hookleftarrow)}$ and later to $v_{(⇤)}$ and $v_{(\leftrightarrow)}$ many phase changes ($\Rightarrow$) are required depending on the distance between the systems S_m and S_n. The third dimension of space and the weave of all the overlapping 4-point spaces between S_m and S_n are not shown.

point travels all the way from S_m to S_n. The information transfer happens through a net or weave of partially overlapping space inversions of all the 4-point spaces between the two systems. From this it follows that actions always are non-local but information transfer between separate systems is always local. It must be kept in mind that every such separation of systems in the universe is an artificial one. This means that the distinction between local and non-local is a subjective one depending on the extent of the artificial separation of the observed systems. These interpretations of action and information, locality and non-locality are the sixth paradigm shift of chirality theory. The questions regarding the velocity of action have been extensively discussed in the Stanford Encyclopedia of Philosophy (2008).

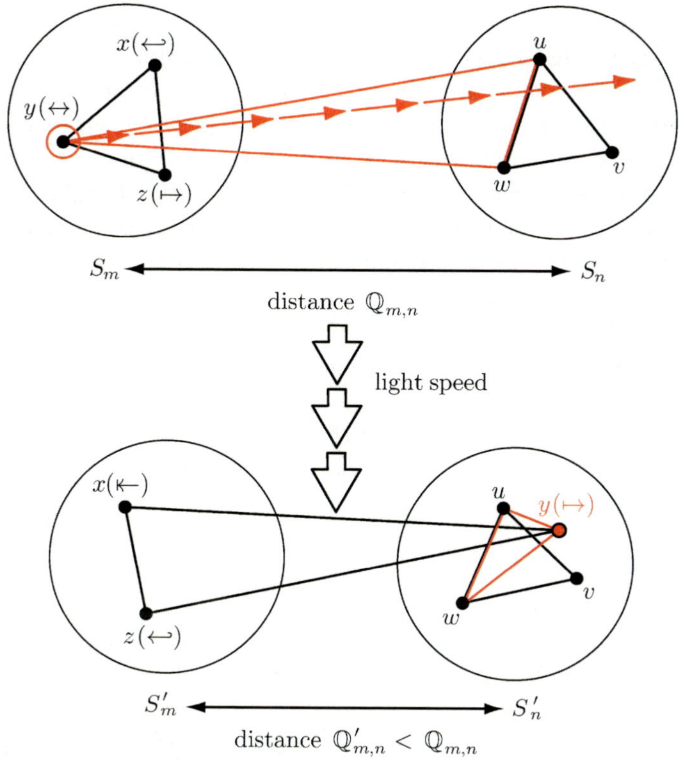

Fig. 28. Information transfer from S_m to S_n with light speed.
Point y (or the wave it has produced respectively) travels with light speed from S_m to S_n whereby both S_m and S_n are changed. The orientation of space (y, u, w) inverts to (y, w, u). The third dimension of space and the weave of all the overlapping 4-point spaces between S_m and S_n are not shown.

8.13.7 The waves of quantum theory

The information or energy transfer through a net of partially overlapping 4-point space inversions might result in a wavelike transfer at light speed as is postulated by quantum theory. An inversion of one of these innumerable 4-point spaces may exert an action in many other overlapping 4-point spaces which share one of their points with the previous 4-point space. According to the axiom of chirality this action produces new inversions of these other 4-point spaces depending on the relative configuration of the spaces and so on. The description of all these numerous inversions is too complicated to be drawn in a figure but the picture probably would resemble that of a wave as in quantum mechanics. But in contrast to quantum mechanics there is no background space for the wave and no continuum with a local neighbourhood of the single points.

Chapter 9

Interaction: Gravitation

A unified theory of all forces will probably require radically new ideas.
Steven Weinberg (1999)

9.1 Interaction in the four-point space

When we speak of interaction we – usually unconsciously – take certain self-evident basic conditions for granted: an interaction always presupposes that one can clearly distinguish between the things which are interacting. That is, they are separate. On the other hand, the interaction means that the two interacting things are connected by this very interaction. That is, they are one. This simultaneous separate-ness and one-ness constitutes the nature of the interaction.[50] If we wish to understand interactions, then we have to have a clear idea about the extent to which the interacting things are separate and the extent to which they are one.

If we look at it in this way, we have already dealt with interactions in the four-point space, namely with the interactions between the four separate points within the one black hole, the four-point space. The four points are separate as (theoretically) countable individuals; they are one through their reciprocal, constantly changing configuration. The interaction between the four points in the four-point space is defined by the axiom of chirality and results in no point being able to leave the black hole because the space inverts beforehand, whereby all four points change their state. The inversion always occurs when one of the four points gets between the other three. In this way, the axiom of chirality

[50] Here also, I refer as I did at the conclusion of Chapter 5 to Goethe's *West-Östlichen Divan* (West Eastern Divan), in which the nature of the pair of lovers is composed of precisely in being both a single and double entity.

effectuates a mutual attraction between the four points. I call this attraction **gravitation F_G**. Gravitation cannot have a repelling action.

The gravitation within the four-point space F_{G4P} is by definition a function *f* of the event frequency $\nu = 1$ within this space:

$$F_{G4P} = f(\nu_{4P}) = f(1),$$

whereby the four-point space becomes a black hole: no point can escape the four-point space with the (according to Section 8.7) radius

$$r_{4P} = 1.$$

9.2 Topology and metric

It is now a matter of deriving a metric from the definitions of distance in Section 8.12.2 or 8.13.5 which corresponds to the metric of the general theory of relativity. Thus, a theory of quantum gravity would emerge. The metric space to be defined is a subset of the topological space [Nakahara (2003) pp. 81 ff.]. Every event is by definition a topological change in the configuration of the four points. The events are the countable quanta of the perception. The requirements for a metric have at least already been met in the simple four-point space, namely distance $AA = 0$, distance $AB = BA$ and distance $AB + BC \geq AC$. The distances must also for spaces with more than four points be so defined that the length contraction corresponds to the function of the mass distribution of the general theory of relativity (Section 6.2.3 and Figure 15) [Hölling (1971a) pp. 157–162].

Now let's suppose that a universe was composed of more than simply one four-point space, for example, of 100 such neutrinos. Each is in a location relative to the other, otherwise it would not belong to the same universe. The 100 neutrinos form a more or less dense network of objects with 400 points, which all move relative to one another, i. e., they change their location as a result of discrete, countable events, whereby their relative topological configuration constantly changes. The points don't really move in a space, and certainly not in a continuum, but only ever relative to the other points. In this way, however, they span a space-like, constantly changing network with locations and distances which must be three-dimensional like the individual four-point space, since with more than three dimensions the locations of the neutrinos would no longer be clearly determinable. The universe is a self-contained, interconnected whole, in which the situation of every part is determined by all the other parts. It is mathematically possible to construct any three-dimensional form, thus the universe as well, from a sufficient number of sufficiently small tetrahedrons [Regge (1961)]. But in the space model of chirality theory different parts can also penetrate each other, since they are only comprised of points, and this penetration results in a dynamic correlation, not just spatially, but temporally as well. The network defines three-dimensional, contacting, discrete modules which change with every event in the universe. These correlations are the subject matter of Chapters 9 to 12. Space and time, according to this theory, are no longer independent entities, but naturally and logically connected to a discrete space-time. It is distinct from the four-dimensional Minkowski space primarily in that it does not form a continuum and that the chirality of both space and time plays a fundamental role.

If one were to succeed in defining the distances in this axiomatic, topological framework mathematically, then the existing paradoxes of the theory of relativity should be solvable fairly easily. Since the distances in chirality theory are a function of countable events, they must be expressible as rational numbers, i. e., as a ratio of counted events. Paradoxes of the theory of relativity which should then be clarified, are, for instance, the Lorentz contraction hypothesis of the rotating annulus, whereby the circumference shortens without its radius becoming shorter, and the slowing of clocks in the twin paradox, which can only be solved on the assumption of negative eigentimes, that is, clocks which occasionally run backwards [Brandes (1995) pp. 83–141]. That would be the first test of chirality theory.

9.3 Volume and the constant of gravitation

The four points of the tetrahedron define in the three-dimensional space a volume V_{4P} with a content of the approximate size $1^3 = 1$. Whether the volume is exactly 1, remains unclear for the time being. A volume of 1 would correspond to a cube with the side length 1. Possibly the smallest volume, however, is rather a sphere with the diameter 1, a tetrahedron with the side length 1 or also a double tetrahedron, since the volume with each inversion is first zero and is then reformed. Obviously, volume 1 refers to the smallest possible volume. Since the inverting or pulsating tetrahedron, as I showed in the previous section, imparts a three-dimensional structure of distance to the universe outside the tetrahedron as well, one can ask what happens to the volume outside of the tetrahedron when the tetrahedron inverts. Since the V_{4P} inverts exactly once per event, the ratio of inverted total volume per number of events n (i.e., per time t) remains constant:

$$\frac{n \cdot V_{4P}}{n \cdot 1} = \frac{1^3}{1} = 1 = G.$$

The constant *G* I call the constant of gravitation. In classical mechanics the constant of gravitation has the dimension distance^3 · time^{-2} · mass^{-1}. A distance^3 is a volume. Since in chirality theory mass is nothing other than a time frequency ν and it therefore has the dimensions time^{-1} in classical mechanics, the dimension of G becomes the variation of volume per time, whereby both volume and time are measured by simple event counting. The constant of gravitation is thereby a dimensionless number, just as Einstein wished for.

9.4 Partial inversion

The account in this section is intuitive. It serves to illustrate the experience that forces or acceleration fields are greater, the shorter the distance between the interacting particles is. However, it is quite conceivable that in a fully formulated theory we could dispense with the picture of a partial inversion. It would then be sufficiently supported purely by event counts: the shorter the distance, the more events the particles should „experience" and the greater their attraction or acceleration should be.

It is possible to turn a glove inside out only partially, for example, just the palm area, just the forefinger or just the tip of the forefinger. Does that also apply to a tetrahedron? It does! The inversion of the lopsided tetrahedron is not a complete one after all, but in a certain respect dependent on the distance r, and this dependency, as shown in Section 8.13, can be expressed purely by event counts. The four-point space is completely inverted with every single event. Following the inversion, nothing distinguishes it from its state before the inversion. It's quite a different matter with the lopsided tetrahedron, where the fifth point *P* forms one of the four corners. The further *P* lies from the four-point space, the lesser the degree of inversion: the angle at *P* becomes more acute, the greater *r* is. (Figure 29).

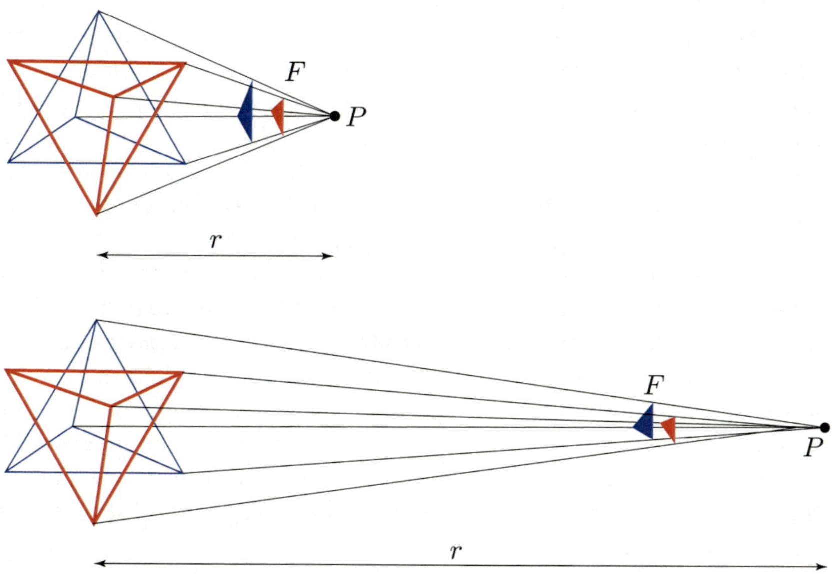

Figure 29. Partial inversion of a four-point space: *The surfaces F are a measure for the degree of the inversion of the lopsided tetrahedrons which contain the point P. F is proportional to* r^{-2}*. When* $r = 0$*, the inversion is complete.*

The inverted triangles *F* are a measure for the degree of the inversion; their size is inversely proportional to r^2. The state of the point *P* does not always change with every inversion, so the states of the lopsided tetrahedron before and after the inversion can be distinguished. The action of the four-point space on the point *P* is, due to the axiom of chirality, always one of attraction and decreases in proportion to r^2, as long as *r* is big in comparison to the radius 1 of the four-point space. At a very small distance *r*, the relation between distance and action – as is the case in the general theory of relativity – must be modified somewhat, since at short distances, the tetrahedron's geometry cannot be neglected. Within the range of the Planck length of 10^{-33} cm the gravitation constant G becomes

$$G \cdot 3^{-\frac{1}{2}} / log\, 2$$

[Smolin (2001) p. 192].

The inversion event can be described mathematically by the external observer as a pulsating, equilateral tetrahedron which according to Section 9.3, constantly changes its volume, thereby interacting with points outside the tetrahedron. The effect is greater, the greater the inverted volume and the greater the frequency of the events is. According to Section 8.7, the results of the events can, however, also be described as a rotation, or as the frequency of a rotation. The measure of a rotation is h, and with this approach, the effect is greater, the greater the rotation per event and the greater the frequency of the events is. The frequency of the events in the four-point space is the same for both approaches, namely one, as long as we state that both h and G equal one by definition. With lopsided tetrahedrons that only partially invert, the frequency must be adjusted accordingly, however, be it conceived as the frequency of volume changes or as rotation frequency. In addition, we will see in Chapter 10 that the frequency is also dependent on the state of the observer; it is therefore a subjective quantity.

9.5 The gravitational field as acceleration field with action

Attraction means that the velocity of a point P travelling in the direction of the object which causes the events increases, or the point accelerates. I interpret velocity as the ratio of the location changes of the point P from the perspective of the presumed observer to the time events of this observer's inner clock. If it has already moved in the direction of the object before the event, then this movement will become faster. If it has moved *away* from the object before the event, then this movement will become slower. If it is at rest before the event, it then begins to move in the direction of the object. The object's events bring about at every location inside and outside the object an **acceleration** $\boldsymbol{a_G}$ of every point which might be located there. This acceleration action of an object I call a **gravitational field**. The gravitational field defines the acceleration for every location, i. e., the acceleration which a point experiences at this location. The gravitation is the result of the volume changes within the object, which always changes the volume outside the object as well. One can also say that the space outside the object is bent by the object's tetrahedron inversions, so that the locations shift accordingly with every event. The gravitation field is proportional to the frequency ν of the inversions G of four-point spaces and proportional to the degree of the individual inversion (i. e., inversely proportional to the square of the distance):

$$a_G = G \cdot \nu \cdot r^{-2} = \nu \cdot r^{-2}.$$

Acceleration a_G is a ratio of three event numbers and thereby a dimensionless number. In classical mechanics acceleration has the same dimensions as

$$\text{distance} \cdot \text{time}^{-2} = a_G = G \cdot \text{mass} \cdot \text{distance}^{-2}$$
$$= (\text{distance}^3 \cdot \text{time}^{-2} \cdot \text{mass}^{-1}) \cdot \text{mass} \cdot \text{distance}^{-2}.$$

The signal of volume change G expands uniformly in all three directions in the three-dimensional space whereby the amount of the volume change remains steady, so that it „thins out" in the course of the expansion. This thinning results in the effect of the volume change decreasing with the square of the distance to the original event. The gravitational field is also a sphere-like wave with the object at its centre and with a wave frequency which corresponds to the object's event frequency. The space has no real existence, no more than time and the object does. The only thing that can claim actual existence is the event. Many successive events then, define space, time and an object. Actually, the events are the real objects, the atoms of Be-ing [Wittgenstein (1963/1994)]. Capra (1991) pp. 261–283, refers to the Heisenberg's S-matrix when he suggests that one no longer proceeds from objects, but from events, and no longer from particles, but from their reactions. Only the entity event is ennobled with the title „real".

9.6 Infinite velocity of virtual gravitation waves

The gravitational field changes at every location in the universe instantaneously in the course of each event, i.e., as soon as the configuration of the points has changed somewhere in the universe. That means that the action of the gravitation spreads out with infinite velocity! One can also say that the action of the gravitation is not a local event; it is non-local. At first glance, that is contrary to the conclusions of the special theory of relativity, according to which there must be a maximum velocity c in nature. However, that which according to Section 8.13.6 can only move at light speed and no faster is a point and not an action. The action does not, however, come about because one or more points from the object causing the field spread out into the space. The points all stay more or less put, for example within the four-point space. Rather, the action is the result of a change in the point configuration, within the four-point space as well as in relation to all the points in the universe. It is the configuration of the points which makes the whole universe a unity. The actions within this unity are instantaneous, without any kind of delay, for the configurations change everywhere immediately.

This finding is consistent with both the theory of relativity and quantum theory. According to the theory of relativity, transfers of information are always bound with energy; according to quantum theory, energy is nothing more than information. Energy, i.e., information, can only be transferred at light speed according to the theory of relativity. Actions, by contrast, are also possible without information transference. They can, therefore, easily spread out faster than at light speed. That occurs for example in the famous thought experiment put forward by Einstein (1935) Podolsky and Rosen (EPR) which, slightly modified, was carried out in practice by John Bell (1964). The EPR experiment can be somewhat superficially and naively illustrated as follows: Socrates is married to Xantippe but lives apart from her; so far apart that they cannot communicate with each other within a useful period of time. Then Socrates dies and his Xantippe instantaneously becomes a widow, but she is temporarily unaware of this. The widowing takes place without delay at superluminal speed, since they are both „entangled" with each other by their marriage. The information concerning the widowing can, however, be transferred at a maximum velocity of light speed. That means that Xantippe initially knows and senses absolutely nothing about the effect, i.e., the action of her husband's death. The results of the EPR experiments are not in themselves contentious, yet the interpretation of these results is debated amongst specialists to this day, always centred on one point: which is correct, EPR or quantum theory [Smolin (2001) pp. 77–87; Görnitz (1999) pp. 166–170]? According to Penrose (2004) pp. 591–594, all particles in the universe are probably entangled, whereby the entanglement (the ability to interlink two quantum particles with practically 100 % certainty) can be dissolved by the measuring process or by an objective process of nature.

In my opinion, both interpretations are right. Their promoters are just talking past each other because they don't share the same understanding of the term reality. For EPR, the actions are real with or without information content, even if they cannot be directly observed. For quantum theorists, only the information is real, whereas the actions remain a part of the theory as long as they are not observable [Hombach (1994)]. According to chirality theory, the actions are real insofar as they permit the theory to predict actions which are at least indirectly observable by certain observers.

Waves which have an action on an object, but which cannot be observed by this object, I call **virtual**. Virtual gravitation waves spread with infinite velocity through the whole universe.

How is an action possible without a transfer of information? An observer sitting in a windowless, freely moving spaceship and who is accelerated through the earth's gravitational field is unable to perceive his own acceleration since he is equally strongly accelerated as his spaceship is. Thus, he receives no information at all through the acceleration field. Nevertheless, the field has an action on him. The action can be perceived by an external observer who observes both the spaceship and the earth and determines that the relative velocity between the two objects changes. The information for this external observer about the action between the earth and the spaceship can be transmitted no faster than at light speed, however. A full description of the EPR problem is beyond the scope of this book, but Genz (2002) pp. 235–246, for instance, provides a vivid account.

Also, according to the formalism of quantum theory, virtual, massless particles, for example virtual photons, which cannot be observed, move at infinite velocity [Feynman (1985) pp. 95 f.].

9.7 The law of gravitation; force

In the previous section I have shown that at a particular location every point is accelerated equally by the gravitational field of an object with the mass m_1, independent of the point's state of motion and independent of the number of points at this location. For example, a kilogram and a single gram of iron are attracted and accelerated equally by the earth's gravitational pull, although the effect is naturally not the same. That becomes immediately clear as soon as one compares the respective holes made in the ground by the falling kilogram and gram of iron. The difference is due to the different masses m_2 of the falling objects: at the same height of the fall, i. e., the same acceleration, the size of the hole in the ground is proportional to the mass m_2 of the object. It makes sense therefore to introduce a new variable which expresses this state of affairs, namely the **force of gravitation F_G** :

$$F_G = a_G \cdot m_2 = \frac{G \cdot \nu_1}{r^2} \cdot \nu_2 = \frac{\nu_1 \cdot \nu_2}{r^2}$$

whereby ν_1 and ν_2 are the frequencies, i. e., the masses of the two attracting objects and $G = 1$. This formula corresponds to the Newton's law of gravitation. Newton's constant of gravitation becomes a pure number and it is one, i. e., it doesn't affect the calculations.

In deriving the law of gravitation, we proceeded from an object which generates a field, which for its part acts on a second object. The law of gravitation itself, however, is symmetrical, in the sense that the two objects with the frequencies ν_1 and ν_2 have the same place value, and affect each other. There is no direction as such in which the action proceeds; rather it's a matter of an action between the two objects. Depending on the observer's position and state of motion, he will ascertain that object 1 attracts object 2 or that object 1 is attracted by object 2. Which is the cause and which is the effect cannot be distinguished in the interaction of virtual particles, since virtual particles move in excess of light speed. Yet each event of the total system $m_1 + m_2$ becomes the cause of a following event.

9.8 The graviton

I call an individual gravitation wave a **graviton**. The gravitational field thus comprises a spatial and temporal sequence of virtual gravitons, which spread out with infinite velocity. With every event, a new virtual graviton comes into existence. The nature of an event is on the one hand the total or partial inversion of a four-point space, and on the other hand, the chiral three-dimensional rotation of this space. The volume change which results from the inversions is measured in units of the constant of gravitation *G*, the three-dimensional rotations of space by contrast in units of Planck's constant *h*. The source of the gravitation is the volume change in the four-point space per event, that means per smallest possible rotation as well. So, it can be measured in units of $G \cdot h^{-1}$. This constant $G \cdot h^{-1}$ is something like a unit charge of the gravitation. If one multiplies $G \cdot h^{-1}$ by the mass (i.e., according to our theory by the frequency of an object), then one obtains the total **gravitational charge L_G** of this object:

$$L_G = \frac{G}{h} \cdot \nu = \nu,$$

whereby *G* is the measure of the gravitational force and *h* is the measure of the rotation in the three-dimensional space. Both constants according to our definition equal one. With every event in the four-point space, the volume 1^3 changes upon the simultaneous rotation by $h = (1 \cdot 1) : (1 \cdot 1)$. If one expresses *G*, *h* and ν via dimensions of distance and time and the mass as frequency, then L_G acquires the dimension of a distance. The meaning of this distance – it concerns half the Schwarzschild radius – will be explained in Section 9.11. The difficulty with every description of gravitation is that upon close observation, the frequency ν of the total system changes a little with every single event as soon as the universe is composed of more than one black hole. So, the gravitational charge is not constant like the electrical charge, for example. This will be clarified in Section 9.11. Moreover, ν is dependent on the observer's state of motion which complicates the description further (see Section 10.9).

If one views the graviton as a particle which spreads out in all directions with a growing distance *r*, then it „thins out" in proportion to $\frac{1}{r^2}$. Its action decreases accordingly. Thus, one can probably do without the picture of the „partial inversions" in Section 9.4.

Since the four-point space rotates with every inversion, the graviton moreover has an angular momentum. As shown in Section 8.9, this is 2.

9.9 Rest mass of the neutrino

Gravitation is a consequence of the inversions of the participating particles. As we have seen, such inversions not only have a frequency, but also a degree, both of which influence the gravitation. In Section 8.2 we saw that there are six different forms of neutrinos of which only two can be described as equilateral tetrahedrons, whereas the other four are lopsided: the inversion of these is less symmetrical, since the apex of the tetrahedron does not pass through the centre of the base but pierces the base plane either over a side or over a corner. In so doing, the tetrahedron becomes lopsided and has a correspondingly greater radius. This radius of the tetrahedron has an influence on the degree of the rotation which it experiences during the inversion, and also thereby on the event frequency with this rotation: the more lopsided the tetrahedron is, the greater the rotational motion and with it, the event frequency. This is because the apex of a lopsided tetrahedron goes through more changes in location per inversion than an equilateral tetrahedron. One can also say that the volume, which the pulsating tetrahedron circumscribes, is the greater, the more lopsided the tetrahedron is. However, the more lopsided the tetrahedron or the higher its frequency is, the greater the resting mass of the particle and with it, the gravitational force which emanates from it. The particle in which the apex of the tetrahedron always penetrates the base above a side, I call a muon neutrino; the particle where the apex of the tetrahedron always pierces the base over a corner, is the tauon neutrino, the neutrino with the greatest event frequency and resting mass, and the strongest gravitational force. However, I am unable to provide a formula for an exact calculation of the relations between the three neutrino masses. From measurements of sun neutrinos and with the help of the so-called seesaw model, the following masses for the three kinds of neutrino – however with very large uncertainties – were calculated: For the ν_e $4 \cdot 10^{-8}$ eV, for the ν_μ $2.4 \cdot 10^{-3}$ eV and for the ν_τ 46 eV [Schmitz (1997) p. 362]. That said, all neutrinos are black holes in the sense of the definition in Section 8.3.

9.10 Potential energy

With an interaction between two neutrinos, that is, an eight-point space, the description of the course of events already becomes quite complicated. Here there are also tetrahedrons whose corners belong to two different neutrinos. In the context of the two individual neutrinos, the states, viewed from outside remain unchanged during the events; the collective state of the two neutrinos together, however, changes with every event: on the one hand, the distance between the two neutrinos shortens, whereas on the other hand, there is an increase in the relative velocity with which the two neutrinos approach each other as a consequence of the gravitation. We are not only dealing with the two rest masses m_{04P}, or their frequencies ν_0 of the two individual neutrinos, but also with the frequencies ν_{12} of all the lopsided tetrahedrons whose corners belong to two different neutrinos 1 and 2. (Beside the two equilateral tetrahedrons of the two four-point spaces, there are 32 lopsided tetrahedrons with three points lying in one four-point space and the fourth point lying in another; and 36 lopsided tetrahedrons with two corners each lying in one of the two four-point spaces. Thus, there are 68 lopsided tetrahedrons in the eight-point space which all overlap and which change somewhat with each event). The frequencies ν_{12} are the frequencies resulting from the interaction between the two particles. These frequencies as well contribute to the **rest mass m_{08P} of the total system** of the eight-point space, and what is more, in the following way.

The space inside the individual neutrino inverts completely with every event. This space, however, is partially within the spaces of all the lopsided tetrahedrons which are comprised of points from two different neutrinos. This total of 68 lopsided tetrahedrons only partially invert per event or not at all, thereby partially offsetting the volume of the inversions of all the single neutrinos. The collective volume of all space inversions is thus reduced by this effect, i. e., the rest mass of the total system decreases and is smaller than the rest mass of the two neutrinos alone. This is a case of holism, i. e., the total system is something other and more than the sum of its parts. The decrease Δm_{12} in the rest mass of the total system, or the frequency $\nu_{12} = \Delta m_{12}$ which corresponds to this decrease is greater, the greater the force F_{G12} between the two particles is. From the perspective of the external presumed observer, as shown in Section 8.7,

ν_{12} concerns event frequencies during rotations, which in accordance with Section 8.10, are measured in units of Planck's constant h. The total rotation of the interaction ν_{12} is the product of two different kinds of rotation, quasi a rotation of the rotation (corresponding to the inversion of an inversion): First, the rotations ν_{04P} of the two neutrinos and second, the rotations ν_{12} between the two neutrinos. One might ask why the total frequency is not simply the sum of the two single turns or single frequencies. The reason is that the 3 dimensions of the four-point space form a black hole, which is self-contained. One cannot add distances from different black holes, if one wants to connect them. A connection is possible only as a product of multiplication. The fact that an interaction must be described mathematically by a product corresponds to the similar procedure in quantum theory during the combining of two objects into only one object: The Hilbert space of the total object is the Kronecker product of the Hilbert spaces of the partial objects. The interaction of the object with other objects is then described by the Hamilton operator of the compound object composed of these objects (Weizsäcker (1999) pp. 24 and 364 f. The radius in the case of the first rotation is 1, whilst the rotation angle with this rotation is at its maximum. The radius of the second kind of rotation is r_{12}; in this case the rotation angle is very small. The measure of the rotation is h. The total frequency, i. e., the product of the two rotations thereby becomes:

$$\nu_{12} = \frac{1}{h} \cdot F_{G12} \cdot r_{12} = \frac{1}{h} \cdot \frac{G \cdot \nu_1 \cdot \nu_2}{r_{12}^2} = \frac{G \cdot \nu_0^2}{h \cdot r_{12}}$$

or

$$h \cdot \nu_{12} = F_{G12} \cdot r_{12} = \frac{G \cdot \nu_0^2}{r_{12}} = - E_{pot}.$$

The product of $F_{G12} \cdot r_{12}$ I call **potential energy E_{pot}**, or more precisely (potential) gravitation energy. It makes sense to define the potential energy as zero when its underlying interaction is zero. That is the case when r_{12} strives towards infinity. In the case of a short r_{12}, E_{pot} is always negative, as long as the forces involved are attractive, since work is required for the mutually attracting particles to be separated. The potential energy can be also calculated by calculating the energy which is released if one shortens the distance between the two particles with the frequencies ν_1 and ν_2 from infinite to a distance of r_1:

$$E_{pot} = \int_{\infty}^{r} \frac{G \cdot m_1 \cdot m_2}{r^2} = \mathrm{d}r = G \cdot m_1 \cdot m_2 \left(\frac{1}{r_\infty} - \frac{1}{r}\right) = -\frac{G \cdot m_1 \cdot m_2}{r} .$$

Since according to chirality theory, the shortening of distances takes place in small, but finite steps, the integral is to be understood as the sum of a finite number of summands. Thus, a small correction could be produced in the final result, which can be neglected over large distances, however.

The mass m_{8P} of the total system is as follows:

$$m_{8P} = 2m_{04P} - \Delta m_{12} = 2\nu_0 - \nu_{12} = 2\nu_0 - \frac{v_0^2}{r_{12}}$$

(As usual $h = 1$ and $G = 1$, whereby the dimensionality in the equation presented is correct.)

The energy of the eight-point space E_{8P} is made up of the mass energy $m_0 = 2h \cdot \nu_0$ of the two neutrinos and the potential energy resulting from the interaction:

$$E_{8p} = 2h \cdot \nu_0 - h \cdot \nu_{12} = h \cdot \left(2\nu_0 - \frac{v_0^2}{r_{12}}\right) .$$

The two kinds of energy can be clearly distinguished: mass energy is the energy from black holes; the potential energy is the energy of an interaction between black holes. The two neutrinos, as black holes, are two separate things; they are one, however, through their interaction. Thus, the eight-point space is simultaneously one and double. Such a clear distinction between mass energy and interaction energy is only possible when one is dealing exclusively with black holes, which in physics is usually not the case. Whether the molecule is a whole particle or whether it consists of atoms and their interactions, whether the atom is a whole particle or whether it is made up of nucleus, electrons and their interactions, whether the nucleus is a whole particle or whether it is composed of protons, neutrons and their interactions, whether the proton is a whole particle or whether it is made up of quarks and their interactions, the gluons, are all distinctions which are left to the arbitrary choices of the physicist. According to how he decides, the interaction energy is counted as mass or measured separately. The divisibility of quarks, electrons and interactions inside black holes will be discussed in Chapters 11 and 12.

The actions of acceleration with each event between the two neutrinos will be considered in Chapter 10. There it will be shown that this acceleration leads to an additional energy term, i. e., kinetic energy.

9.11 Schwarzschild radius

Not only the points inside the four-point space are attracted by the inversions, but according to Section 9.4, all the remaining points outside the four-point space are as well. The inversions within the four-point space, i. e., inside a black hole, I call first-order inversions; the inversions of the lopsided tetrahedrons whose volume lies partly outside the black hole, I call second-order inversions. In the previous section, I have shown that the first-order inversions are partially offset by the second-order inversions. The potential energy of the total system is thereby reduced. Now if the system's total mass increases, i. e., if several or a great number of black holes are present and, moreover, are sufficiently close to each other then it could happen that the second-order inversions fully offset the first-order inversions. The potential energy in this case reaches the same magnitude as the rest mass of a particle, only negative.

An equal amount of energy is thus required to separate the particles again as it does to create a particle out of nothing. The system has become indivisible, i. e., it has become a new black hole. Such a hole I call a second-order black hole. Now, the total frequency or mass of the second-order black hole is made up of the frequency ν_0 of inversions in the first-order black hole and the frequency of second-order inversions, which must offset the former. Consequently, the collective frequency is $2\nu_0$. The gravitational charge L_G, which corresponds to this frequency, is

$$L_G = \frac{2G \cdot \nu_0}{h} = 2\nu_0 = \mathscr{R} .$$

As already explained in Section 9.8, L_G has the dimension of a length, if one expresses G, h and ν_0 by means of dimensions from classical physics and the mass as frequency. L_G is the radius where the mass m with the frequency ν_0 becomes a black hole. I call this radius **Schwarzschild radius** $\mathscr{R}$.

It is to be noted that in chirality theory $h = c^2$ whereby $\mathscr{R}$ can just as well be expressed as in the general theory of relativity in line with Section 6.2.2

$$\mathscr{R} = \frac{2Gm}{c^2} .$$

In a particle, which is at the Schwarzschild radius of a black hole, the internal inversions of the particle are fully offset by the external partial inversions which the black hole effectuates in this particle.

9.12 Internal and external properties of an object

The distinction between black holes of the first and second order corresponds to the distinctions which Whitehead made between extension and change of location, inwardness and outwardness, substance and configuration, thing (point) and relation, oscillation (nexus) and event, corpuscle and wave. For these pairs of opposites, one could use the term complementarity. The principle of complementarity in this context means that inside and outside are not only two different aspects which are linked to each other and are opposites in any given cognitive process, but that the two aspects also always block or impair each other in some way with their alternating complementation [Wiehl (2000) pp. 320–373].

A good deal of attention was paid by Lewis (1998) to the philosophical characteristics of the internal and external properties of a physical object. Two things are the same or duplicates of each other precisely when they have the same internal properties. All external, i. e., relative properties, can be as different as they like. The difficult problem that this raises is distinguishing between internal and external properties, since one inevitably gets caught up in circular reasoning: a duplicate is the consequence of identical internal properties; internal properties, however, are precisely those which remain the same in the duplicates. What the properties of the object are really like can only be determined by the external observer. He measures for example, a spin $+\frac{1}{2}$. This spin can only be measured by comparing it with an external spin defined by the observer. Therefore, internal properties also are always relative, which means that in a certain sense, they are external properties as well. The observer is free to choose where exactly he wishes to set the boundary between internal and external properties, what he regards as a part and what as a whole. However, the new possibility arises in chirality theory of setting between internal and external properties an objective, absolute boundary, which is independent of any observer, namely at the Schwarzschild radius of the black holes.

The smallest possible black hole is the neutrino (see Section 8.3). Perhaps it is the only possible first-order black hole at all. Of particular interest now is the question concerning the smallest possible second-order black hole, i. e., a black hole composed of black holes. It will be answered in Chapter 11 in the treatment of electromagnetism.

Chapter 10

The Real Observer: Mechanics

> We often think that when we have finished our study of one we know all about two, because „two“ is „one and one“. We forget that we have still to make a study of „and“.
>
> Arthur Stanley Eddington (1882–1944)

10.1 Perception by a real observer

It is entirely possible that my transcendental consciousness makes observations and perceives things. Indeed, these are simple to describe by means of a presumed observer. They are, however, subjective and thereby only conditionally scientific. Physics is concerned with objectifiable perceptions, for which it relies on real observers who can exchange information amongst themselves about their perceptions. In this way, physics becomes objective and thus more scientific. So, it is time to replace our presumed observer with a real one. Real means that he himself can be described by the same physical theories by which he describes the rest of the world. Thus, the perception becomes an interaction between the observer and the observed object. As we have seen, interactions are always something mutual, i. e., during the observation, the subject and object influence each other. Not only the subject and the object change thereby, but also the relation between the two. To be able to speak of perception, information must flow from the object to the subject and then be stored in the subject in such a way that it is recallable and available to the conscious mind.

Probably the simplest possible measuring instrument that fulfils these conditions was described by Arnulf Schlüter (1993). It consists of a single atom with a single electron of interest to us which can exist in four quantum states. Thus, the flyby of a charged fast particle can be measured, i. e., counted. To prepare for the measurement, the atom is brought into its ground state 1. The ground state of a quantum mechanical system is its lowest energy state. An excited state is any state with energy greater than the ground state. If a particle now flies past, the electron can, under this force effect, convert to the excited state 4 of the atom. If, under the emission of a photon, it falls to the metastable state 2, the flyby is thus registered. The atom is thereby a document for this process.

Naturally, a document must be able to be read, and repeatedly at that. This occurs here via the so-called resonance fluorescence. The electron is irradiated with a single photon, raising it from level 2 to level 3. From there it falls back immediately and in so doing, in turn emits a photon of the same frequency. This photon can now travel in all spatial directions and is thereby almost always easy to distinguish from the radiated one. As is the case with every proper document, this reading process can also be repeated at will. (Figure 30).

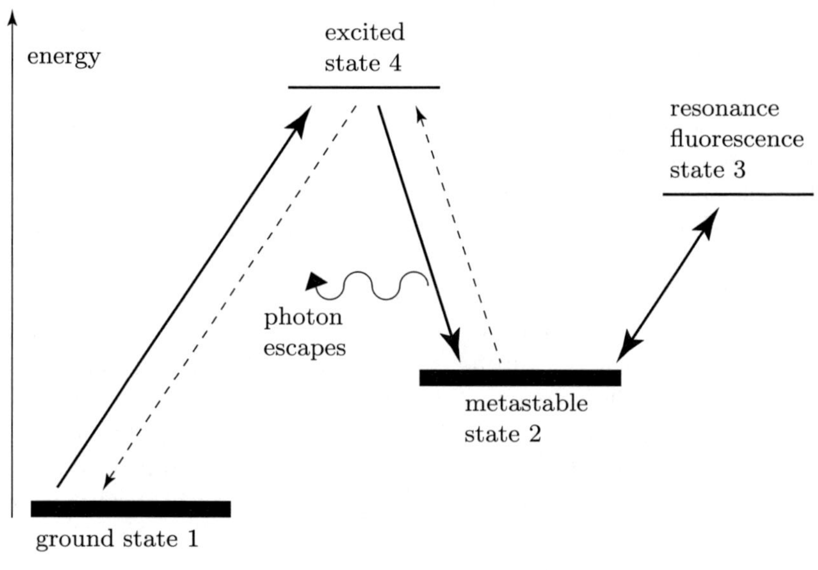

Figure 30. An atom as a measuring instrument: *The arrows indicate possible trajectories upwards: by energy uptake downwards: by the emission of a photon (the boldness of the line indicates the respective probability).*

The remarkable thing about this example is that no macroscopic measuring instrument is assumed. The whole process seems to take place completely in the quantum world; an observer is not mentioned. The only condition is that the radiated photon is not reflected back and that no other incoming photons are to be expected, i. e., that the process is not reversible. Rather, the radiated photon must disappear irrevocably into the universe, which is only possible if the universe is dark. Only by this cosmological constraint, by the photon's definite disappearance, are the observer and object definitively separated following the measurement process.

In physical practice, all sensory information transfers are photons, i. e., electromagnetic interactions. Other interactions, such as gravitation or the so-called weak and strong interactions for example, we can never perceive with our senses; they are purely theoretical. That doesn't mean, however, that also these other interactions cannot transfer information and be thereby suitable for explaining certain aspects of real observers. This also makes sense above all because the theory of gravitation is much easier to describe than the complicated electromagnetic interactions. To do this, it is not necessary to construct the concrete model of a gravitation wave observer, as Schlüter did for the photon observer; fortunately, it is sufficient to specify a few conditions to which such a real observer is subject:

1. For the „real observer of gravitation waves", the axiom of chirality applies (see Section 7.5).
2. The observer is able to register the information of individual gravitation waves. This information consists of the energy $h \cdot \nu$ of the waves, i. e., a dimensionless ratio, as well as a rotational direction (spin).
3. The observer is in a location relative to the object (see Section 8.13.2).
4. The observer has a distance from the object (see Section 8.13.5).
5. The observer has an inner clock, which functions just like a pulsating or rotating four-point space. These pulse beats or numbers of rotations can also be counted.
6. The observer can store the received information.
7. The stored information can be retrieved by a second observer without it being lost to the first observer. The information transfer between the two observers can likewise take place via gravitation waves.
8. The observer has a consciousness, which distinguishes him from the measuring instrument.

10.2 Time dilation by a black hole

We now compare the frequencies ν_A and ν_B of the inner clocks of two real observers B_A and B_B, which are at the distances $r_B = \infty$ and r_A respectively away from a black hole. The two real clocks are assumed to be models of four-point spaces. So, they are calibrated such that their frequencies, in line with Section 8.6, correspond precisely to a four-point space. At a great distance from the black hole, the frequency, i. e., the clock's running speed, is practically uninfluenced by it. At a distance of one Schwarzschild radius $\mathscr{R}$, however, the first-order inversions inside the four-point space, i. e., the clock, according to Section 9.11, are completely compensated by the many external partial second-order inversions resulting from the interaction between the clock and the black hole: the clock stands still! Since not only the observer's inner clock, but all clocks at a distance of $\mathscr{R}$ stand still, the observer B_A is unable to determine whether time runs more slowly or even stops altogether by comparing the clocks or any other events in his vicinity. He will not be able to tell until he compares his clock with that of observer B_B.

When $r_A = \mathscr{R}$, then from the perspective of B_B, $\nu_A = 0$; when $r_A \gg \mathscr{R}$, $\nu_A = \nu_B$. What does B_B actually see exactly when he looks in the direction of B_A? What comprises the frequency ν_A? It involves the frequency of a volume inversion, i. e., the change of a three-dimensional space. However, B_B is unable to observe the volume inversion of A as such directly. Rather, he simply sees that the configuration of points changes at B_A. This change results in a three-dimensional rotation of object A, which in accordance with Section 8.7, can be described by rotations through two angles, φ and ψ. It concerns the rotation of a rotation, i. e., a rotation squared. The two angles φ and ψ change simultaneously and periodically, indeed in such a way that the total rotation h always remains the same. When φ is big, then ψ is small and vice versa. The rotation, or frequency, observed by B_B is really a frequency squared, namely $\nu_A{}^2$. The two rotations, or frequencies, cannot be added, since they are in two different, mutually independent space planes. They can therefore only be mathematically joined by multiplication. B_B then compares this frequency $\nu_A{}^2$ with the frequency $\nu_B{}^2$ of his own inner clock. Without a black hole in the vicinity of B_A, ν_A and ν_B would not differ. The black hole, however, has the effect that the inverted volume, or the three-dimensional rotation with B_A is partially compensated by the second-order inversion via the black hole. This compensation is greater, the greater the Schwarzschild radius $\mathcal{R}$ (or the mass or frequency) of the black hole, and the smaller the distance r between the black hole and B_A is. Consequently:

$$\nu_A^2 = \nu_B^2 - \nu_B^2 \cdot \frac{\mathcal{R}}{r} .$$

Thus, we have

$$\nu_A = \nu_B \sqrt{1 - \frac{\mathcal{R}}{r}} .$$

Between the clock times t_A and t_B of the two observers, the relationship

$$t_A = t_B \sqrt{1 - \frac{\mathcal{R}}{r}}$$

applies similarly.

When $r = \mathcal{R}$ then $t_A = 0$; when $r \gg \mathcal{R}$ then $t_A = t_B\,(1 - \frac{\mathcal{R}}{2r})$; when $r = \infty$ then $t_A = t_B$ (due to the so-called Taylor series).

The considerations presented are not of a strict mathematical derivation; rather they are intuitive and presume that in the event that a transcendental observing consciousness existed, its activity would slow down in exactly the same way as its real inner clock, i. e., that conscious time and real time run synchronously. The existence of such a consciousness, however, is not presupposed.

10.3 Length contraction by a black hole

According to Section 8.12.2, the relationship $l = c \cdot t$ exists between distance l and time t. A consequence of this is that the time dilation automatically leads to a corresponding shortening of distance or length. If a standard length l_B is brought from the infinite into the proximity of a black hole, then, viewed by an infinitely distant observer B_B it is shortened to the length l_A

$$l_A = l_B \sqrt{1 - \mathcal{R}/r} .$$

The observer B_A notices nothing of the shortening. At the Schwarzschild radius, however, all lengths viewed by B_B become zero. In the course of this, it is immaterial whether the lengths are measured radially or tangentially to the black hole. In layman's terms, one could say the space is bent by this length contraction. Since according to chirality theory space doesn't really exist, this manner of expression is rather misleading, however.

10.4 Red shift

A consequence of the length contraction is that the wavelengths of non-virtual waves, which are sent outwards from the direction of a black hole, become longer from the perspective of a very distant stationary observer B_B, and that the frequency of these waves is lower than the frequency which B_A measures in the proximity of the black hole. This applies equally to all waves which travel at light speed, i. e., to non-virtual light waves as to non-virtual gravitation waves. The phenomenon is called red shift. Since according to Section 8.11 the wave energy is $E = h\nu$, B_B measures a lower energy than B_A. Energy and time are relative and depend upon the observer's point of view [Einstein (1922) pp. 88 ff.].

10.5 Deflexion of light by gravitation

According to Section 8.11, $E = m_0c^2 = h\nu$. So, a wave with the energy $h\nu$ also has something like a mass m amounting to $m = \frac{h\nu}{c^2}$. If it involves a standing wave, such as in the case of a four-point space, then that is the rest mass m_0 of the particle. However, if the wave moves at light speed like a real light or gravitation wave, then the event frequency ν no longer corresponds to a rest mass, but to a „moving mass m" with the energy $h\nu$. Also, this mass, or its event frequency, interacts with the black hole or with any other heavy mass, since all events are, after all, total or partial inversions of space. The interaction is stronger, the nearer the wave passes to the black hole and the higher the wave frequency is. The gravitational force between the black hole and the wave results in the wave being drawn towards the black hole, i. e., it is diverted from its course.

Virtual waves, according to Section 9.8, move with infinite velocity, rather than at light speed. Thus, their relativistic mass becomes zero, and consequently, their energy is also zero. Therefore, they cannot be attracted by a black hole.

10.6 Particle motion

During the previous considerations, we have always presumed that all particles and their observers rest, i.e., that their mutual configuration does not change. Up until now, the only things which ever moved have been the individual points within the particles at light speed, and the virtual gravitation waves with infinite velocity. Slower velocities did not come up. With the waves, no points shifted between the particles. In this sense, the waves were only virtual. Particles or black holes and waves define, as we have seen in Sections 9.2 to 9.11, a three-dimensional space, indeed solely by means of the relation „between“ and by event counts. In this space, the interactions always effectuate a mutual acceleration of the particles (Section 9.7). Also, when the particles are possibly at rest initially, inevitably they already start moving with any event and the rate of motion changes with every further event. If the observer is real instead of presumed, naturally that applies to him as well. These accelerations and movements change the relative frequencies of the particles and the observers' clocks. Relative means that the observed frequencies depend on the relative movement between the observer and the object. In quantum mechanics, the motion of a particle is described mathematically as the „beat-note“ of the wave function ψ of this particle. The beat-note moves as frequency from location to location with a velocity which corresponds to the classical velocity of the particle [Feynman (1966a) 7–4f.]. The reason for this is that with a relative movement between two objects, the distance changes, or – with circular movements – it is the relative direction of movement that changes, which is connected in each case to partial space inversions. These external second-order space inversions partially or wholly compensate the object's internal first-order space inversions. The process is comparable to the effect black holes have on particles in their vicinity: the external space inversions by the black hole partially or wholly compensate the internal inversions of the particle. Since the observer's clock has been calibrated as the frequency of the internal inversions of a four-point space, in accordance with each movement it will measure a time which corresponds to the frequency of a four-point space at rest, or which stands still when the relative velocity reaches light speed.

The relative velocity v between two objects is always between zero and lightspeed c.

10.7 Time dilatation by motion

An object A moves at the speed v towards the resting observer B. The observer feels an eigentime t_B, which he measures by means of a four-point space. In the object A there is also a four-point space functioning as a clock, which measures A's time. The points within the four-point space move with each inversion at light speed c, in each case over the diameter d of these spaces (Figure 31).

a) Observer B, at rest

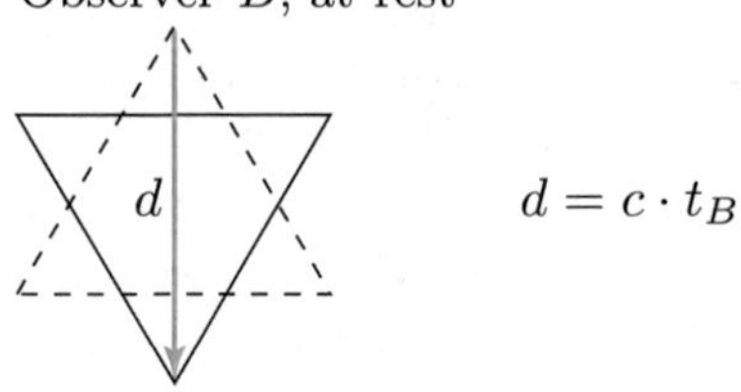

b) Object clock A, moving with velocity v

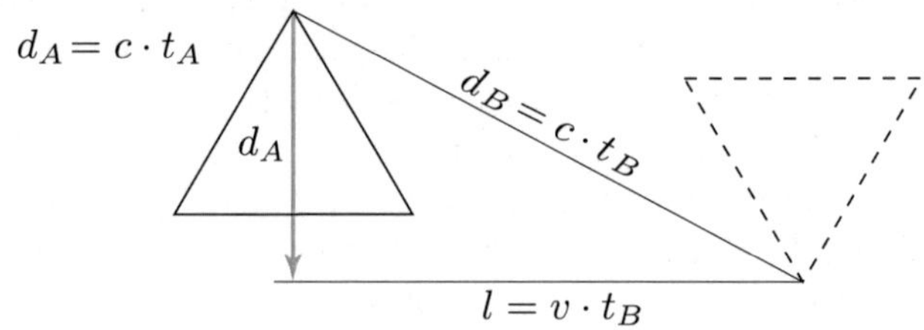

Pythagoras:

$$d_A^2 + l^2 = d_B^2$$

$$c^2 t_A^2 + v^2 t_B^2 = c^2 t_B^2$$

$$t_A^2 = \frac{t_B^2 (c^2 - v^2)}{c^2} = t_B^2 \left(1 - \frac{v^2}{c^2}\right); \quad t_A = t_B \sqrt{1 - \frac{v^2}{c^2}}.$$

Figure 31. Time dilation of a moved clock

A will have travelled the length l when the observer B measures the time t_B. The length l, the diameter d_B of the observer's clock and the path d_A of a point of the object clock form a right-angled triangle, in which Pythagoras' theorem applies:

$$d_A{}^2 + l^2 = d_B{}^2$$

Thereby $d_B = ct_B$, $l = vt_B$ and $d_A = ct_A$, that is

$$(ct_A)^2 + (vt_B)^2 = (ct_B)^2.$$

The solution according to t_A yields for time dilation the well-known equation of the special theory of relativity

$$t_A = t_B\sqrt{1 - \tfrac{v^2}{c^2}}\,.$$

The clocks of objects which are moving from the resting observer's perspective, run more slowly than resting clocks and stand completely still when an object moves at light speed.

10.8 Length contraction by motion

Between length l and time t, according to Section 8.12.2, there exists the relationship $l = ct$. Thus, the relationship for the length contraction of a moved object l_A in comparison to a resting object l_B can be derived from the formula

$$l_A = l_B\sqrt{1 - \tfrac{v^2}{c^2}}\,.$$

Distances in moved objects are, seen from the perspective of the resting observer, shorter than in resting objects and shrink right down to distance zero when an object moves at light speed.

10.9 Kinetic energy

In Section 8.11 I defined the energy E of a resting four-point space as

$$E = m_o c^2 = h\nu = \tfrac{h \cdot n}{t_B}\,.$$

n is the number of events counted by the resting presumed observer B and t_B is the time measured by the observer's clock. This definition only applies to resting masses m_0. If the mass A is moving, then the frequency

ν_A measured by the resting observer *B* changes. The resting observer *B* counts *n* events of the moved object *A* in the time t_A, which the moved clock *A* shows. The moved object *A* counts the same *n* events with its (in its view resting) clock *A* and measures therefore the „rest time“ t_B. In accordance with Section 10.7 thereby,

$$t_A = t_B \sqrt{1 - \frac{v^2}{c^2}} \, .$$

Thus

$$\frac{t_A}{t_B} = \sqrt{1 - \frac{v^2}{c^2}} \, .$$

The rest mass m_0 is, in accordance with Section 8.11

$$m_0 = \frac{h}{c^2} \cdot \nu_B \, .$$

The relativistic mass m_A is

$$m_{\mathrm{A}} = \frac{h}{c^2} \cdot \nu_A \, .$$

Consequently,

$$\frac{m_0}{m_A} = \frac{\nu_B}{\nu_A} = \sqrt{1 - \frac{v^2}{c^2}} \, ,$$

or

$$m_A = \frac{m_0}{\sqrt{1 - \frac{v^2}{c^2}}} \, .$$

Thus, the energy E_A of the moved object *A* measured by the resting observer becomes

$$E_A = h \cdot \nu_A = m_A c^2 = \frac{m_0 c^2}{\sqrt{1 - \frac{v^2}{c^2}}} \, .$$

This equation can also be written as

$$E_A = m_0 c^2 + m_A c^2 \left[\frac{1}{\sqrt{1 - \frac{v^2}{c^2}}} - 1 \right] .$$

That is the famous conclusion of Einstein that mass is equal to energy [Møller (1969) pp. 77–82].

At first sight, the right-hand side of this equation appears to be complicating things unnecessarily. It is, however, very useful since in this way, a distinction can be made between the energy of the rest mass $E_0 = m_0 c^2$

and the additional energy which results from the motion, the so-called **kinetic energy E_{kin}**

$$E_{kin} = m_0 c^2 \left[\frac{1}{\sqrt{1 - \frac{v^2}{c^2}}} - 1 \right] .$$

When $v = c$ then $E_{kin} = \infty$, when $v = 0$ then $E_{kin} = 0$. Since there is no infinity in nature, no object with mass, i. e., with $m_0 \neq 0$, can ever reach the speed of light. If the speed of the object *A* is very low in comparison to light speed, i. e., $v \ll c$, then the equation is simplified to the well-known expression for kinetic energy in classical mechanics.

$$E_{kin} = \tfrac{1}{2} m v^2 .$$

The mass m_A of the moved objects, as result of the movement, becomes heavier in comparison to the the rest mass:

$$m_A = m_0 \frac{1}{\sqrt{1 - \frac{v^2}{c^2}}} .$$

The total energy E_{tot} of a system, then, is made up of the mass energy, the potential energy and the kinetic energy, whereby the boundary between the mass energy and the potential energy – as described in Section 9.12 – can be arbitrarily set by physicists in the way that it most simply suits their purpose:

$$E_{tot} = E_0 + E_{pot} + E_{kin} .$$

10.10 Energy conservation

The kinetic energy and with it, the total energy, depends on the relative motion between observer and object. Observers in different states of motion measure quite different total energies for the same object, just as they have also measured different times and distances. Energy is not something absolute, but relative. This is so because the observer, who relative to the object is at rest, counts fewer events, or experiences fewer inversions of space than the observer, who relative to the object, moves. If the object accelerates in the gravitational field of another object, then the frequency of all of the total system's events counted by the same observer remains unchanged, however; the kinetic energy increases at the expense of the potential energy between the two objects, whilst the rest mass remains constant. This holds true as long as the observer does not change his state of motion. As in the four-point space, time in complicated systems is also a direct consequence of the periodic events experienced by the observer, whereby the measure with the event counts is Planck's constant. Every event has an event as its cause and in turn, brings about another event. In the process, the number of observed inversions of space per time unit cannot change, because there is no cause for such a change.

10.11 Momentum conservation

Events are not only countable in time; they also take place, as we have seen, in a location relative to the other points in the space. If the objects move, then the movement always has a direction in the three-dimensional space, a space defined by all the universe's points. In principle, this also applies to rotations, at least to the individual points with the individual events. Every moving point of an object A pulls this in a certain direction. In precisely this direction, the next event is thereby also shifted, and what is more, with precisely the velocity v_A, with which the object, in which the event has taken place, shifts. The greater the mass m_A of this object is, the greater the number of events which are shifted in this way. Since, as explained in the previous section, the total amount of the observed inversions of space does not change for the observer, as long as his state of motion remains unchanged, the quantities $p_A = m_A \cdot v_A$ must also remain constant, indeed not only in number, but also in their direction. I call this quantity **momentum $\boldsymbol{p_A}$**:

$$p_A = m_A \cdot v_A = \frac{m_0 \cdot v_a}{\sqrt{1 - \frac{v_A^2}{c^2}}} .$$

Like kinetic energy, momentum is also not an absolute; its quantity is dependent on the state of motion v_B of the observer and thus relative. The momentum, however, remains constant for the observer, as long as his state of motion does not change. When $v_A = 0$ then also $p_A = 0$, when $v_A = c$ then $p_A = \infty$, when $v_A \ll c$ then $p_A = m_0 \cdot v_A$.

In the event that $v_A = 0$, all events proceed periodically in time. The translation in all directions in space is zero. The mass thereby becomes quasi a momentum in the time direction [Russell (1925/1958) pp. 93 f.].

10.12 Centripetal force

If an object A with the frequency ν_A and the velocity $v_A \ll c$ flies past an object D with the frequency ν_D at a distance r_{AD}, then A and D are attracted, in accordance with Section 9.7, with the force

$$F_{AD} = \frac{v_A \cdot v_D}{r_{AD}^2} .$$

If the observer is with object D and is at rest relative to it, then he will see how the object A is diverted from its original path by gravitation in the direction of D. This so-called centripetal acceleration a_A of A in the direction of D amounts to

$$a_A = \frac{G \cdot v_D}{r_{AD}^2} .$$

If A is so strongly attracted by D, i. e., accelerated, that A flies around D on an orbit $2\pi r_{AD}$ with the period t_A, the velocity $v_A = \frac{2\pi r_{AD}}{t_A}$ and the angular frequency $\omega = \frac{2\pi}{t_A} = \frac{v_A}{r_{AD}}$, then

$$a_A = r_{AD} \cdot \left(\frac{2\pi}{t_A}\right)^2 = r_{AD} \cdot \omega^2 = \frac{v_A^2}{r_{AD}} = \frac{G \cdot \nu_D}{r_{AD}^2} .$$

If object D with the event frequency ν_D forces object A with the velocity v_A into an orbit, the distance r_{AD} will be

$$r_{AD} = \frac{G \cdot \nu_D}{v_A^2} .$$

At this distance, centrifugal and centripetal forces cancel each other out exactly. The distance r_{AD} is a function of the mass (or frequency) of D and the velocity of A; it is independent, however, of A's mass.

Also, on any given curved paths $a_A = \frac{v^2}{r_{AD}}$ when r_{AD} is the path's radius of curvature in the place envisaged. Movements along any given path can always be divided into tangential and normal components of velocity and acceleration. For more details, the literature on mechanics can be referred to.

10.13. Centrifugal force reversal in the vicinity of a black hole

Let us now consider the case where the object A circles a black hole D at a high velocity $v_A \simeq c$. The question arises as to what the minimum distance $r_{AD\min}$ must be so that A does not fall into the black hole. In principle, the same equations apply to such a relativistic object as in the previous section, simply with the addition that for time, the eigentime of A must be used in each case. That is the time which a clock carried by A would show as seen by a far-removed resting observer. This time depends, as shown in the previous section, on the velocity v_A as well as the distance r_{AD} from the black hole. Clifford Will (1989) calculated that

$$r_{AD\,\min} \;=\; \frac{4G \cdot \nu_D}{c^2} \;=\; 2\mathscr{R} \;.$$

A generally understandable, although not quite precise illustration is also given by Sexl et al. (1990), pp. 77 ff. Their simplified calculation produces the result $r_{AD\min} = 3\mathscr{R}$ instead of the correct $2\mathscr{R}$. That is, double the Schwarzschild radius is the shortest possible distance where a mass particle can still circle a black hole without falling into it. Even at a smaller distance r_{AD} from the black hole, A can still be perceived by the greatly-distant observer, though the orbit becomes unstable.

If A is a wave with the velocity c, then according to Section 10.5 this wave can also be forced into an orbit around the black hole, that is, when

$$r_{AD} \;=\; \frac{3G \cdot \nu_D}{c^2} \;=\; \tfrac{3}{2}\,\mathscr{R} \;.$$

That means, at a distance of $1\frac{1}{2}$ Schwarzschild radii, a wave with the velocity c will orbit a black hole forever [Abramowicz et al. (1990)].

Of interest now is the behavior of centrifugal force in the space between 1 and $1\frac{1}{2}\,\mathscr{R}$, i.e., in a section close to the black hole where there could, however, very well be observable objects. It is evident that in this area, the centrifugal force draws the orbiting object A not outwards, but inwards! The faster A circles the black hole, the more strongly it is drawn towards the black hole, into which it then inevitably falls. This effect can be explained qualitatively in chirality theory as follows:

The far-removed real observer sees A as an object whose inner clock almost stands still due to the high velocity and the proximity to the black hole. The two effects of the velocity and the proximity to the black hole are to be multiplied by one another. If, for example, the time measured by the inner clock runs only $\frac{1}{5}$ as fast because of the high speed and this slowed down time is slowed down further by the proximity to the black hole to $\frac{1}{3}$ of that, then the time measured by the observer is slowed down to $\frac{1}{15}$. The diameter of *A* has also become very small. However, a standard length on the surface of the black hole being circled would be even smaller in the observer's view, i. e., zero. The velocity, i. e., the relationship of the length to the eigentime of *A*, also in the observer's view, remains unchanged and high, however. The observer sees and measures not only the eigentime of *A*'s inner clock which becomes ever slower with the shorter distance, but also the unchanged rapid period of orbit around the black hole. Now, not only does *A* as a whole move, but according to chirality theory, the individual points within *A* also criss-cross between each other at light speed. (In the process, the individual point is not observable, however.) When the velocity v_A approaches light speed, a stage is reached where the movements of the individual points within *A* begin to compete with the movement of *A* around the black hole: at a small r_{AD}, the points of the object *A* distribute themselves more or less evenly around the black hole, although as seen by the observer, they still form a single particle outside the black hole.

It can happen that the space inversion within *A* inverts itself around the black hole in such a way that inside and outside are exchanged, not only within the four-point spaces of *A*, but also in the relation of *A* to the black hole. If, however, the relationship of inside and outside reverses with this relation, then that means that the original outward-acting centrifugal force also reverses and now pulls inwards. Whether one now says that in this process *A* has fallen into the black hole, or rather the reverse, that the black hole has actually penetrated the particle *A*, is a question of how one looks at it.

Such an effect, even today, is still surprising for many physicists. It was discovered and described in mathematically exact terms by Abramowicz, who dared to make statements like „space appears to be turned inside out“ or „inward and outward are not absolute concepts; they are relative in spaces warped by strong gravitational fields [Abramowicz (1993)].“ In the terminology of chirality theory I call that an inversion of space. The phenomenon has important consequences for astrophysics. It explains how the clouds of gas which circle a black hole in the centre of galaxies, the so-called accreation disc, supply this centre with energy. The viscosity within the gas cloud makes its rotation more rigid, i. e., it normally slows down the rotation within and accelerates it without. Thus, the torsional moment is shifted outwards from inside the cloud. Calculations show, however, that in the proximity of black holes the effect reverses, so that the torsional moment is transferred inwards from the outside [Anderson et al. (1988)]. This is only explainable if outside and inside change places, i. e., if the space in the proximity of a galaxy centre inverts itself.

10.14 The Big Bang is (was) not a Bang

In Section 8.4, I established concerning the four-point space: that which the presumed observer outside a black hole perceives as a length of time, is a distance for the observer inside the black hole. For the object itself, there is no difference between space and time; there are only events. From this I concluded in Section 8.9 that the question whether a particle is a fermion or a boson depends likewise on the point of view of the observer and is thus relative: what the observer outside of a black hole perceives as a rotating fermion is for the internal observer the translation of a boson. A theoretical cause for this new form of relativity is the axiom of chirality.

This state of affairs generally applies not only to the four-point space, but to black holes generally. The boundary where space becomes time and vice versa is the Schwarzschild radius. Here the rotation changes into a translation and the translation becomes a rotation. Such a rotation is always three-dimensional, which according to Section 8.7, can be described by means of two angles which, in dependence of each other, change from event to event. The translation, however, is the progressive movement of an object from event to event, whereby the distance covered is measured by counting the events in the three-dimensional space which revolves around this stretch of distance. This rotation of the space comes about because with each event, a point pair flies past the object at light speed, traversing the translation axis.

Both time and space have a one- and a three-dimensional chiral aspect: the chirality of time shows up as time direction, its one-dimensionality is the time axis and the three-dimensionality is expressed by the two angles which change with every event. The chirality of the space shows up in the three-dimensional handedness with the contrast of the outside and inside of the inverted glove; the one-dimensionality of space is represented by the translation and the three-dimensionality by the space which turns about the translation axis with every event. In comparison to the simple four-point space, larger black holes are additionally accompanied by the innumerable partial inversions of the overlapping lopsided four-point spaces, yet that makes no fundamental difference to the observations made above. These partial inversions simply mean that always when a relationship between two or more event counts is involved, the result can be not only a whole, but also any rational number.

A further striking analogy between space and time is the time dilatation and the contraction of length in the general and special theories of relativity, which I have illustrated in Table 1, Chapter 6.3. Could this perhaps involve one and the same effect which is simply being described from another observer's perspective? For instance, the observer on the inside of a black hole describes the time dilatation as a consequence of the translation at high velocity by means of the special theory of relativity, whereas the observer on the outside of the black hole formulates the analogous process as the phenomenon of a rotation or gravitation in the general theory of relativity?

All the symmetries or analogies of the situation outside and inside a black hole can lead one to suspect that perhaps things on the inside of a black hole look quite similar to the outside: there is a chiral three-dimensional space, a one-dimensional time with a direction, and fermions and bosons. And how does the boundary between outside and inside, i. e., the surface at the Schwarzschild radius, look from the outside and from the inside? The black hole, seen from outside, is a location in space. If it is correct that space becomes time at the Schwarzschild radius, then this location in space at the Schwarzschild radius must become a location in time, a point in time. What the external observer perceives as a spherical plane with a Schwarzschild radius, is for the observer on the inside of a black hole a point in time: it's called a Big Bang. The surface or the horizon of a black hole is an objective property, i. e., it is independent of the location and state of the observer. The same applies to the Big Bang. Whether these two boundaries are perceived as a surface or as a Big Bang, however, is relative and depends on the location of the observer. Is he inside or outside the black hole? How can a property be objective and nevertheless dependent on the observer? This is possible if the different observers fundamentally cannot communicate with one another, and precisely that is the case here. The passage of time at the Schwarzschild radius, as it is seen by the external observer, creates the impression that matter in motion disappears into the black hole. Seen from within, the passage of time becomes an expanding three-dimensional space, the expanding universe. The fact that, seen from outside, no matter can escape the black hole, corresponds to the speed of light seen from the inside, which cannot be exceeded. Only things that can travel faster than the speed of light, i. e., the so-called virtual waves, can leave the black hole. They carry the gravitational action of the black hole outwards, but not any information about the internal structure of the black hole.

What does the inside of a black hole look like? Have a look around your kitchen or gaze up at the starlit sky: that's what it looks like inside a black hole! Our universe is the inside of a black hole. This idea is not completely new. Cosmologists certainly speculate about whether a developing black hole could be the birth of a new universe. The cosmos would then consist of many universes; it would be a „multiverse" [Smolin (2001) pp. 192 ff. and 198].

The fermions which fall into the black hole are „smeared" all across the entire space inside by the partial inversions of the new environment in the black hole: they become Bosons. The bosons which fall in from outside at light speed, however, were still not yet in a fixed location outside the black hole and their inner clock has remained still since the fixed point in time of their creation. At the Schwarzschild radius they become fermions at a fixed location relative to the observer, and their inner clock begins to run. In Figure 32 I attempt to illustrate the complex state of affairs at the Schwarzschild radius with a simple diagram, by exchanging the red time axis with the blue space axis at the crossover point from outside to the inside. Objects seen from the outside fall into the black hole at quite different points in time, (possibly separated by billions of years) but at a defined location. When seen from inside the black hole, they all appear simultaneously at a fixed point in time. Following this Big Bang, they form a new, expanding space. From outside, nothing further can be seen of this; it can only be determined that the surface, which forms a barrier between inside and outside, has grown somewhat.

Was the Big Bang a bang? As perceived by the observer inside the black hole, it might look that way. Since it is relative, however, whether we perceive an event as taking place in time or in space, the Big Bang is not a bang as such, but can just as easily be seen as the surface of an expanding black hole. What is happening here is a compensation of first order inversions by second order inversions. The Big Bang is a temporal and spatial overlay of events in time and space. In the process, the symmetry between space and time is reversed. So there „was" not a Big Bang, there „is" a Big Bang, in the state of something existent.

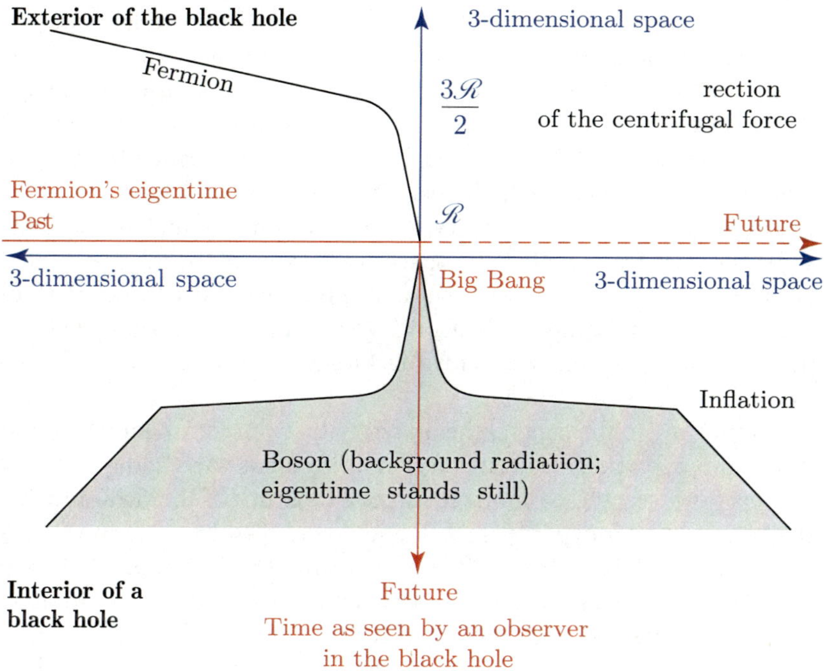

Figure 32. The Big Bang as a black hole: *The fate of a fermion which falls into a black hole. blue = space; red = time; $\mathscr{R}$ = Schwarzschild radius*

10.15 Inflation

If one tries to stretch the analogy between the situation within and without the black hole somewhat further still, then there should be a time inside the black hole which corresponds to radius, where the direction of the centrifugal force reverses. I suspect that there actually is such a phenomenon. Most cosmologists today are of the opinion that the area of our universe expanded in the time from about 10^{-36} to 10^{-32} seconds after the Big Bang with superluminal velocity at a factor of around 10^{50} to about the size of a soccer ball. Among other things, this explains why the background radiation of approximately 3° Kelvin, as a relic of the Big Bang, is spread out so nice and evenly over the whole universe. Due to the expansion, the substance, which at the Big Bang was still extremely hot, cooled down and condensed into particles of matter [Guth et al. (1989); Fritzsch (2000) pp. 382–384]. It is important to keep in mind that according to this inflation theory, it was not a case of matter dispersing throughout space, but that the space itself exploded [Lineweaver et al. (2005)]. That is, of course, only possible if the space actually exists. If this effect is portrayed graphically as in Figure 32, then an analogy between the inflation theory and the centrifugal force reversal does actually seem to show up.

According to the general theory of relativity an astronaut who disappears into a black hole would feel nothing of it. According to newer models however, the cosmonaut in the proximity of the Schwarzschild radius would hit a hyperdense wall and would dissolve in gamma radiation [Musser (2003)]. If one traces the fermion in Figure 32 on its way into the black hole, then the following can be observed: The fermion has a defined location at each measured time. Drawn by the gravitations of the black hole, it moves towards it with an accelerating speed which, however, always remains below the speed of light. Its eigentime runs, measured by the outside observer, ever more slowly, due to the time dilation. If the fermion does not steer exactly towards the centre of the black hole, a centrifugal force operates in addition to the gravitation, which has the effect that the fermion can only approach the black hole at a reduced velocity. At a distance of $\frac{3}{2}\mathscr{R}$ from the centre of the black hole, the direction of the centrifugal force reverses, however, and the fermion falls with increasing acceleration into the black hole. When it reaches the

Schwarzschild radius R, its eigentime stands still and in principle, it can no longer be perceived by an observer outside of the black hole. The only thing that the observer can still establish is that the surface, the mass, the angular momentum and the electrical charge of the black hole have increased somewhat.

Viewed by an observer inside the black hole, the particle appears suddenly as a boson at the beginning of (all) time. This particle travels at light speed and its inner clock stands still. From this observer's perspective, however, depending on his location, it has already been travelling for perhaps billions of years since the Big Bang. Shortly after the appearance in the black hole inflation commences, whereupon which the boson's location expands exponentially. What does this mean? Since in chirality theory the space does not exist *per se*, but is a purely theoretical consequence of the events, the nature of these events must fundamentally change upon inflation, just as outside of the black hole the nature of the centrifugal force changed. With the centrifugal force, that was the reversal of a direction in space. Accordingly, with the inflation, the phenomenon should be a reversal of the direction in time: past and future will be switched shortly after the Big Bang, just as outside and inside will be switched in the case of the centrifugal force reversal. The reason for this time reversal seems to be an interaction of the observed boson with the cosmos outside the universe at the time of the Big Bang. Such an interaction apparently has as a consequence that an observer in the future, who retraces the journey of the boson, arrives at a point near the Big Bang where time, and thus causality, reverses. On this last short section of the journey between the Big Bang and the inflation, from the perspective of this backward-looking observer, the boson comes not from the past, but from the future. The observer cannot, in principle, dig any further into the past.

As long as the boson is not observed and thereby destroyed by the observation, the boson does not have a specific location, but is „smeared" over an expanding area on the inside of the black hole. In so doing, it enters into a relationship with all the other particles in this area: it becomes dependent on their configuration. The boson is part of the total situation in the area coated by it. When it is perceived the boson disappears everywhere in the entire area simultaneously. The point where the boson can be observed is affected by all the relations to all the other particles which it has „experienced“ on its way from the Big Bang to the observer.

10.16 Real gravitation waves as precondition for every observation

An object is real when it can be perceived. Upon perception, at least part of the information which the object carries is transferred to the observer. If the in-falling fermion was a neutrino, then it will become a graviton inside the black hole, indeed a real one. I wish to leave open for the time being the question as to whether there is a chaotic intermediate state when there is a change from the neutrino to the graviton at the Schwarzschild radius, since after all, there time and distance become zero from the perspective of the external observer, while the observer on the inside of the black hole cannot learn anything of the time before the Big Bang. In the case of a chaos at the Schwarzschild radius the quantity of information would be conserved for the system, but it would not be accessible for an observer, however, irrespective of whether he was now inside or outside the black hole. In contrast to the virtual graviton, the real graviton is perceivable because essentially, it did not previously exist for the observer. Thus, for the observer, something changed and this change is the perceivable. The information of the graviton, which upon perception is transferred to the observer, from the observer's perspective arose quasi out of nothing with the Big Bang. According to chirality theory, the cause of the emergence was an interaction, namely the coinciding of the fermion with the black hole. Because also the neutrino is a black hole, albeit only a mini-hole, then one can say that the graviton, as a real particle, resulted from the fusion of two black holes.

Since our graviton developed with the Big Bang, it cannot move faster than time and space, which were born along with it: from the observer's perspective, time was born; from the graviton's perspective, space. Just as time moves into the future for the observer, three-dimensional space expands for the graviton. The ratio of the two quantities is the **speed of light**. It is the ratio of the number of the real graviton's space events to the number of time events of the observer's clock. The graviton cannot fly out into a space which for it, doesn't even exist.

The **frequency** of the time events or inversions which come from the fermion are, in principle, conserved as a frequency, except that now space and time are exchanged: instead of the graviton having a fixed location as the neutrino once did, in which it moved towards the future, it now has a stationary eigentime and on account of this expands out into the space. Seen by the observer, this increase in the graviton area corresponds to the flow of time into this observer's future. The fermion's temporal rotation events have become the gravitons spatial translation events. The frequency of the graviton is the information which it carries or, conventionally expressed, it has energy. This is so small, however, that it could not be perceived or even measured directly with measuring instruments available today [Fritzsch (2000) pp. 287–300; Smolin (2001) p. 251; Sexl et al. (1990) pp. 101 f.]. Only since 2016 gravitational waves can be directly perceived [Abbott et al. (2016)].

Up to now we have only been concerned with the newly-emerged graviton inside the black hole. Yet real gravitation waves emerge in the space outside the black hole as well. This happens whenever a moving mass changes its form. For example, one can regard the system fermion + black hole (F+BH) as a single, moving object which continually deforms. The attraction between the fermion and the black hole is a consequence of the energy-less virtual gravitons which are exchanged between the two. The (F+BH), for its part, also attracts all the other particles outside of the black hole by virtual gravitons. The deformation of the (F+BH) results from of a translation of points and can therefore take place at a maximum of light speed. Naturally, this alters the gravitation action between the (F+BH) and the other particles, unless the deformation is spherically symmetric. Although the gravitation action is transmitted by virtual gravitons with infinite velocity, the *change* of this gravitation action, however, only proceeds at the speed of light, i. e., by information-carrying, real gravitation waves. In this way gravitons can transmit information and energy from one object to another. That is precisely the precondition for observation being possible at all. So, the actual things can never themselves be observed, but only their deformation, and these also can only be observed if they are not spherically symmetric. If the real gravitation wave arrives at a measuring instrument, then it deforms it and in so doing, disappears completely. The deformation of the measuring instrument is the document of the observation. If the measuring instrument has a consciousness, then it is an observer.

What is that, a **deformation**? The term is all very descriptive and comfortable, but not precise. In chirality theory, all points in the universe are in constant motion. All these motions are described as frequencies of events, which for their part, are rotations or translations, i. e., are time or space events. A deformation means that the relationship of such frequencies with and between the objects observed changes. Only such deformations or frequency *changes* are perceivable and can be measured or counted, but never the actual form. Moreover, all sequences of events are chiral; they have a direction, be it forwards or backwards in time, be it rotating to the left or to the right in space. This orientation of the chirality also changes during the deformation. Thus, each deformation is accompanied by a spin change as well. The spin itself is not perceivable, but only ever the spin *change*. The difference between virtual and real gravitons is that real gravitons themselves carry frequency and spin, and transport these from one object to another or to the observer. Thus, they transmit not only an action, but also information. The transportation of spin and frequency, and of energy and information, can occur at a maximum of light speed [Møller (1969) p. 321].

Since the real gravitons have energy, they have also a relativistic mass. Due to this mass they attract each other. The interaction between gravitons can be observed indirectly: the attraction between the sun and the earth affects the attraction between the earth and the moon and thereby the moon's orbit. Real gravitons have a gravitational effect on each other [Goldman et al. (1988)].

10.17 Duality of particles and waves

So, what is a graviton, a wave or a particle?

The answer depends on the state of motion, location and theory of the observer. Let's take for example the gravitation between the earth and the moon. For the observer, there are three fundamentally different perspectives: Firstly, he can rest on one of the two celestial bodies, – let's say the earth. Secondly, he can move from the earth to the moon; let's say, to examine an extreme a state as possible, at the same speed as the graviton. And thirdly, he can be far beyond the earth-moon system and look at it as an individual object which deforms.

The observer resting on earth describes the virtual graviton as a wave which has its origin on earth and spreads out spherically from here with infinite velocity over the entire three-dimensional space. Its target is the moon, which is accelerated. Conversely, the moon for its part accelerates the earth. The acceleration is mutual, or put another way, relative. If the area between earth and moon has a structure as the consequence of further celestial bodies and waves between them, then the information of this structure is added to the properties of the gravitation wave, which as long as it is travelling, is not in a specific location, but „is smeared" over all of space. Since the graviton is the consequence of the inversion or variation in the volume of a four-point space which spreads out over the whole of its surrounding area, this volume is increased, made smaller or distorted by the existing spatial structure. Depending on which of these is the case, the frequency and/or the direction and the destination of the wave changes. The whole process from the observer's perspective is non-local; rather, it is a succession of two events, first on earth, later on the moon. That which happens in between, i.e., the wave *per se* and its changes in volume, direction or frequency, is not directly perceivable for the observer. That which this observer describes is thus the relative acceleration between earth and moon, the cause of which he interprets as a virtual wave, which is intrinsically non-perceivable.

A presumed observer, who accompanies the wave from the earth to the moon, sees the graviton not as wave, but as an individual event which he interprets as a four-point particle. 1 to 3 of the 4 points of the particle are on earth, the remainder is on the moon. Thus, the starting point and the destination of the graviton are clearly fixed from the outset. The configuration of the points on the earth and the moon changes at a stroke. Since, however, the four-point space (with points both on earth as well as on the moon) is very lopsided, other points, which lend an order structure to the space between the two widely separated objects, affect the event because the arrangement of all points in the universe is interconnected: no point can move without its location in relation to all other points of the universe changing somewhat. Thus, the frequency of the virtual graviton also changes. This presumed situation would naturally be impossible in reality, since a real observer, due to the special relativity theory, can never move with infinite velocity. For such a presumed observer, the virtual graviton is a virtual particle. Its particle character consists of the fact that the thing has individuality and is quantised. It is not local, however, as would be right and proper for a decent particle. From this point of view, cause and effect of the gravitation can no longer be kept apart. That is always the case in situations which are explained by means of virtual particles, which move with infinite velocity. The effect is mutual. The third observer sees the earth/moon system as an individual object. It is a rotating, form-changing object, which in addition to the virtual gravitons from the earth and the moon, radiates further real gravitons. The latter are a consequence of the relative acceleration between the earth and the moon, and can in principle be observed, because they deform the observer. The third observer will describe the observed deformation of the earth-moon system with a theory of virtual gravitation waves and the effect of the system on himself in all probability as a real gravitation particle. What the observer really believes he sees is first and foremost a question of the theory which he has devised for the description of his observations.

The characteristics of the gravitons in chirality theory are comparable to von Weizsäckers Ur's (Ur's = plural of Ur). The Ur's are wave functions in cosmic space. The universe is comprised of about 10^{120} Ur's and a nucleon of 10^{40} Ur's. Although the individual Ur's are local, the action of their momentum is non-local and is distributed over the whole universe. The three-dimensional space can be deduced from the Ur entity. The agreement of Ur theory with modern cosmological data is striking in some respects [Lyre (2003)].

10.18 Heisenberg uncertainty principle

The observer perceives the deformations and, in chirality theory, that is the changes in frequency. The measure of it is Planck's constant h, the angular momentum of the simple four-point space. The rotation of this four-point space is, as it were, the primordial event. Alternatively, one can regard it as the inversion of a tetrahedron or as a rotation through two mutually dependent angles. These angle changes give the event its frequency ν. As explained in Section 10.16, it is not the frequency itself which is measured, however, but the change in frequency $\Delta\nu$. The definition of energy, in accordance with Section 8.11, is

$E = h \cdot \nu$ and since h is a constant, this automatically yields

$$\Delta E = h \cdot \Delta\nu .$$

One now measures ν by counting the number of rotations during the time t. In order to measure ν accurately, one would have to wait for several periods. Then, however, t is no longer precisely determined. Thus, we have

$$\Delta\nu = \frac{1}{2\pi \cdot \Delta t} .$$

and

$$\Delta E \cdot \Delta t = \frac{h}{2\pi} = \hbar .$$

This equation is called the **Heisenberg uncertainty principle** [Penrose (2004) pp. 521–524]. If the observer wants to know exactly when he has measured a particle, then the inaccuracy of the time Δt must be as little as possible. The consequence of this, however, is that the inaccuracy of the energy ΔE at time t becomes correspondingly greater. If, on the other hand, the energy is accurately measured, then the uncertainty of the time will be greater.

The Heisenberg uncertainty principle applies similarly to location r and momentum p:

$$\Delta p \cdot \Delta r = \hbar .$$

Here, Δr is the distance perpendicular to the path of the object with the moment p. If, for example, one sends a photon or a bullet through a split with the width Δr, then the moment of the photon or the bullet can only be measured with an accuracy of $\Delta p = \frac{\hbar}{\Delta r}$. The deeper reason for the uncertainty is that whilst t and r are one-dimensional, the event itself is three-dimensional. One might well be familiar with the frequency of the four-point space; however, one never knows where the four individual points are. This brings with it an uncertainty with space and time which can never be overcome. The question whether the uncertainty is inherent to nature or merely a consequence of perception or perception theory, has been, and continues to be, discussed endlessly by philosophers and physicists. The discussion culminates in Einstein's dictum „God does not play dice". In reality Einstein is supposed to have said „The old bloke doesn't play dice" [Einstein (1926/1969)]. That does not, of course, exclude the possibility that Einstein called the old bloke „God" on another occasion. Bohr is said to have replied with: „It does not depend on whether God plays dice or not, but on whether we know what we mean when we *say* God does or doesn't play dice." [Weizsäcker (1986) p. 509]. In chirality theory, the question is: is it only the events which exist, or are the individual points also real? Since the individual points cannot be perceived, but only the change in a whole „point region", in each case comprising four points, it is evident that physical quantities cannot be measured with absolute precision and that a measure of uncertainty always remains. The answer to the question of reality depends on the understanding of the term reality. If nature is precisely that which is perceivable, then the uncertainty is also real according to chirality theory. If, on the other hand, one believes that behind observed nature, there is another real, divine one, then the uncertainty is not real: it is unreal and God does indeed not play dice. Atmanspacher and Primas (2003) differentiate between empirical reality, which, according to Bohr, only regards what we know about nature as real, and ontological reality, which according to Einstein, regards nature as that which is real, even if that is in principle not perceivable.

In chirality theory reality is defined as the entirety of all those things which can, in principle, be empirically – directly or indirectly – perceived (Whitehead 1925/1939) whereas existence is an ontological property of anything that is either real or – if not real – must have a Be-ing in nature (see Section 1.2). Thus, single points are not real but they exist whereas the four-point space is a model of reality.

10.19 Pauli's Ring *i*

Wolfgang Pauli (1995) dreamed occasionally about a Chinese lady, who probably embodied the duality of Be-ing. In his dream, „The piano lesson", she pulled a ring from her finger, allowed it to float in air and lectured Pauli:

„You probably know the ring from your school of mathematics. It is the *Ring of i.*"

Pauli nodded and said „The *i* makes a pair of the void and the one. At the same time, it is a quarter rotation of the whole ring."

The Chinese lady „It makes the instinctive or compulsive, the intellectual or rational, the spiritual or extra-sensory of which you spoke an integrated or monadic whole, which numbers cannot portray without the *i*."

Pauli „The ring with the *i* is the unity beyond particles and waves and at the same time, the operation which produces one of the two."

The Chinese lady, „It is the atom, the indivisible …"

Pauli, „It makes time a static picture."

The Chinese lady „It is marriage and at the same time the central realm,[51] which one cannot enter alone, but only as a pair."

Then, from the centre of the ring, the voice of the *Master*[52] spoke, transformed, to the Chinese lady, „Be merciful."

Pauli was unwilling to discuss the dream with his psychologist Marie-Louise von Franz. No doubt Pauli's fantasy concerns the central issue of the relationship between the rational and the irrational, the empirically perceivable – represented by the real number line, and the non-perceivable – possibly the unconscious or the transcendental. Put briefly: the question about the duality of Be-ing in its different forms.

In mathematics, the imaginary number is $i = \sqrt{-1}$ (see Figure 33). Euler proved that $e^{i\pi} = -1$. Furthermore, the function $e^{ni\varphi}$ describes a helical wave, whereby $0 \leq \varphi \leq 2\pi$ and n is a whole number (Figure 34, p. 255). In theoretical physics, particularly in quantum theory, such waves are used for describing states. I assume that, for example the state of the neutrino, as it is described in Section 8.7 by means of the two polar coordinates φ and ψ, can be formulated mathematically as such a wave. Both angles φ and ψ change continuously, but not independently of each other. As a product they form a three-dimensional space. The knowledge of only the one angle makes no physical sense thereby; only the product of both angles, i.e., the spin, is observable. Similarly, the quantum mechanical wave function ψ (which has nothing to do with the angle ψ) is not an empirically perceivable size, but only its product ψ^2. Characteris-

[51] In the German original, „Reich der Mitte"; this can also be translated as „middle kingdom" and is very likely a pun on the Chinese term for their own country.

[52] The *master* is a central figure in Pauli's dreams. As Pauli (1992) remarks in a letter to Emma Jung, the master is a mental light figure of superior knowledge. He „is in a certain sense an antiscientist, whereby science is to be understood here as the scientific perspective in particular."

tic of wave functions is the fact that something always changes whilst something else remains unchanged. Usually, not every value which the mathematical wave model indicates can also be empirically verified, especially not when the value itself is imaginary. Yet the wave also repeatedly assumes a real value. On top of this, the *ring i* is chiral, which Pauli apparently overlooked. That is, it turns anti-clockwise. The direction of rotation is not absolute, however, but is always based on a convention. If φ is a function of time, then the time direction is not absolute, but is only established by means of an agreement reached by physicists. Viewed thus, the *Ring i* does indeed make time a static picture: it does not flow; it is simply there. Hawking (2001) pp. 58–69 and 90, has also suggested a cosmological model with a two-dimensional imaginary time, according to which the universe was neither causally created, nor will it end, but according to which it simply exists.

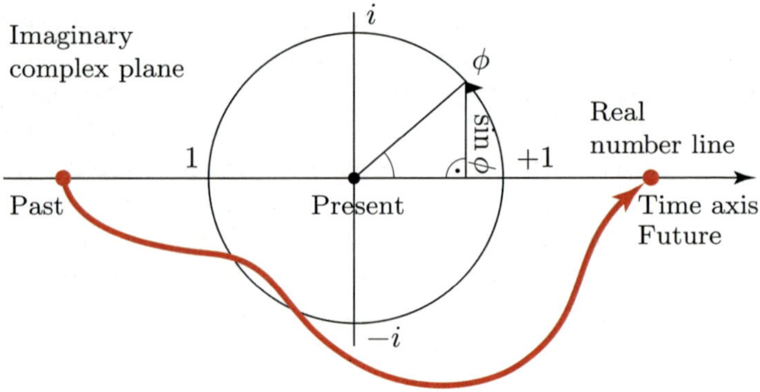

Figure 33. The ring i: *Via the imaginary complex plane, the past and the future can be connected without touching the present.*

Via the imaginary complex plane, the past and the future can moreover be directly connected, which via the temporal real number line would be impossible due to the causality principle. On the real number lines, the path from the past to the future always passes through the zero point, that is, through the present. Locations as well, which in space are too far apart to be connected by signals moving at the speed of light, are nevertheless directly connected via the range of imaginary numbers. If the synchroni-

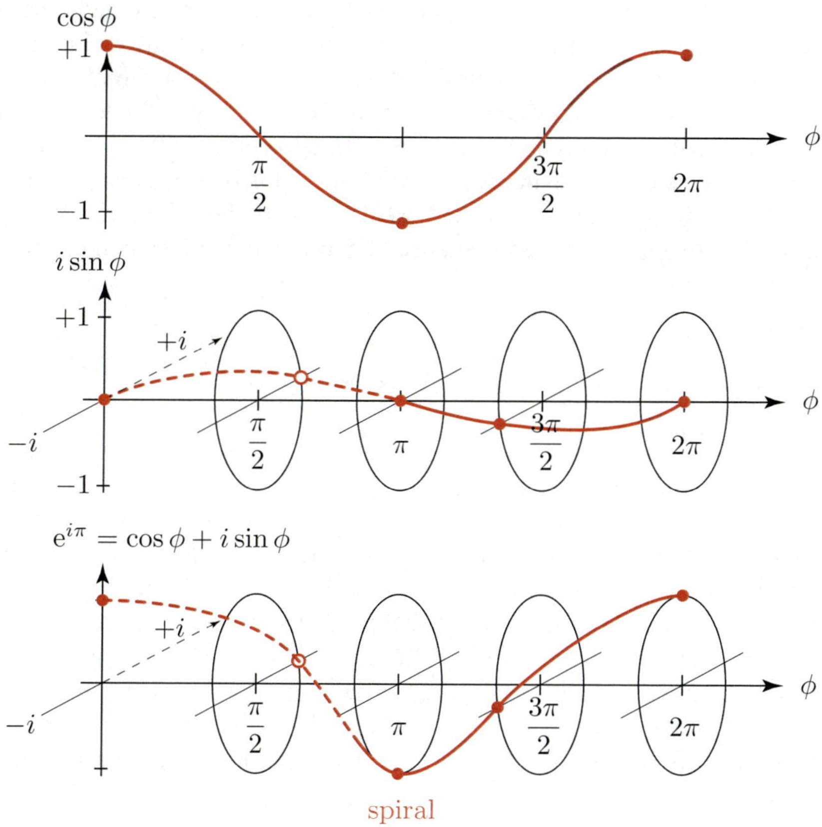

Figure 34.: $e^{i\pi}$

city theory of C. G. Jung (see Section 3.1.5) can ever be described by a mathematical theory, then this would have to be with the help of a mathematically imaginary wave. This could circumvent the causality principle.

If one compares all the characteristics of the ring i with the chirality theory of the gravitation wave, then the analogies are immediately apparent: the chirality theory's event is Pauli's unity of particle and wave and simultaneously the operation which always produces one of the two. Whether the event in chirality theory is a particle or a wave depends on the observer: It is a pure particle when it concerns a black hole. It is a pure wave when the event is perceived as a gravitation field. Certain aspects of events, for example rotation and periodicity, can be described mathematically by the *ring i*. The event is the atom, the indivisible entity of Be-ing. The *ring i* connects the rational with the extra-sensory, that which cannot be perceived empirically, namely the countable events with the four-point space and with the individual points. For the perception of the event, however, a marriage is always needed, i. e., an interaction between two partners.

Long ago, Aristotle (1987) vol. VIII, 1 (251b), described the circulation as an original, directed sequence of countable events which produce time.

10.20 Theory of everything (TOE)?

Most cosmologists and theoretical physicists are convinced that there must be a „theory of everything", the TOE, which describes the development of the universe from the Big Bang at Time Zero at least up to Planck time of approximately 10^{-43} seconds and in which the four known interactions gravitation, electromagnetism, strong and weak interaction can be described in common [Genz (1994) pp. 324 ff.; Schmitz (1997) 384 ff.]. The TOE must apply at all times. It is thus independent of time. That means in particular that the point in time of the Big Bang must not appear in the law [Genz (2002) p. 250]. I suspect that chirality theory, as I have described it in Chapters 7 to 9, is in essence the TOE, or at least the basis of it. All other scientific theories should be reducible to the TOE, in particular quantum theory, the theories of relativity, the Grand Unified Theory (GUT), the theories of the electrically weak and strong interactions, electromagnetism, atomic theory and thermodynamics, as well as cosmology, chemistry and the theory of life and its evolution. For its part, I have derived the TOE from only one theorem of metaphysics: the laws of nature are as they must be in order for perception to be possible.

Chapter 11

Electrodynamics

Time, it is a peculiar thing.
If one simply lives,
It is absolutely nothing at all. But then all at once,
One feels nothing besides it: It is all around us,
it is also inside us.

Hugo von Hofmannsthal[53]

11.1 Black mini-holes within black mini-holes

If our universe and the Big Bang are interpreted as a black hole, then the black holes, which the cosmologists perceive owing to their gravitational effect, are black holes within the black hole.[54] There is nothing to preclude there being further black holes within these black holes. As far as I'm aware, no-one has yet posed the question whether this is possible and how such holes in the holes might act on the outside world. If physicists and cosmologists arrive at such an idea at all, then they probably assume that this question makes no sense, because no information about the inside of a black hole can reach the outside in any case, so any given hypotheses are fundamentally untestable. Thus, they are outside the realm of physics.

Black holes are as still waters, from whose motionless surface no information springs. But still waters are deep. What is hiding their depths?

[53] Hugo von Hofmannsthal: Der Rosenkavalier. Die Zeit, die ist ein sonderbar Ding. Wenn man so hinlebt, ist sie rein gar nichts. Aber dann auf einmal, da spürt man nichts als sie: Sie ist um uns herum, sie ist auch in uns drinnen.

[54] The black holes in the universe proven to exist by cosmologists have masses within the range of a few solar masses up to approximately 10^8 solar masses in the centres of galaxies [Klapdor-Kleingrothaus et al. (1997)].

Presumably the definition of the black hole (see Section 8.3) could be extended to include spaces with more than 4 points. The neutrino model is the smallest possible black hole. An extension of the notion of the black hole might be the case where four black mini-holes together form during a second symmetry breaking a new black hole the same way as the neutrino model had been formed by 4 single points. For this possibility $4 \cdot 4 = 16$ points would be required. Such a system might be stable due to its high symmetry. It could be interpreted as an electron or positron (Section 11.5). Until now no such electron structures have been experimentally observed. This is no surprise because it is impossible to observe in the universe or produce in the laboratory temperatures and densities as they must have existed during the Big Bang. The action of these new particles could be as well attractive as repulsive depending on the relative orientation of such black mini-holes.

Also, other elementary particles such as the 72 different quarks and antiquarks, or mesons and nucleons might be formed in a similar way according to dynamical rules that all can be derived from the axiom of chirality. Such an extension of chirality theory would result in the standard model of the *Grand Unified Theory* (GUT) (Chapter 12).

In this section the dimensionality of nature itself is described. This means that also the dimensionality of the black holes must be included. Every symmetry breaking that leads to a new level of black holes also gives rise to 3 more space dimensions. This is because only 3 space dimensions are observable and no information can be transferred from the inside to the outside of a black hole. Therefore, there are in nature a 1-dimensional time +3 dimensions of space +3 concealed dimensions of black mini-holes +3 concealed dimensions of black mini-holes within the black mini-holes = 10 dimensions, the same as is postulated by some string theories. The 10 dimensions cannot be observed empirically but they must exist. In both chirality and string theory 6 of these dimensions are „concealed". In chirality theory, this is due to the black mini-holes from which no point and no information can escape.

Even the universe itself could be the interior of a black hole. The not perceivable, chaotically moving single points in such a universe then could be interpreted as the so called *dark matter*, which contributes about 27 % to the total mass of the universe.

11.2 Interaction of black mini-holes

Those who consider the problem carefully and without preconceptions quickly come to the conclusion that the question of holes within holes is not as trivial as it may appear at first glance. The situation is portrayed schematically in Figure 35 (p. 260). At the Schwarzschild radius, the three-dimensional chiral space becomes a one-dimensional linear time and vice-versa. If inside the black hole there is a further black hole with a Schwarzschild radius, then there the linear time becomes a chiral space again, and the space changes back to a linear time. In which direction does this time run? Into the future or into the past? In chirality theory, time in itself does not actually exist; only the countable chiral events with their frequencies exist. The frequency is the ratio of the number of events per time unit of the presumed observer's inner clock. The frequency doesn't actually tell us anything about the time direction. The events as a whole are chiral, but the theory doesn't say whether the orientation is now to the left and time runs into the future, or whether the orientation is to the right and the time thus runs into the past. In particular, one knows nothing about the direction of time during a transformation of space into time. All the same, seen from outside, at the Schwarzschild radius all lengths are zero and the clocks stand still, so it is quite conceivable that the direction of time reverses. However, that is probably only possible when at the Schwarzschild radius, the orientation of the space turns around at the same time, since only thus does the black hole's angular momentum remain unchanged, which is required by the classical theories and observation as well. The space-time symmetry as a whole cannot change. It can thus be accepted quite plausibly that upon penetrating a black hole, the probability that the re-emerging time – seen from the outside – runs into the past, is just as great as the probability that it runs into the future. The only question is whether this effect is perceivable from the outside. According to the conventional theories, only three characteristics of black holes are perceivable: the gravitation (and accordingly, the mass and the Schwarzschild radius), the angular momentum and the electrical charge. Electrical charges attract each other when they have different signs and they repel each other when they have the same sign. The theory of chirality, as it has been formulated so far, knows only attractive interactions, and with these, it is not possible to

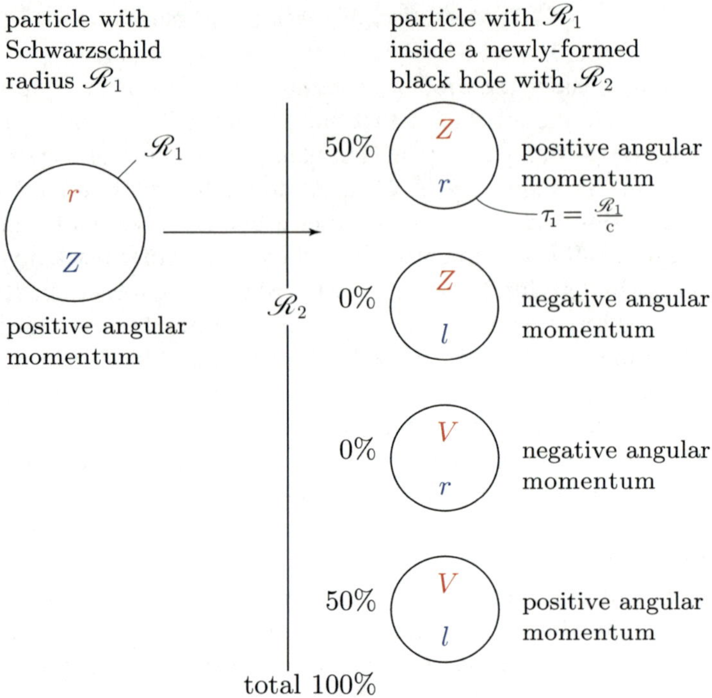

Figure 35. Transformation of temporal into spatial orientation at the Schwarzschild radius $\mathscr{R}$

Inside the black hole $\mathscr{R}_2$, the Schwarzschild radius $\mathscr{R}_1$, seen from outside, becomes a „Schwarzschild time" $\tau = \pm\frac{\mathscr{R}_1}{c}$
r = *right (turning)*
l = *left (turning)*
Z = *time runs towards the future*
V = *time runs towards the past*
red = outside $\mathscr{R}_2$ spatial, inside temporal
blue = outside $\mathscr{R}_2$ temporal, inside spatial

explain repulsion. In 2001, as I lay sleeping one beautiful night, it suddenly occurred to me that a repulsion is basically nothing more than an attraction, with which time, as seen by the observer, runs backwards. How can time run backwards? Such a situation can result if a chiral particle gets into a black hole. There the chiral space orientation becomes a time direction, which can just as well be directed backwards as forwards.

This leads directly to the hypothesis that repulsive interactions can occur if objects are black holes consisting of black holes. The conception that time can in a certain sense also run backwards and that the frequencies accordingly become positive or negative, is not new. It plays an important role in the formalism of quantum theory [Feynman (1985) pp. 95 ff.].

Electrons and quarks are the smallest independent particles with an electrical charge. In most theories they are treated as point charges, although that cannot be the whole truth, because one point cannot have a geometrical structure and therefore no direction of rotation or spin. During an experiment at the German particle accelerator in Hamburg anomalies are supposed to have arisen with the collision of high-energy electrons and positrons which can be best explained by the assumption that electrons possess an internal structure [Maddox (1998) pp. 88 and 387]. Could the tiny electrons be black holes, and what is more, ones which consist of further black holes? There is an experimental indication that electrons are in a certain sense black holes: The „gyromagnetic moment" of the electron is precisely the size it would have to be for a black hole with an electron charge and mass [Penrose (2004) p. 832]. That cannot be ruled out *a priori*, since according to chirality theory, even the much lighter neutrinos are black holes. If the electrons actually are black holes, then it would also explain why they can only be observed as point charges, because first of all, no information about the internal structure of a black hole can make it to the outside, and secondly, due to the electron's small mass, its Schwarzschild radius is much too small for its extension to be measurable. The radius of the electron is less than 10^{-17} cm [Klapdor-Kleingrothaus et al. (1997) p. 107]. This Schwarzschild radius is even much shorter than the Planck length, so for this reason also, the extension of the electron would, in principle, not be perceivable.[55]

[55] The Planck units are derived directly from the natural constants. Their sizes in the MKS system are as follows: Planck length $L_P = \sqrt{Gh/c^3} = 1.6 \cdot 10^{-35}$ m; Planck time $T_P = \sqrt{Gh/c^5}$. $= 5.3 \cdot 10^{-44}$ s; Planck mass (that is the mass of a black hole with the radius L_P) $M_P = hc/G = 2.5 \cdot 10^{-8}$ kg; Planck entropy $S_P = k = 1.38\ 10^{-23} J/°K$ (*k* is the Boltzmann constant). The meaning of the Planck units for the theory of physics is still unclear, but most theoretical physicists assume that the physical laws break down at shorter lengths and times than the Planck units [Finkelstein (1996) pp. 165 f.].

The electrical charges are never directly perceivable, but only ever their interaction with other charges. The presence of charges can be concluded from the resulting attraction or repulsion. The same applies to the mass: It can likewise never be perceived directly, but only its interaction with another mass, and even then, only when as the result of this interaction real gravitons appear. For the perception of electrical charges, one thus always needs at least two electrically charged objects A and B, from which four possible constellations can result: A^+B^+, A^-B^-, A^-B^+ and A^+B^-. As with gravitation, the frequencies of the two interacting objects must be multiplied with one another to determine the interaction's strength. It is new, however, that with the frequencies not only the value, but also the time direction is to be considered in such a way that in an A^+B^+ and A^-B^- situation, the two particles repel each other, whereas with A^-B^+ and A^+B^-, they attract each other.

The interaction between two black holes which for their part consist of black holes is represented schematically in Figure 36. In principle, it can be described in the same way as the gravitational interaction between the points of two four-point spaces. With the four-point spaces, the interaction is a consequence of the inversion of all lopsided tetrahedrons, which are formed from points belonging to the two four-point spaces. The strength of the interaction is proportional to the product of the frequencies of the two four-point spaces and inversely proportional to the square of the distance between the two spaces. With the black holes, the interaction is a consequence of the inversion of all lopsided tetrahedrons which are formed from the small black holes in the two large black holes. The strength of the interaction here is proportional to the product of the frequencies of the small black holes in the large black hole and inversely proportional to the square of the distance between the large black holes. Within the same black hole, the clock times of all small black holes run into the same direction. If the time of the small black holes in the second large black hole runs in this direction as well, then the product of the frequencies will be positive and the black holes repel each other. However, if the time of the small black holes in the second large black hole runs in opposite direction, then the product of the frequencies will be negative and the black holes will attract each other. Negative frequencies in this sense were introduced by Dirac (1930) in his description of the electron and its antiparticle, the positron. Richard Feynman interpreted antiparticles as particles in backwards-running time [Penrose (2004) pp. 609–626, 639 f.]. There is no such effect with gravitation, since the individual, structureless points do not have internal clocks whose time could run in different directions. Here, therefore, there is no repulsion.

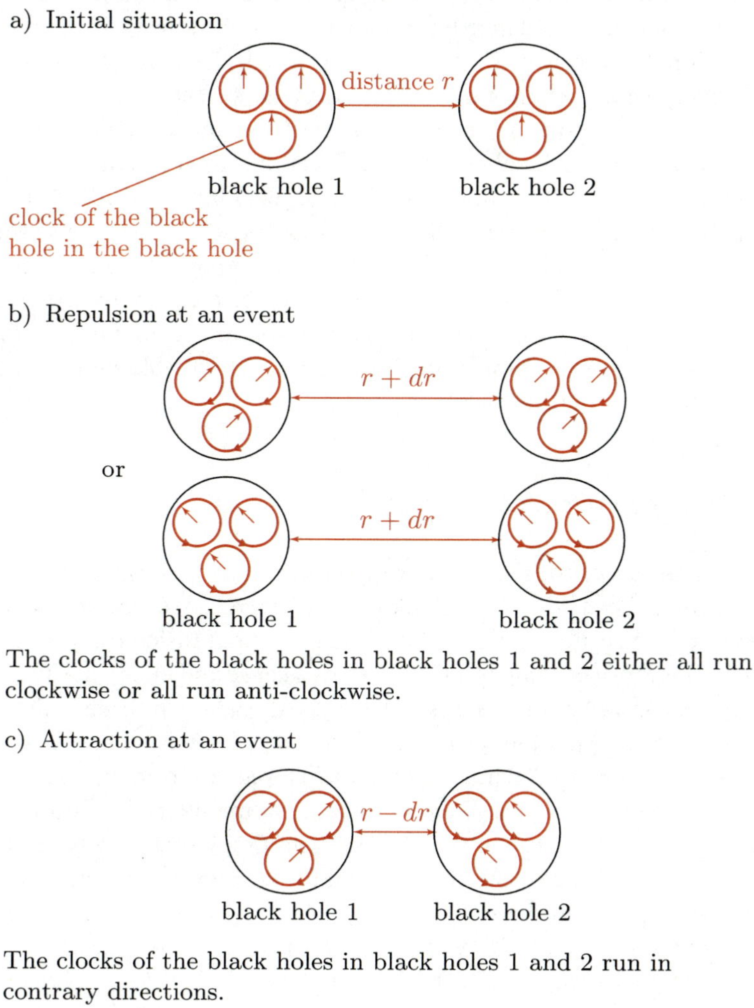

Figure 36. Interaction between black holes in black holes

11.3 The arrow of time

If it is correct that the time direction is the cause of the electromagnetic forces, one should know what time direction actually means. The arrow of time is rather seldom a topic among physicists. On the contrary, one assumes that in accordance with common sense, time exists, it is one-dimensional and it runs from the past, through the present into the future. Everywhere, where time occurs in physics, seems to occur or appears as a parameter in a formula, it has these three characteristics: it exists, is one-dimensional and it has a direction. However, the entity time has completely different characteristics and meanings depending on context, about which we usually have little awareness. It makes sense therefore, to take a closer look at the meanings and contexts of importance to us [Hölling (1971b)]. According to Mehlberg, the arrow of time only ever has a meaning relative to a (subjective) observer. Mehlberg (1980) pp. 152–202, gives a comprehensive and understandable description of the entire arrow of time problem.

11.3.1 Time in our consciousness

Linear time is a condition *a priori* for consciousness, for thinking and for cognizance. Consciousness, thinking and cognizance consist of a sequence of – possibly transcendental – acts of the transfer and processing of information according to the pattern of a cause and effect. In each case, the cause comes before the effect, whereby the arrow of time is defined. During the information processing, the information received is connected and compared with existing and stored information. There are only ever memories about the past. Nobody remembers the future [Carnap (1928/1998) pp. 110 and 229]. The temporal structure which differentiates past and future is a fundamental form of every perception, thus a condition *a priori* in the Kantian sense [Bauberger (2003)]. Time in this sense is measured in chirality theory with the help of the internal clock of the (presumed) observer, i. e., of the ego, the Self.

If the conscious mind, in an act of extrasensory perception, should nevertheless remember the future, then this, as explained in Section 10.19, would probably have to be described by a theory with imaginary,

i. e., two-dimensional time. Such a theory could violate the causality principle. The act of perception would be, however, only conditionally conscious and not controllable by the will of the subject. It simply happens.

11.3.2 Theology

Concepts of time play a fundamental role in theology. Creation, final judgement, transmigration of souls and Karma, repentance and atonement, life after death or before birth are all impossible without time. Eternity, the precise meaning of which remains essentially unknown, may or may not be timeless, depending on the belief system. What does it mean when we say, „His time has come“? Does the soul of the deceased live in a timeless world without causality or does it live on indefinitely [Kather (2003) pp. 182 ff.]? For the questions posed by this book, it is not necessary for me to go into greater detail with such problems. I am aware, however, that there are natural scientists and philosophers, whose theories are not free of such thoughts. See for example Augustinus (396–400 AD/1946).

11.3.3 Classical mechanics

Classical mechanics are deterministic: each cause has a clear effect, which can be precalculated accurately if one knows the initial condition and the cause, i. e., the forces at work. Formally, that means that wherever time makes an appearance in an equation, the direction or the sign of this time can be reversed without rendering the processes described impossible. All processes can run forwards just as well as backwards. There is no arrow of time.

11.3.4 The theory of relativity

The same basically applies to the theories of relativity. In the special theory of relativity, determinism can be formulated as the fact that initial data in any given synchronous space specify the behaviour of the entire space-time, both for the future and for the past. One does not, by the way, even need the initial data of the entire space for the calculation, but only those of a finite, limited area. The reason for this is that information cannot be transmitted faster than the speed of light. This distinguishes the theory of relativity from Newton's mechanics, where the forces work instantaneously.

In the general theory of relativity, determinism is a substantially more complicated affair. First of all, the calculations are difficult, since space itself, or its curvature, continuously changes locally. Secondly, it can happen that even knowing all the initial data for the entire space is insufficient for calculating the future. This is the case if black holes are involved. Apart from this limitation, determinism also applies in general relativity theory [Penrose (1989) pp. 211 ff.]. Thus, in both theories of relativity, the past does not fundamentally differ from the future. There is no arrow of time.

11.3.5 Quantum theory

Quantum theory, in its current form, is deterministic in certain sense as well, despite the Heisenberg uncertainty principle. With the help of the Schrödinger equation, the past can be calculated from a state at a certain time just as precisely or, for that matter, as imprecisely as the future. One only ever knows the exact value of a variable at a certain time and at a certain place, however, once one has measured it, one can only speak about future and past values in the form of probabilities, according to quantum theory. These probabilities are then precisely determined, i.e., calculable. Quantum theory says nothing about what happens exactly during a measurement or during another, similar act. It simply states that the wave function of the Schrödinger equation „collapses". From the time of this collapse, the new probable value of the new past or the new future, nothing more, can then be calculated with the help of a new Schrödinger equation from the concrete measured value. The act of measuring is an interaction between the measuring instrument as a macro system and the measured object as a microsystem [Atmanspacher (1995)]. Quantum theory does not say anything about the boundary between macro and micro. Penrose (2004) pp. 787–868, assumes that this boundary is crossed upon reaching Planck energy. It corresponds to the energy or mass of a black hole with the Planck length as its radius, i.e., $2.5 \cdot 10^{-8}$ kg. The act of measuring is irreversible and thereby has a direction in the time. Penrose therefore calls for a new quantum theory which also applies to the act of measuring and where the arrow of time then appears. Penrose (1989) pp. 296 ff. and 367 ff., like Albert Einstein, considers today's quantum theory to be incorrect, or at least incomplete. He calls for radically new ideas about the nature of space-time geometry, something along the lines of a non-local theory of quantum gravitation, in which the Planck mass plays a central role. At present, there is no such theory. Could chirality theory meet these conditions?

In the Section 9.8, I described the gravitons: real gravitons carry information and move at light speed. Their inner clock, and thus time „from the perspective of the graviton", stands still. Virtual gravitons, however, carry no information and act instantaneously. They move at infinite velocity. For an external observer as well, no time passes during its effect. With a timeless procedure of this kind it is doubtful at least to speak of a direction of time which is zero. A cause and effect can no longer be distinguished here. One speaks therefore of an interaction.

There is, however, a mathematical reason which nevertheless suggests an arrow of time for quantum theory. The variables of the quantum theory change with the imaginary number of $i = \sqrt{-1}$. Since i changes upon the reversal of time, i should actually be a variable in quantum theory, not a constant [Finkelstein (1996) p. 448]. Thus, a fundamental difference between past and future would be introduced into quantum theory.

11.3.6 The second law of thermodynamics

Very few processes in daily life can proceed just as well backwards as forward. A scrambled egg cannot be unscrambled and „decooked", the yolk and albumen cannot go back in the shell and the egg cannot (easily) be put back in the chicken. That is so, although all important physical laws for describing particles and interactions in eggs do not know an arrow of time. That applies both to the laws of classical mechanics and the theories of relativity, as well as to quantum mechanics. To date, there have been no experiments by which the theories of relativity or quantum mechanics would have been violated. From where then, does this arrow of time come?

Naturally, we know that the arrow of time plays a central role in thermodynamics. Disorder is more probable than order. If one has two boxes, one filled with one hundred red balls, the other with one hundred black balls, and blindly takes one ball at a time from one box and puts it into the other one, then one will soon find red and black balls in both boxes, approximately half and half in each case. If one continues playing with this for long enough, then at some point the most improbable case will arise, whereby all one hundred red balls are once again in the same box. Yet the greater the number of balls, the more seldom this case will be. If, for example, one is prepared to wait until all air molecules just happen to be in the same corner of the room, then one would have to wait around many times longer than the current age of the universe. The world thus progresses from a highly improbable state of order to a more probable one. The arrow of time of thermodynamics is the result of a special, improbable initial condition. From the high state of order of this initial condition, each closed system progresses to a condition with a lower state of order. As physicists put it, the entropy, i. e., the disorder, increases. Defined somewhat more precisely, entropy is the *potential* information, negative entropy, on the other hand, is the *current* information of a system. Every single air molecule has many more places in the whole room where it can reside than in just one corner. For all air molecules combined, the possibility of relative locations therefore grows exponentially. The potential information of the air in the whole room is much greater than if it is compressed in a corner. In accordance with the second law of thermodynamics, the entropy of a closed system increases until it reaches a maximum state of disorder [Weizsäcker (1986) pp. 119–162; Penrose (1989) pp. 302–314].

Clearly, entropy has also increased in the past, indeed ever since the universe began to exist. Ludwig Boltzmann recognized in 1896 „that the universe, … or at least a very large part of it which surrounds us, started from a very improbable state, and is still in an improbable state." His statistical theory was violently attacked by prominent physicists such as Loschmidt, Mach, Ostwald, Poincaré and Zermelo: Statistics cannot be part of a law of nature. The mockery and contempt drove Boltzmann as a martyr of his ideas to suicide [Popper (1998) pp. 181–183]. During the development of the universe, the Milky Way, the solar system and life on earth, entropy increased constantly. Proof of this can be found, or at least assumed, in documents from the past, for example in the cosmic background radiation as a document of the Big Bang. The entropy will also increase in the future, because this process leads from the less probable to the more probable state. However, if we did not have such documents of the past, if there were no memory, then there would also be a state of higher entropy for the past than we find today, the more probable. The state of the universe with its nevertheless highly improbable order, from today's perspective, is much more likely to have arisen from a state of disorder by chance than from one in the past with a much higher and accordingly more improbable order still. That it is not so is an assumption we make based on the documents of the past. We have explained nothing thereby, however.

This arrow of time argumentation from the point of view of thermodynamics cannot satisfy the physicist fully. The laws of thermodynamics are not laws of nature in the strict sense, because they avoid making statements about the details of the observed objects. They only describe statistical average values of an enormous number of single objects, for example of air molecules, about whose individual characteristics one actually knows little. Accordingly, the statements about the future of the individual objects are very vague. As long as the fundamental contradiction between thermodynamics and the other, much more precise theories concerning the arrow of time continues, a need of clarification remains. Penrose hopes such a clarification will be forthcoming in a new theory of quantum gravitation. Can chirality theory with its interpretation of the Big Bang as the formation of a new black hole and of the elementary particles as black miniholes be such a quantum gravitation theory?

11.3.7 Cosmology

So, is the arrow of time a consequence of the Big Bang? The entropy was indeed extremely low at the Big Bang, because the universe's entire energy was concentrated at a very particular time in a very small area. According to the current cosmological theories, nothing precise can be said concerning the extension of this area. Perhaps the universe was the size of a grapefruit; perhaps had it a diameter of a Planck length, i.e., 10^{-35} m, perhaps the universe was a point. In any case, nearly all cosmologists speak of a singularity.[56] Some imagine that space and time only began to exist at all with this singularity, which is why it makes no sense to them to inquire into the cause of the singularity. After all, a cause would have to have existed in time before the singularity. That is not possible, since time did not yet exist, and with no time, there was no „yet". Others say that the singularity was a vacuum fluctuation [Tryon (1973)], i.e., pure chance. Hawking (2001) pp. 58–69 and 90, believes that there was no singularity at all in reality; it only looks that way from today's point of view. In reality one must imagine time in two dimensions like a surface of a sphere, on which the Big Bang is something like the North Pole. Whoever stands at the pole, sees nothing special there at all [Penrose (2004) pp. 769–772; Genz (1994) pp. 320 ff.].

Even if one admits the possibility that the Big Bang had no cause, one would nevertheless like to have an explanation for the Big Bang, in particular since there obviously must have been a highly improbable state. So far, my explanation in Section 10.14 still seems the most plausible to me: the entire matter of the universe, seen from outside, is at every point in time in a tightly confined location, namely in a black hole. The order looked at from the outside is indeed high, yet the information about this order is not accessible for an external observer. For him, the entropy is therefore very high. Seen from within, this tightly confined location is a point in time, namely the Big Bang, and on the inside the information about the configuration of the points is accessible everywhere. The entropy is thus very low when seen from the inside. It would be unsatisfactory for me, if one had to explain the second law of thermodynamics and the arrow of time solely with the improbable state of the Big Bang.

[56] *Singularity* is a mathematical term which, somewhat simplified, can be defined as follows: a is called a singularity of the function *f* if in each environment of *a* there are locations in which *f* is holomorphic; however, there is no (not even a miniscule) environment of *a*, in which *f* is analytically continuable from each holomorphic location in any given way to *a*. Holomorphy is defined as complex differentiability, i.e., is a consequence of the continuity of a mathematical space. [Reinhardt (1994) pp. 425 ff.].

In this context, the question vis-à-vis the end of the universe always arises. If the universe is a closed system, so that over time it pulls together up until the „big crunch", is this final state then not simply another Big Bang with an extremely deep entropy? Penrose (2004) pp. 707, 727–734, 766 ff., demonstrated that there may well be a certain similarity between the Big Bang and a big crunch, but that the entropy constantly increases the entire time from the Big Bang up to the big crunch. This applies in particular if, with a big crunch, a black hole forms. Black holes have an enormous entropy, a huge amount of information which is not accessible to the outside observer. Today, the majority of the entropy of the universe probably already consists of the entropy of its black holes. A black hole with the mass of our universe would have the tremendous entropy of approximately 10^{123}.[57] A big crunch therefore, is not particularly improbable. The extremely improbable Big Bang cannot be explained thereby in any case [Penrose (1989) pp. 302–345]. If the universe is an open system, however, because the mass is not sufficient to pull it all together in a big crunch, then it expands into all eternity. Because of the increasing volume, the entropy will likewise constantly increase thereby, and ultimately the universe will die a cold death.

11.3.8 Kaon decay

There is also, however, very much a phenomenon with an arrow of time in the microrealm of quantum theory! This concerns kaon decays:

$$K^0 \rightarrow \pi^+ + \pi^- \; ; \; K^0 \rightarrow \pi^0 + \pi^0 \; ; \; K^0 \rightarrow \pi^+ + \pi^- + \pi^0 \; ;$$

$$K^0 \rightarrow \pi^0 + \pi^0 + \pi^0 \; ; \; K^0 \rightarrow \mu^+ + \mu^-$$

and in particular

$$K^0 \rightarrow \pi^- + e^+ + \nu \text{ and } K^0 \rightarrow \pi^+ + e^- + \text{anti-}\nu.$$

In the arrow of time discussion, kaon decay, oddly enough, barely rates a mention, although the phenomenon has been known since 1964 [Chris-

[57] The number 10^{123} states the entropy of the universe as a black hole in so-called natural units, with which the Boltzmann constant k = 10-23 Joule/°K equals one.

tenson et al. (1964)]. The reason for this perhaps is that there is no simple experiment whereby the symmetry of the time is clearly violated. Such an experiment would be a particle decay or a particle transformation which is not reversible.

A problem with the kaon decay is that the electrically neutral Kaon K^0 with spin zero is not a particle with clear properties, but that as a meson it consists of a quark and an antiquark, which for their part constantly change their properties and spontaneously turn into one another. The same applies in certain respects to the pions π which arise from the decay. From the perspective of quantum theory, all these different kaon states do not exist in a temporal sequence, rather they are simultaneously overlaid or oscillating with one another, whereby every possible kaon state has a certain probability of being found during a possible measurement. The K^0, according to current theories, is a combination of K^0 and anti-K^0, which are constantly turning into, or oscillating with, one another. The K^0 differs from its antiparticle by the opposing electrical charges of the two quarks which make up the K^0. As to how the spin is apportioned with these two quarks remains open; the only thing that is clear is that the sum of the spins is zero. Since K^0 and anti-K^0 cannot decay into the same particles, they can be distinguished from each other by the decay products. From this, conclusions can then be drawn about the percental portion of the K^0 and anti-K^0 in the composition of the two states, or about the transformation of the K^0 in anti-K^0 and vice versa. Thus, it is shown that the transformation $K^0 \rightarrow$ anti-K^0 is clearly 0.66 % less probable than the transformation anti-$K^0 \rightarrow K^0$. In other words: the kaons sense the arrow of time [Wolschin (1999)]. I have already referred to it in Chapter 4.7: with the Kaon decay, apparently not only are space parity and charge symmetry violated, but time symmetry as well. In all this, however, nothing is said about the direction in which time effectively runs during the transformation of the K^0, towards the past or towards the future. I will return to this in Section 12.10 (Kaon decay).

11.3.9 The arrow of time and the axiom of chirality

The axiom of chirality is composed entirely of arrows:

$$(\leftarrowtail) \Rightarrow (\leftrightarrow) \Rightarrow (\mapsto) \Rightarrow (\hookleftarrow) \Rightarrow (\leftarrowtail) \Rightarrow$$

The arrows in brackets signify the state of a point in relation to the other points in a system. The broad arrows in between signify the phase transition from one state to the next. In this axiom, all four entities – space, time, interaction and substance (or frequency) – are concealed, as I explained in Chapter 8. The arrows in brackets were interpreted as relative locations or changes in location. From these follows the chirality of space, i. e., the spin. The arrows for the phase transitions are an expression of time, which runs in the direction of the arrows and is thus chiral. The interaction forces a point to reverse whenever another point is between the remaining ones. The frequency comes about when the event count, i. e., the number of phase transitions is compared with the observer's inner clock. Also, the observer's time, that of the perceiving consciousness, is chiral according to Section 11.3.1. As soon as an event is observed, we are then dealing with at least two different times, the eigentime of the object and the eigentime of the observer. The number of events of these two clocks can be compared easily. That does not necessarily have to mean, however, that these two times, that of the observer and the object, also have the same direction, because there is no such thing as absolute time. Not only the measure, but also the direction of time in an object is relative and can only be determined in relation to a second clock. It is obviously important to always know which time is being referred to, that of the object or that of the observer, and above all, whether these two times have the same or opposite directions. If the direction is opposite, then an attraction becomes a repulsion. These considerations indicate that the orientations of space, time and possibly electrical charge as well, are closely connected. An observer who observes a spin of +1 if his time runs in the same direction as that of the observed object, will measure a spin of –1 if his time runs in the opposite direction. Naturally, such an effect cannot be demonstrated with presumed observers, but only with real observers.

11.4 The arrow of time and the arrow of space

At the Schwarzschild radius, one-dimensional, linear time becomes three-dimensional space. How does the arrow of time change in the process? Is it simply lost or does it take on a new significance? And how would one have to imagine a three-dimensional arrow? In Section 10.14, the Big Bang was described as the inside of a black hole, in which space expands from time zero in all three spatial dimensions. Just as the future is not simply the opposite of the past, neither is the direction outwards simply the opposite of the direction inwards. Space can expand outwards without limit; inwards, however, it can contract no further than to a point. The future is, in principle, open-ended, whereas the past always ends with the Big Bang. So, the arrow of space corresponding to the arrow of time is the outwards-directed radius of a growing spherical plane. The interior of the sphere corresponds to the past, the exterior to the future. The surface of the sphere is the present. Just as the theories of relativity and the laws of quantum theory know no arrow of time, these laws of nature are also independent of location and direction. Exceptions to this principle are at most possible in cases where black holes play a direct role, such as in the case of the Big Bang or the orientation of spin.

11.5 Basic conditions for an electron model

If one tried, on the basis of chirality theory, to construct a model of the electron in such a way that the observations of an experimental physicist could be explained, the model would have to have the following characteristics:

1. In order to be stable, the electron must, like the neutrino, periodically, ideally after each individual event, have the same structure as before the event. Thus, it fulfils the condition for a black hole: no point can leave the black hole.
2. It must have a spin of $\pm\frac{1}{2}$.
3. Like the neutrino, it must come in three variants or „flavours" with different energy values, in order to explain the existence of muons and tauons.

4. All these variants must each exist with positive and negative electrical charges, with the characteristic that electrical charges with same sign repel each other. Thus altogether, there should be $2 \cdot 3 \cdot 2 = 12$ different kinds of electron – more precisely „charged leptons“ – analogous to the 6 different neutrinos.
5. The electrons with spin $+\frac{1}{2}$ must differ fundamentally from the electrons with spin $-\frac{1}{2}$, so that the parity violations can be explained by the so-called weak interaction: only the *CPT*-symmetry is not violated (see Sections 4.7 and 12.10).
6. In the three-dimensional electron there should be an internal axis which corresponds to the North-South Pole axis.
7. If possible, an extension of the electron model should lead to a model for the $2 \cdot 2 \cdot 2 \cdot 3 \cdot 3 = 72$ different quarks (with spin $\pm\frac{1}{2}$; electrical charge $\pm\frac{1}{3}$ or $\pm\frac{2}{3}$; 3 colours and 3 flavours), which do not exist as single particles, but only in the compound which fulfils condition Nr. 1, be it as a pair comprising a quark and an antiquark (mesons) or as a trio in three different „colours“ (nucleons) [Close (1989) pp. 396–424]. Quarks do not have to be black holes: individual points can change between the compound's two or three quarks, but after this change, the group as a whole should have the same structure again.

Naturally, we also want to proceed here according to the rule of Ockham's razor, the principle of parsimony for all theories. If we are looking for a model for black holes in black holes, it surely makes sense to have an analogous to that with the neutrino model, the model for the simplest and smallest black hole. Therefore, I wish to assume for now that the black holes in the electron's interior could all be described in the same way as the six different neutrinos. This is also obvious because among other things, neutrinos develop with the decay of muons and tauons. For the presumed external observer, the neutrino models were four-point spaces with a frequency and a direction of rotation or spin as their only characteristics. The real observer can only perceive them as point objects, since no information about the internal structure of a black hole can reach the outside. A new particle should now be constructed from a number of such four-point spaces, so that again, the emerging object becomes a black hole. The simplest way to do this is obviously the repetition of the construction of the neutrino model: four neutrinos together should form a superordinate four-point space, in which they move analogously to the individual points in the neutrino's four-point space. I now want to show that such a model actually has all seven required characteristics. In the course of the description I proceed similarly to as I did with the gravitation: I begin with the model of an individual electron. Then I place a second electron beside it and examine the interaction between the two electrons. Only at the end do I introduce the real observer of this two-electron system and explain the interaction between observer and two-electron system, i. e., the observation.

11.6 Model of the individual electron

The electron should thus be a black hole which is composed of four neutrinos. This is the simplest possible model of a black hole consisting of black holes. The neutrinos move relative to each other in accordance with the axiom of chirality, just like the four points in the four-point space. From this motion, a black hole with an additional contribution of $\frac{1}{2}$ to the spin of the electron results. Every change in the arrangement between any of the four points within the black hole is an event. Since 16 points within a total of 1820 four-point spaces move relative to one another, there are many more events inside an electron than inside a neutrino. Moreover, most of these four-point spaces are lopsided, like the four-point spaces of the muon and the tauon neutrino. Therefore, the frequency of the internal events and with it the **mass of the electron** is very much greater than that of the neutrino.

Only if the rhythm or the phase of the inversions of all neutrinos and of the electron as a whole is the same, is the **state after each event the same as it was before the event**, i. e., back to how it was. The standard measure for this frequency is the event in the four-point space, exactly as with gravitation. That means that the neutrinos within the electron move at light speed. Since no information is transferred with this movement, this is not a violation of the special theory of relativity.

It is scarcely possible to provide an accurate and correct pictorial representation of the electron's structure, first of all because one would need more than the three spatial dimensions available and secondly, because the significance of these dimensions would be interpreted as a time or place, depending upon the location of the viewer. Nevertheless, I make an attempt, with which I draw the three-dimensional, chiral four-point spaces, i. e., the neutrinos as two-dimensional, chiral triangles, which are connected to the closed space of a spherical plane, from which they can no longer escape (Figure 37).

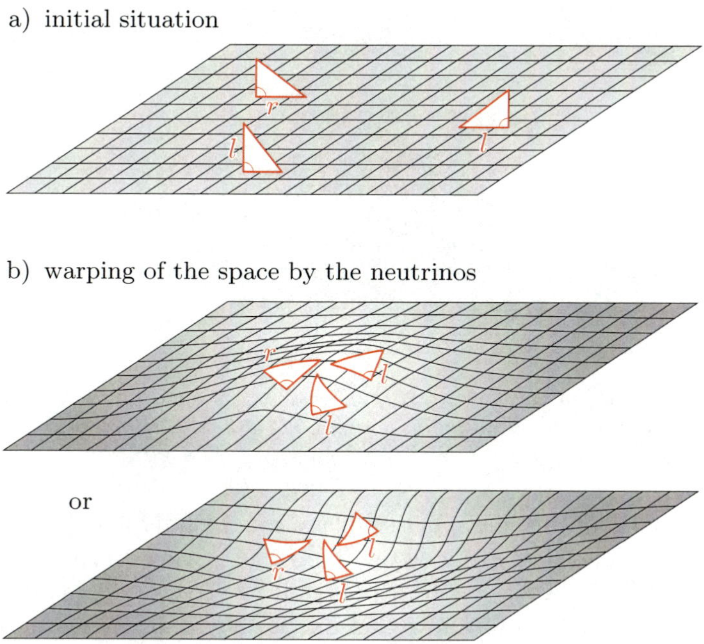

Figure 37. Model of the electron, Part a) and b): *The formation of an electron or a positron out of 4 neutrinos. The neutrinos are represented symbolically as chiral triangles instead of tetrahedrons. The space is represented as a two-dimensional surface, which warps into a new black hole, represented as a spherical plane. Due to the missing third dimension, instead of four neutrinos, only three are required, which form a new chiral pattern on the spherical plane.*

c) new black hole

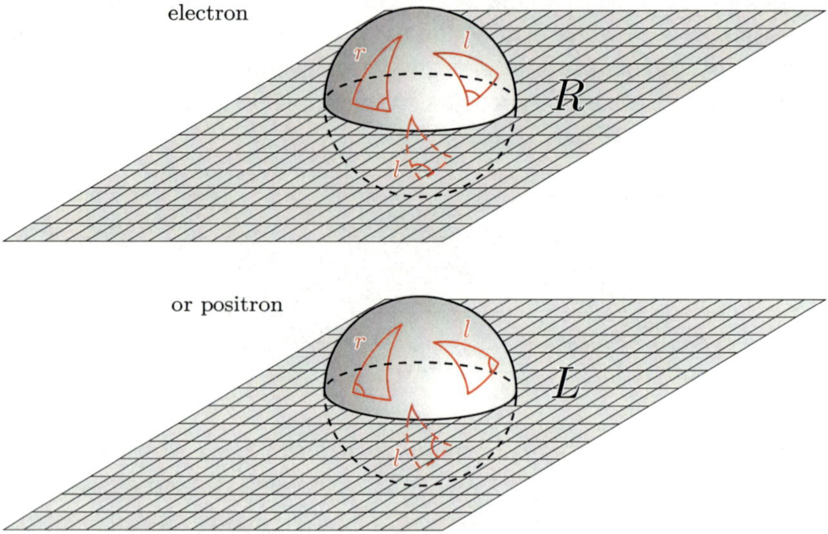

Figure 37. Model of the electron, Part c): *Observed from the outside the rotation of the chiral triangles in the positionis reversed. But since the spin is the same in both particles, the time arrow of the positron is opposite.*

In the course of this, the spherical plane is to be interpreted as a newly-formed black hole. In effect, two additional, new space dimensions develop with the spherical plane, which can only be perceived from the outside in one dimension, however, in that one is able to differentiate between inside and outside. Depending upon the curvature of the spherical plane, the orientation of the triangles is then seen differently from the outside. Looking more closely, one sees that there are two different possibilities for the formation of the spherical plane. They differ in the opposing orientation of the triangles on the spherical plane; this depends on which side of the triangles becomes the outside and which becomes the inside as a result of the curvature. Outside and inside thus define the chirality of the spherical plane. The chirality of the neutrinos can be interpreted as the chirality of a (four-point) space and the chirality of the spherical plane as the chirality of time. In addition, the movement of the neutrinos on the spherical plane causes a further chirality. In this way, a relationship between the chirality of the neutrinos and the chirality of the new black hole results. By means of the chirality of the spherical plane I define now the sign of an **electrical charge** as follows: if, as seen from the outside, the orientation of the triangles remains unchanged with the formation of the spherical plane, then the charge of the new black hole is positive; if the orientation of the triangles changes, then the charge is negative.

The spin of the neutrinos does not change with the formation of the electrons, irrespective of whether a positive or a negative charge develops in the process. Since, however, the direction of rotation of the triangles (the models for the neutrinos) differs depending upon the curvature of the spherical plane (the model for the electron), the time must run in accordance with the curvature in one direction or the other. It would be misleading to say that in one case, time runs forwards and in the other, it runs backwards. There are simply two opposing directions, and only an observer can decide what, seen from his perspective, is forwards and what is backwards.

For some considerations it is more appropriate, if not necessarily more descriptive, if one speaks of the spherical surface's spin, rather than its chirality. This spin is produced by the relative motion of the four neutrinos, just as the spin of the four-point space is a consequence of the relative motion of the four points. The total spin of the electron is made up of the spins of the four neutrinos and the orbital spin, which is produced by the relative motion of the four neutrinos. All five spins have a value of $\pm\frac{1}{2}$. The total spin of the electron is the sum of these five spins and can thus amount to $\pm\frac{5}{2}$, $\pm\frac{3}{2}$ and $\pm\frac{1}{2}$. Naturally only the electrons with spin $\pm\frac{1}{2}$ are elctrons in the conventional sense. Higher spins should be possible, however. The fact that no one has found such a spin in nature up to now could be due to the fact that such electrons with spin $> \pm\frac{1}{2}$ are not stable.

For the rest of the discussion it is appropriate to clearly differentiate between the orbital spins due to the relative motion of the neutrinos, and the spins of the neutrinos themselves. I call the former *R* or *L* (for *R*ight and *L*eft, or spin $+\frac{1}{2}$ and $-\frac{1}{2}$, or electrically negative and electrically positive), and the latter *r* or *l* (for *r*ight or *l*eft, or spin $+\frac{1}{2}$ and $-\frac{1}{2}$). Writing the charges in this way makes particular sense, because the electrical charge is by definition a direct consequence of the chirality of the spherical plane. *R* means that the electrical charge is negative, whereas *L* is always positive. This applies, however, only as long as one is dealing with whole charges. With the charges $\pm\frac{1}{3}$ and $\pm\frac{2}{3}$, as they occur in the quarks, the situation is moere complicated. The charge $+\frac{2}{3}$ is L, $-\frac{2}{3}$ is R (see Sectieon 12.5).

The electron in the conventional sense with spin $+\frac{1}{2}$ has the spin structure *Rrrll*, the elctron with spin $-\frac{1}{2}$ has the corresponding structure *Rrlll*. One can see immediately that the structure of the electron with positive spin differs fundamentally from that of the electron with negative spin. The former has two levorotatory and two dextrorotatory neutrinos, the latter, however, has three levorotatory and only one dextrorotatory. It is not surprising therefore, that the two electrons have fundamentally different physical characteristics: the electron with negative spin is less symmetrical with respect to the neutrinos and is subject to the so-called **weak interaction**, while the electron with positive spin is not affected by this. It seems that the weak interaction has to do something with the exchange of neutrinos (see Section 12.9).

The four points of the four-point space are completely structureless. A relation between the internal structure of the points and that of the four-point space is therefore impossible. It is different with the electron. Here both the structure of the neutrinos and that of the newly-formed black hole are chiral, be it spatial or temporal. The space and time directions of the different structures adjust themselves to one another. The consequence of this is an internal direction within the electron. By definition, it runs from the **North to the South Pole.** In quantum field theory one calls this mutual adjustment between an external and an internal space symmetry breaking. The external gauge field tries to rotate the phase of the local internal field [Moriyasu (1983) pp. 86 ff.] (see also Section 14.4.).

With the muon, one of the four neutrinos is a muon neutrino, with the tauon, a tauon neutrino.[58] The frequency of the neutrino inversions is the same as in the electron. Since the individual points in the muon must, however, move further per inversion than in the electron, the frequency of all internal events, i.e., the frequency of the inversions of all lopsided

[58] According to chirality theory it would not be impossible that several muon neutrinos are in the same charged lepton, but such particles have not been observed, probably because their lifespan is too short.

tetrahedrons, and thus the mass of the muon, is greater than that of the electron. In the tauon, this effect is still more pronounced. With muon and tauon, the condition that the state should be the same again after each event as before the event is no longer met. Since muon and tauon neutrinos must in each case go through four phases with their movement within the muon or the tauon until they are back in their initial state, this also applies to the muon and the tauon themselves. Therefore, the muon and the tauon are less stable than the electron.

11.7 Attraction and repulsion

If we assume that an electron is a black hole consisting of black holes, then the repulsion between two electrons is the outcome of an interaction between two black holes which both consist of black holes. In order to make it from the heart of one electron, i. e., from the interior of a neutrino, to the heart of the other one therefore, no less than four Schwarzschild radii have to be overcome; places or points in time, where time becomes space and space becomes time, and where all information about internal structures is lost. The electrons themselves do not feel the interaction at all, just as little as an astronaut in a windowless spaceship can feel the diversion of his spaceship by virtual gravitons. The mutual repulsion can only be observed by an outside observer who measures the distance between the two electrons and finds that this becomes greater at an ever-increasing speed. So, we are dealing with five arrows of time, two each for the inside of the two electrons and for the neutrinos within these electrons and one for the observer.

Due to their mass, an attractive gravitation force F_G naturally always exists between electrons or positrons. It is small in comparison to the electrical force and will be disregarded in the following considerations. The attraction or potential energy between an electron and a positron is, due to the gravitation, about $2 \cdot 10^{33}$ times weaker than their electrical attraction. The **electrical force F_E** comes about – in principle just like the gravitation – via an electrical field, only that we now have four neutrinos instead of the four points in the lopsided tetrahedrons. These tetrahedrons, formed in each case from four neutrinos from two different electrons or positrons, only ever partly invert themselves. As already stated in

Section 9.9 there are 68 such lopsided tetrahedrons between two neutrinos. The same number applies to the lopsided tetrahedrons between two electrons or positrons. The force F_E at work between electrons and positrons is analogous to the gravitation force F_G

$$F_E = e_1 \cdot e_2 \cdot r^{-2}.$$

e_1 and e_2 are the electrical charges of the electron or the positron, r is the distance between the two charges. Unlike the situation with the masses, which are the cause of the gravitation, the electrical charges can be positive or negative, depending on how the black hole is formed from the four neutrinos. Two questions arise: Why do equally curved black holes, i. e., with the same charge repel each other? And why is the potential electrical energy or the electrical force between the electrons and positrons so much greater than it is with the gravitation? Ultimately, both are a consequence of the fact that in the electrical interaction we are not dealing with one such as there is between individual points, but with one between black holes, i. e., the neutrinos, which themselves have an internal structure.

e_1 and e_2 cannot be directly perceived or measured, only their product can.[59] This can be positive or negative, depending on whether they have the same sign or not. If it is positive, then the signs of e_1 and e_2 are the same and their eigentimes run in the same direction. If two clocks run in the same direction, then the relative positions of the two hands always stays the same. Thus, no periodic events ensue between the two clocks. If $e_1 \cdot e_2$ is negative, however, then the eigentimes of the two particles run in opposing directions. This results in additional periodic events in the system of the two particles, like two clocks whose hands turn in opposing directions. Such clock hands periodically point once in the same direction, and then once again in the other. However, as seen by the resting, external observer, the total number of the events in the system of the two particles is always the same. If he sees more periodic time events, then he perceives fewer non-periodic space events as a result, and vice versa. Fewer space events means that the distance between the two particles is get-

[59] The product of $e_1 \cdot e_2$ is mathematically correct for the description of the relation between e_1 and e_2, because e_1 and e_2 concern separate black holes whose spatial dimensions cannot be interconnected by addition.

ting shorter. They attract each other. On the other hand, with more space events in the course of the process, the particles will repel each other and the distance between them grows larger. In Figure 38 (, p. 284), this situation is portrayed schematically. However, only one of the 68 possible lopsided tetrahedrons has been portrayed.

The electron-plus-positron system can also be represented graphically as a picture of two cogs which turn in opposite directions (Figure 39, p. 284). If the negative charge always turns clockwise and the positive charge turns anti-clockwise, then the two cogs interlink beautifully. If, however, one is dealing with two negative charges, i. e., cogs which both turn in the same direction, then the two cogs block each other and one can imagine that they must repel each other. It is clear that this explanation of attraction and repulsion is intuitive and not very mathematical. A mathematically correct model is yet to be found.

11.8 The measure of electrical charge: the factor $\frac{1}{137}$

The electrical charge is an intrinsic property of the electron. It can be derived from the formula

$$F_E = \frac{e_1 \cdot e_2}{r^2}$$

that $e_1 \cdot e_2$ must have the same dimension as the product of the two natural constants c and h. Measurement [Georgi (1989b) pp. 425 ff.] shows that

$$e_1 \cdot e_2 = e^2 = c \cdot h \cdot \frac{1}{137} \, .$$

The factor $\frac{1}{137}$ is a dimensionless number. It is a measure of the strength of the electromagnetic interaction. It is called the coupling constant of electrical interaction or also Sommerfeld fine-structure constant α. According to the Grand Unified Theory *GUT*, which explains electromagnetic, weak and strong interactions with a unified theory, α is not constant, however, but at distances shorter than 10^{-16} cm increases slightly until at about 10^{-29} cm it unites with the coupling constants for the weak and strong interactions. The exact value of α at greater distances amounts to 0,0073 = 1/137,03599976. At a very short distance or high energy, e. g., when two electrons collide at practically light speed, the value of α rises to

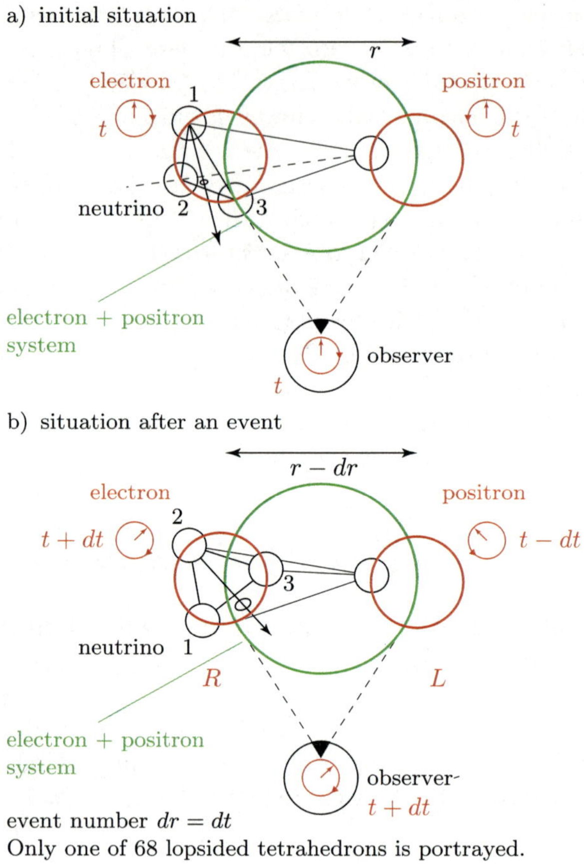

Figure 38. Attraction between electron and positron

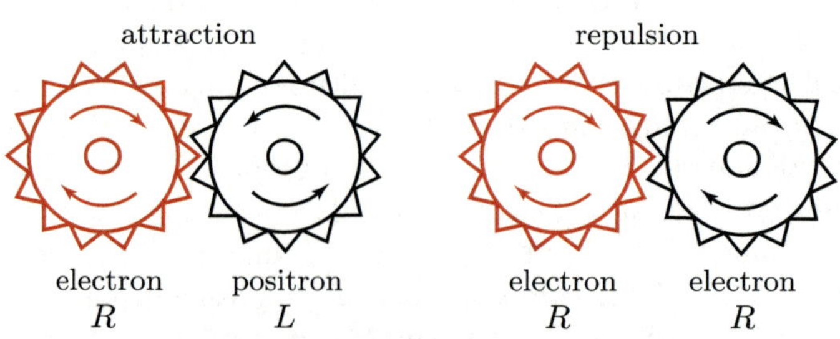

Figure 39. Cog model of electrical attraction and repulsion

0,0078 [Collins (2004)]. Although there are repeatedly reports that the constant α must have changed in the course of the development of the universe, such reports have never been able to be confirmed [Barrow et al. (2005)]. Since the „constant“ α is dependent on the energy, i.e., the distance of the interacting particles and thus their density, it would be no surprise if under extreme conditions, e.g., shortly after the Big Bang or in the vicinity of black holes, it had a somewhat different value. So far, there is no explanation for the number α. For Pauli, the explanation of the coupling constant would be the most important touchstone of field theory, yet he was unsuccessful in finding this explanation. It unsettled him, however, to be tended to in number 137 of all rooms in Zurich's Red Cross hospital, where he died shortly thereafter [Enz (1995) pp. 21 ff.]. In the history of modern physics there have been mountains of papers about the value of α, which, however, are all rendered invalid by the continual experiments producing specifications of ever further decimal places. Quantum electrodynamics interprets the coupling constant as the probability that a real electron emits a real photon and absorbs it again [Feynman (1985) pp. 125 f.]. I won't presume to suggest a further explanation for the value of α. Even so, I would like to give an indication as to where to look for the explanation.

Figure 38 shows that the electromagnetic interaction is a consequence of the inversion of 68 lopsided tetrahedrons whose corners are formed by neutrinos in two different electrons or positrons. In accordance with the axiom of chirality, two tetrahedron corners are in motion with each event, while two others are each in one definite location. The neutrinos have the spin $\frac{\hbar}{2}$ and within the electron they move at light speed c. $\frac{\hbar}{2}$ is the measure for the internal motion of the neutrinos and c is the measure for their motion relative to the other neutrinos, whereby the relative direction of all these movements would probably still have to be considered here. $c \cdot \frac{\hbar}{2}$ is the measure for the movement in the entire system which for its part effectuates the inversions. In addition to the consideration of all these basic conditions, the factor $\frac{1}{137}$ would have probably to be explained with tetrahedron geometry. In chirality theory h and c are described as frequencies of counted events:

$$h = (1 \cdot 1) : (1 \cdot 1) = 1^2 : 1^2 = 1$$
$$c = 1 : 1 = 1$$
$$h \cdot c = 1^3 : 1^3 = 1$$
$$e^2 = (1^3 : 1^3) \cdot \tfrac{1}{137} = \tfrac{1}{137}$$
$$e = \pm \sqrt{\tfrac{1}{137}}.$$

Also, the electrical charge *e* is thus a dimensionless number which can be expressed by the ratio of counted event numbers.

In this way, the model for the individual electron fulfils the required conditions 1 to 6 in accordance with Section 11.5. Condition 7 will be looked at in Chapter 12.

11.9 Virtual and real photons

The „particles" which transfer the interaction between electrically charged particles are called photons. They can, in principle, be described analogously to the gravitons in the theory of gravitation. As is the case with gravitons, it is rather questionable to speak of a photon as a particle. Actually, it is simply a model for the difference between two states, be they perceivable or otherwise. Therefore, we can forgo a detailed description of the photons. The distinction between real and virtual photons follows the same path as with the graviton. **Real photons** transmit information and move at the speed of light. **Virtual photons** transmit no information, but only an action or a momentum. They move with infinite velocity. Only the real photons are ever perceivable. They develop when the form of an electrically charged object changes, e.g., when the distance between the atomic nucleus and an electron becomes shorter. Thus, the event frequency of the object always changes as well. The difference in the **frequency** before and after such an event is transmitted as information from the object to the measuring instrument. If the measuring instrument has a consciousness, then it is an observer. As with gravitons, one can also interpret photons alternatively as **particles** or **waves**. Since object, subject and the space between them stand in relation to the entire remaining universe and its configuration, and since with each change in the object's form this configuration likewise changes a little, the information connected with this change is added to the photon. It is then most appropriately described as a wave with its **interferences**. Real photons can – like real gravitons – interact with one another.

Naturally there are also divergences between gravitons and photons. Since the electrically charged particles have a substantially more complicated internal structure than the simple four-point spaces, the photon, as the difference between such charged states, must likewise have a more complicated structure than the graviton. In particular, the internal north-south direction within the charged particles has as a consequence that the photon, as the difference between these charged states, must likewise have an internal direction. This can run in all three space dimensions, thus also in the photon's direction of motion. That leads to the different possible **polarisation directions** of the photons. The photons have a spin. Since they are the difference between the states before and after the event, and since the charged states, as fermions, have a spin of $\pm\frac{1}{2}$, the photons must have a spin of ± 1. More detailed descriptions of photons can be found in any book about quantum theory.

It can nevertheless make sense to regard the photon as a particle, i. e., if one views it from the perspective of outside of the universe. Seen by an observer outside of the black hole „universe", space and time are exchanged. Instead of being the unlimited expansion of the photon with an eigentime zero, the photon for this observer is a particle during an unlimited time at a location with zero expansion, i. e., a fermion.

11.10 Magnetism

The magnetism follows automatically from the connection of the theory of the electromagnetic interaction described above on the one hand and the special theory of relativity on the other. If a charged particle moves relative to an observer, then from the observer's perspective, the spherically symmetric electrical field is compressed by the relativistic length contraction perpendicular to the direction of motion, so that the strength of the electrical field there increases. Thus, the electrical field exerts a stronger force on other charges (Figure 40) [Sexl et al. (1987) pp. 83 and 171].

This additional attractive or repulsive force is the magnetic force. The magnetic moment of the electron comes from its internal axis, which it has – in contrast to the neutrino. Within the electron the paths of the neutrinos and the paths of the individual points within these neutrinos can align themselves to each other, whereby a special direction or axis develops (see Section 11.6).

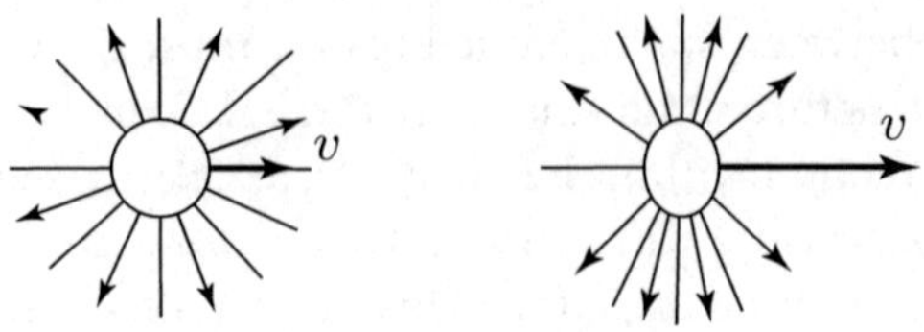

***Figure 40. The electrical field of a particle moving with velocity** v: The spherically symmetric electrical field of a slow particle is deformed at a higher velocity v. The lines of electrical flux are compressed in the plane perpendicular to the direction of motion so that the strength of the electrical field there increases. In so doing, the electrical field exerts a stronger force on other charges.*

Chapter 12

Strong and Weak Interactions

In jeden Quark begräbt er seine Nase.
Johann Wolfgang Goethe[60]

12.1 (Spontaneous) symmetry breaking

In Section 11.5 and Figure 37 I described how an additional chirality results from the curvature of the originally Euclidean space in which there is a chiral object. Owing to this curvature, a distinction can be made between outside and inside. This new chirality concerns the electrical charge, which can be positive or negative. For the external observer, here positive and negative are not simply the opposite of each other in an overall symmetrical situation. Positive and negative are fundamentally different, and consequently have different physical properties. They are different because inside and outside are not direct opposites of each other in a total system which is intrinsically symmetrical. If a symmetrical situation spontaneously becomes asymmetrical in such a way, then I call it **symmetry breaking**. The consequence of the symmetry breaking is the electromagnetic interaction.

The symmetry breaking thus comes about spontaneously through the formation of black holes. However, in chirality theory black holes have already formed spontaneously in two other constellations, i. e., as individual points coalesced from chaos into four-point spaces and at the

[60] There are numerous translations in English of Goethe's *Faust* which includes the Prologue in Heaven. Here Mephistopheles is speaking scornfully of mankind's quest for knowledge, saying „He buries his nose in every quark." The pun doesn't translate, but quark in this context, following common German idiom, is rendered to mean something like „bit of dirt". (Quark in the context of physics was first used in the 1960's.)

Big Bang. These two processes also are accompanied by symmetry breakings: from the completely symmetrical chaos of the points[61], space and time have developed with the four-point spaces, with rotation and translation, left and right, inside and outside. The symmetry of the chaos was broken, and with gravitation, the first interaction developed. If one interprets the expanding universe with its Big Bang as the inside of a black hole, then a symmetry breaking, a new asymmetry, should be ascertainable here as well. Here also, there is an outside and an inside, whereby we can only perceive the inside. As consequence of the Big Bang, time got a direction and became chiral. I suppose this is the reason why in the universe there is almost exclusively matter and practically no antimatter. Matter and antimatter differ by the fact that they have opposite electrical charges and that their internal arrows of time point in opposite directions.

The phenomenon of the symmetry breaking arises whenever a system gets out of balance and begins to organise itself [Nicolis (1989)]. Prime examples of this are snowflakes, which form from cooled-down water droplets. In biology and chemistry as well, such examples of equilibrium disturbances are frequent. An example from physics is the magnetisation of an iron bar. Characteristic of the symmetry breaking is the relativity of the symmetry: a system is not symmetrical or non-symmetrical *per se*. Whether it is perceived as symmetrical or non-symmetrical depends on the location of the observer and on the scale against which he measures it. An observer inside the magnetised iron bar will perceive it as asymmetrical, unless he is in the exact centre of the bar. An observer outside of the bar sees that the bar is, as a whole, very much symmetrical. A pile of cooking salt looks symmetrical to the cook. Whoever resides in a grain of salt, however, ascertains that three directions in his space, the salt crystal, are distinguished. He measures on a shorter scale than the cook does, and for him, the rotational symmetry is broken. The cook can pick out the grain of salt, turn it around in space as he chooses and the observer in the grain of salt along with it. But for the latter, the space remains asymmetrical. The world for a third observer between the cook and the inside of the grain of salt is the most complicated. He sees both

[61] The term „completely symmetrical" means that in the chaos there is no place, no direction in space and no time direction which are distinguished in any way from all the other places and directions in space or time.

a certain symmetry and asymmetry and has difficulty reconciling these. Here therefore, one speaks of chiral symmetry. Such an intermediate state is illustrated in Figure 37b. The space is neither Euclidean, nor is it totally warped into a black hole. The symmetry is only partly broken. This is also the case with the phenomenon of the so-called weak interaction, which becomes effective with new symmetry characteristics at distances of approximately 10^{-16} cm [Georgi (1989b) pp. 435 f.]. And something quite similar happens at distances of approximately 10^{-29} cm, where the strong interaction develops with new characteristics, likewise as consequence of a symmetry breaking [Genz (1994) p. 324]. In chirality theory, however, neither of the two are complete breakings of symmetry, i.e., no new black holes develop in the black holes. But the organisation in the relative motion of black holes, i.e., the neutrinos, becomes so great at these short distances that new asymmetries arise which split the past interactions up into more complicated, apparently new interactions. As with the snowflake, the system consisting of a neutrino plasma had to cool down before this new order became possible.

12.2 A quark model

For particles to have a certain stability and be perceivable as objects with specific characteristics, they do not necessarily have to regain the same internal structure after every single event, yet they must still do so periodically. Therefore, they need not be black holes. Mesons, according to the standard theory, are composed of a quark and an antiquark. They have integral electrical charges, an integral spin and a mass which is greater than that of the electron. Since the electrical charge, as we have seen, is a consequence of the periodic motion of four neutrinos, it is to be expected that also the meson can be described by a model of neutrinos moving periodically amongst one another. The mass of the lightest meson, i.e., the electrically neutral π^0-meson, is about 264 times heavier than that of the electron, which is why probably more than just four neutrinos are involved in the structure of the meson. Let's try it once with six!

Electrical charges result when a neutrino periodically travels between three others, whereby a quasi positive or negative curved space forms. Thus, four neutrinos are required for each charge. A total of six neutrinos means that two of the six neutrinos belong to two different charges at the same time. The result is a meson whose six neutrinos form two triangles which interlink in such a way that the (↔)-neutrino of one triangle always forms the corner of another triangle. I call the two entangled triangles quarks (Figure 41). Quarks can thus only occur entangled with other quarks, never in isolation. This principle is called **quark confinement**. The quark triangles in Figure 41 are drawn curved, so that one can differentiate between outside and inside and the triangles with their own direction of rotation in the three-dimensional space are chiral. Thus, here inside and outside are merely characteristics of the mathematical model, not of the quarks themselves. Similarly, the direction of rotation in the quark model is not based on the effective rotation of a triangle but is meant to express that the quark has a chiral orientation.

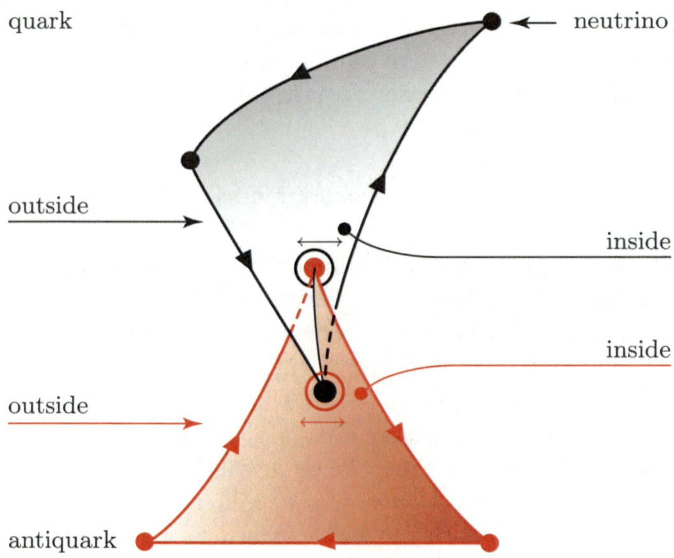

Figure 41. Meson composed of a quark and an antiquark: *Mesons are composed of a chiral quark which is interlocked with a chiral antiquark. Quark and antiquark have opposing orientations. In the example illustrated, the orientation of the quark, viewed from outside, is clockwise; that of the antiquark is anti-clockwise. Every quark is composed of four neutrinos. The two (↔)-neutrinos belong to both the quark and antiquark.*

Before I analyse the structure and parameters of this quark model, it is advisable, to remind oneself of the uses and dangers of such models. Certainly, as a support for our imagination and as a basis of discussion, they are very convenient. Yet, they are still only ever **models** of nature and never nature itself. Unfortunately, a model will always be inferior to nature itself. For example, my drawings are only possible in the two-dimensional space of the paper. According to chirality theory, however, there is no such space, rather the drawn points are themselves the space and this has no background. The paper is a continuum; there is no such thing according to chirality theory, because a continuum is fundamentally not perceivable. The drawing is static. According to chirality theory, however, nothing is motionless; there are only events. The drawing of events can only be done in a makeshift way with arrows, which symbolize the time or the direction of motion. Or one draws, as in film, a temporal sequence of static pictures. A black hole is a three-dimensional space that is warped and bent into a four-dimensional one. Thus, to portray black holes in black holes, one needs many more than the two dimensions available on paper. Furthermore, these dimensions are to be interpreted depending upon the point of view of the observer, either as three-dimensional space or as linear time, which cannot really be portrayed anyway. Often such drawn geometrical models are overinterpreted because the theoretician is no longer aware of the extent to which his model deviates from observed nature. He is no longer practising physics, but only mathematics and from his mathematics, he draws the wrong conclusions about nature. One must tread carefully, therefore!

12.3 Quark parameters

Quarks can be described by the following five parameters, which we have already partially met with the neutrino and with the electron.

12.3.1 Rotational direction of the triangles (2 spin directions)

The spin is depicted in Figure 42. The direction of rotation of the curved triangles represents the chiral orientation of the three-dimensional space inside a black hole. It is the result of the fact that we can differentiate between left and right. The result of this chirality is the spin, which can be levo- (l) or dextrorotatory (r), negative (–) or positive (+). In the electron and in the meson, the orbital spin (L or R) is added to the spin of the individual neutrinos. The total spin of the quark or the meson is the sum of the spins of its particles on the one hand and the orbital spin on the other. Thus, the spin of the quarks is $\pm\ (\frac{1}{2}\ + n)$ and the spin of the mesons, each consisting of one quark and an antiquark, $\pm\ (0 + n)$, whereby n is a positive integer.

Depending upon the observer's point of view, the direction of spin can also be interpreted as a time direction. Looked at in this way, a particle with negative spin corresponds to a similar particle with positive spin, with which time runs in the opposite direction. Such particles are called **antiparticles.** The idea of portraying an antiparticle as a particle with which time runs in opposite direction also comes up with Capra (1991) pp. 181 f. The neutrino rotates levorotatory, i. e., its spin is $-\frac{1}{2}$ whereas the antineutrino is dextrorotatory with spin $+\frac{1}{2}$.

With charged leptons and quarks, the situation is more complicated. To the spin of the neutrinos (r or l), the orbital spin of its chiral, relative motion (R or L) is added. The total spin of the charged leptons and the quarks is the sum of all partial spins r, l, R and L. Here we are dealing with black holes in black holes, thus with two different levels, to which different time directions also apply as well. Notably, on each level 2 time directions are possible, i. e., in total $2 \cdot 2 = 4$ time directions. This corresponds to a two-dimensional time, which could be represented mathematically by complex numbers. However, only one of the two times is directly perceivable by the real observer. The other time is concealed in the black hole and is not directly accessible. More about this in Section 12.3.3! I remind the reader in connection to this that Hawking (2001) pp. 58–69 and 90, has suggested such an imaginary time with his cosmological model. He may have come to this idea along a completely different path, but his theory also has something to do with black holes.

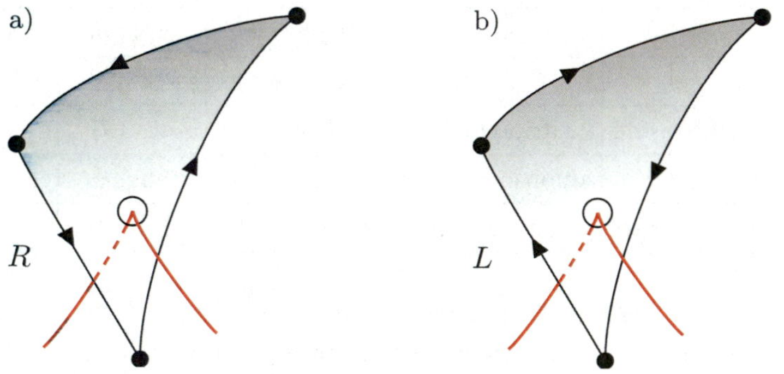

Figure 42. Orientation of the curved triangles (spin): *A quark is composed of 4 neutrinos with spin (r or l) which move along a chiral path (R or L). The total spin is the sum of the neutrinos' spin and the orbital spin.*

12.3.2 Topology of the events (3 flavours)

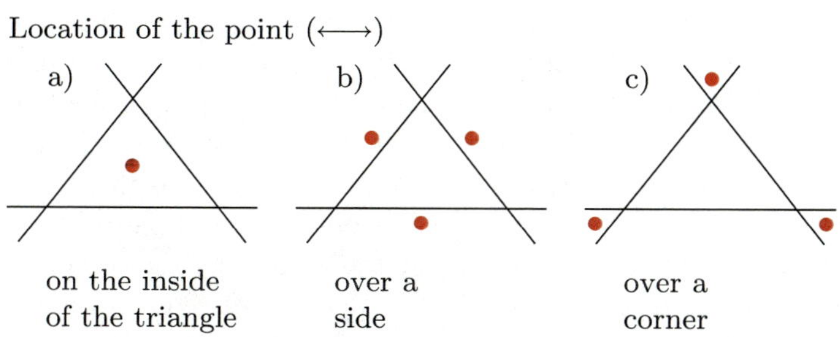

Figure 43. Topology of the event (flavor)

The (↔) point of a neutrino can be either inside the triangle, over a side or over a corner. The three variants are depicted in Figure 43. Corresponding to these, there are three different flavours. The two parameters spin and flavour we already find with the simple four-point space, the neutrino. Accordingly, there are $2 \cdot 3 = 6$ different neutrino/antineutrinos. The excited flavour states of the charged leptons and quarks occur when in these particles one of the 4 neutrinos is not in the flavour ground state. The different flavour states of the leptons and quarks have imaginative names at times, as can be seen in the following table:

	neutrino	lepton with	quark with charge	
		charge –1	$-\frac{1}{3}$	$+\frac{2}{3}$
$(\leftrightarrow)$ on the inside	neutrino (ν_e)	electron (e)	down (d)	up (u)
$(\leftrightarrow)$ over a side	muon-neutrino (ν_μ)	muon (μ)	strange (s)	charm (c)
$(\leftrightarrow)$ over a corner	tauon-neutrino (ν_τ)	tauon (τ)	bottom (b)	top (t)

Each of these particles has a corresponding antiparticle (anti-ν_e or anti-u etc.). As we have already seen in Section 9.9, the mass of a particle increases with its flavour from „$(\leftrightarrow)$ on the inside" to „$(\leftrightarrow)$ over a side" to „$(\leftrightarrow)$ over a corner".

12.3.3 Time direction (2 signs of the electrical charge)

In the axiom of chirality $\Rightarrow (\mapsto) \Rightarrow (\hookleftarrow) \Rightarrow (|\!\leftarrow) \Rightarrow (\leftrightarrow)$ time is symbolised by the arrow $\Rightarrow$. Time runs phase by phase into the future. A quark's time, however, can also run in the direction opposite to the observer's time. The quark is then an antiquark. That can be represented by an opposing phase arrow $\Leftarrow$, which points to the past. Thus, the axiom of chirality becomes $\Leftarrow (\mapsto) \Leftarrow (\hookleftarrow) \Leftarrow (|\!\leftarrow) \Leftarrow (\leftrightarrow)$. In this way, a new parameter for quarks emerges, which is portrayed in Figure 44 (p. 297).

Depending upon the time direction, i. e., depending upon the orientation of the orbital spin of the neutrino paths, the sequence of events of the charged particle, seen from the outside, is different and its electrical charge (as described in Chapter 11.7) has an outward-working positive or negative effect. If the „curvature" is complete, then a new black hole develops with a new time. That is the case with all charged leptons, e. g., with the electron. With each event, the electron – apart from the internal distribution of the neutrino spins – arrives back in its initial state. One can also say, the electron turns a full –360° with each event. For these cases we define the value of the electrical charge as –1. If the electron's space curvature is opposing, then the charged lepton turns +360°, its charge becomes positive and one calls it positron.

Orbital spin, time direction and electrical charge are not independent of each other. This concerns not three parameters, but only two of them.

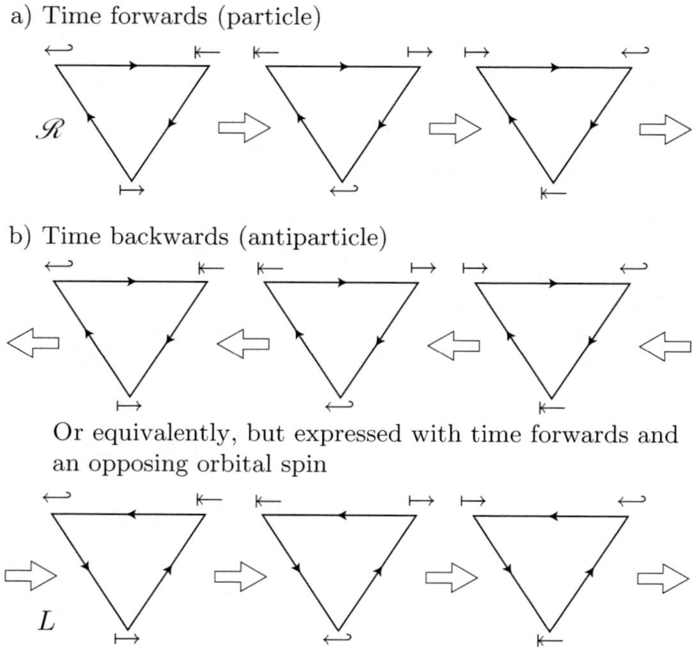

Figure 44. Time direction (orbital spin, particle or antiparticle)

If, for example, the orbital spin is *R* and the time direction is positive, then the electrical charge automatically becomes negative. Since physicists are used to only considering the – always positive – time direction of the observer as a basis, the two parameters are usually understood not as time direction, but as spin and charge direction. We find the three parameters described so far with all charged leptons. Accordingly, there are $2 \cdot 3 \cdot 2 = 12$ different charged leptons.

12.3.4 Phase of the triangle rotation (3 colours)

With the triangles, which symbolize the quarks, the three corners differ by their motion states (↦), (↩) or (⇤). The (↔) states, which are always also part of the other quark, thus change in three phases from (↦) to (↩) to (⇤). In chirality theory, these three states of the quark correspond to the colours in the standard model. I define them, based on the $(\leftrightarrow)$ state of the neutrino as follows (Figure 45):

State of the $(\leftrightarrow)$ neutrino of a quark	colour of this quark
$\Rightarrow (\mapsto) \Rightarrow$	red
$\Rightarrow (\hookleftarrow) \Rightarrow$	blue
$\Rightarrow (\mapsfrom) \Rightarrow$	green

As we have seen with the spin and the electrical charge, the arrow of time can also show an opposing direction with the colour. Thus, not only the orbital spin and the electrical charge turn into their opposites, but the positive colour becomes also the negative anti-colour:

State of the $(\leftrightarrow)$ neutrino of the antiquark	colour of this antiquark
$\Leftarrow (\mapsto) \Leftarrow$	antired
$\Leftarrow (\hookleftarrow) \Leftarrow$	antiblue
$\Leftarrow (\mapsfrom) \Leftarrow$	antigreen

Since the sign of the orbital spin, the sign of the electrical charge, and the sign of the colour on the same time level in the same black hole turn around with each other during the time reversal, the anti-colours are not new parameters which entail further quark variants, but – like the electrical charge – are simply a consequence of the positive or negative time direction. A parameter is just the time direction, positive or negative.

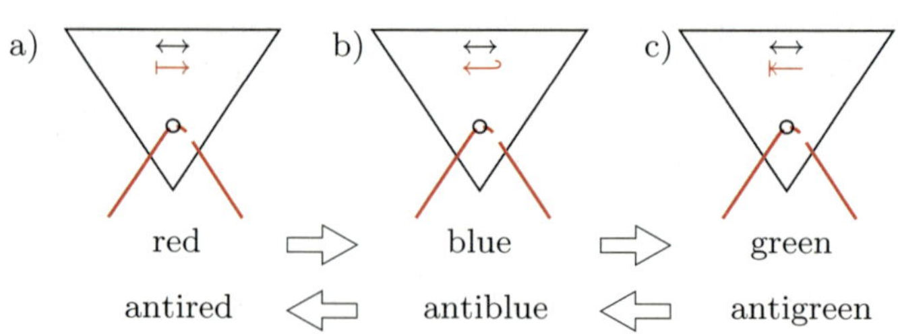

Figure 45. Phase of the triangle rotation (colour): *When time runs backwards, the colour becomes an anticolour and the particle becomes an antiparticle.*

12.3.5 Curvature of space: angle of rotation of the triangles (2 possible values of the electrical charge)

As already pointed out in 12.3.3, each event of a charged lepton can be described as a rotation through 360°, whereby the electrical charge resulting from it was defined as ± 1. Owing to the three colours, curvatures (or rotations), which are only fractions of 360° and accordingly effectuate smaller electrical charges, can now be defined for the quarks. Since the curvature no longer causes a complete 360°-rotation per event as with the charged lepton, the quarks are not black holes, but at most „nearly" black holes.

How high do the electrical charges of the quarks become? If one draws the quark as an equilateral triangle, then there are six possibilities for shifting the quark into its original state by rotation of the triangle in the plane around the ($\leftrightarrow$) point, i. e., by periodic events: one can rotate the triangle 3 times through 120°, 3 times through 240° or once through 360°, and in each case, alternatively clockwise or anti-clockwise. If, seen from the outside, one rotates the triangle clockwise ↻ through 360°, then the lepton, as a black hole with orbital spin *R* carries a negative charge, but if it is rotated anti-clockwise ↺, then it has a positive charge. The rotation through 360° corresponds to the complete inversion as it occurs with electrons or positrons. The charge is then ± 1. If the curvature is not complete, i. e., only a „nearly" black hole is present, then the electrical charge becomes lower than 1. With a small space curvature corresponding to a rotation through $\pm 120°$ per event, the charge is $\pm\frac{1}{3}$. In the case of medium curvature corresponding to a rotation through $\pm 240°$ per event, the charge is $\pm\frac{2}{3}$. The situation is depicted in Figure 46. These are not complete inversions so consequently, no new black holes develop. The quark – in contrast to the lepton – always needs three events in order to arrive back in its initial state.

The quarks change colour with each of these three events, indeed from perspective of the external observer, in the following sequence:

Curvature (Rotation per event)	Name	Charge	Colour
–120°	*d*-quark	$-\frac{1}{3}$	red ⇒ blue ⇒ green ⇒ red
+240°	*u*-quark	$+\frac{2}{3}$	red ⇒ blue ⇒ green ⇒ red

Additionally, there are the corresponding antiquarks, with which time runs in the opposing direction:

+120°	anti-*d*-quark	$+\frac{1}{3}$	antired ⇒ antiblue ⇒ antigreen ⇒ antired

or put differently but equivalently: red ⇐ blue ⇐ green ⇐ red

–240°	anti-*u*-quark	$-\frac{2}{3}$	antired ⇒ antiblue ⇒ antigreen ⇒ antired

or put differently but equivalently: red ⇐ blue ⇐ green ⇐ red

For the quark there are thus the following free parameters: 2 spin directions, 3 flavours, 2 time directions or signs of the electrical charges, 3 colours and 2 possible curvatures according to the angles of rotation for the triangles, i. e. values of the electrical charges. Accordingly, there are $2 \cdot 3 \cdot 2 \cdot 3 \cdot 2 = 72$ different quarks and antiquarks.

Usually, however, physicists speak only of 6 or 12 different quarks, because they do not take into account that there are 2 spin directions, nor that although the 3 colours are not directly perceivable, they are nevertheless present somehow, nor that each quark has a corresponding antiquark.

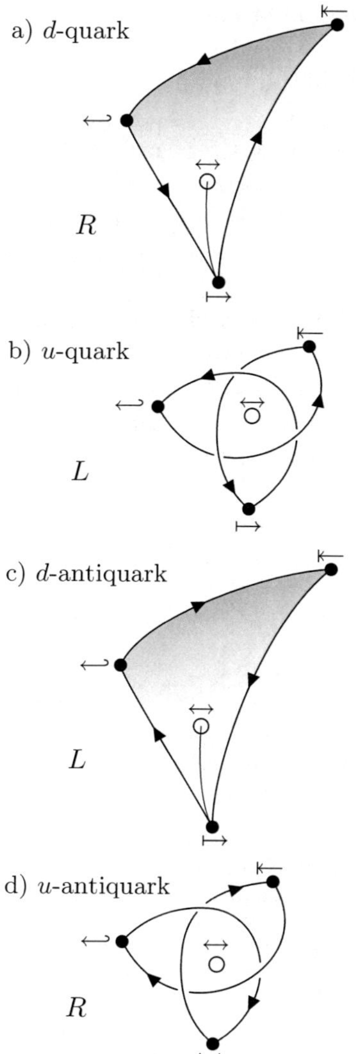

a) Weak curvature corresponding to a rotation through $-120°$ per event. Electrical charge $= -\frac{1}{3}$

b) Medium curvature corresponding to a rotation through $+\ 240°$ per event. Electrical charge $= +\ \frac{2}{3}$

c) Weak curvature corresponding to a rotation through $+120°$ per event. Electrical charge $= +\ \frac{1}{3}$

d) Medium curvature corresponding to a rotation through $-240°$ per event. Electrical charge $= -\ \frac{2}{3}$

Figure 46. Curvature of space corresponding to the rotation of the triangle per event (value of the electrical charge)

12.4 Rules for the combination of quarks

To begin with, I will once more define the most important symbols.

⇒ Time runs – as seen by the external observer – forwards, that is, in the same direction as the time of this observer's inner clock. The sequence of events is (⇤) ⇒ (↔) ⇒ (↦) ⇒ (↩) ⇒ (⇤), or red ⇒ blue ⇒ green ⇒ red.

⇐ Time runs – as seen by the external observer – backwards, that is, in the opposite direction from the time of this observer's inner clock. The sequence of events is (⇤) ⇐ (↔) ⇐ (↦) ⇐ (↩) ⇐ (⇤), or red ⇐ blue ⇐ green ⇐ red. One can also portray the same situation thus: antired ⇒ antiblue ⇒ antigreen ⇒ antired.

r The spin of the internal black hole, i.e., of the neutrino, is dextrorotatory, or $+\frac{1}{2}$.

l The spin of the internal black hole, i.e., of the neutrino, is levorotatory, or $-\frac{1}{2}$.

R The orbital spin of the external black hole, i.e., the neutrino path is dextrorotatory or $+\frac{1}{2}$. The electrical charge is negative.

L The orbital spin of the external black hole, i.e., the neutrino path, is levorotatory or $-\frac{1}{2}$. The electrical charge is positive.

colour	state of (↔)	colour	
⇒ (↦) ⇒	red	⇐ (↦) ⇐	antired
⇒ (↩) ⇒	blue	⇐ (↩) ⇐	antiblue
⇒ (⇤) ⇒	green	⇐ (⇤) ⇐	antigreen

The following rules apply to the combination or entanglement of quarks:

12.4.1 Per (partial) electrical charge four neutrinos are required.

12.4.2 The phases of all quarks must agree. Only thus can the new construct return after three phases to the initial state and become a particle in the process.

12.4.3 The newly constructed particle must be without colour. A relatively easily understood mathematical explanation of this condition by gauge theory can be found in Moriyasu (1983), pp. 121–127. Only with such symmetrical multiple events does the relative motion of the at least six neutrinos involved remain ordered. This way, no neutrino can escape from the particle; the neutrinos remain „imprisoned“ in the particle, similar to in a black hole. Particles are colourless if their coloured phases are offset by anti-coloured phases, or if all three colours occur at the same time and change with each event.

12.4.4 The more symmetrical the particle, the more stable it becomes. Hence it follows that the total spin, the flavour-state and the electrical charge should be as small as possible.

12.5 Meson Structure

The mesons can now be constructed according to these four rules. 36 quarks and 36 antiquarks are available for this in each instance, which in itself yields $36 \cdot 36 = 1296$ possibilities. In reality, however, the number of the meson types is much larger, because *two* quarks and *two* antiquarks can also be connected to form a meson comprising four quarks. The relative motion of all these quarks, according to the special theory of relativity, additionally produces magnetic effects, indeed both electromagnetic as well as „colour magnetic“, which lend a fine structure to the meson spectrum. Furthermore, there can theoretically also be states with greater spins and electrical charges which are improbable, however, since they would require very high levels of energy. Even so, excited meson states often occur, with which the orbital movements of the neutrinos or the quarks exhibit higher frequencies. It remains requisite for a meson state that the meson, after a periodic number of events, arrives back in its initial state. The professors do not require of their physics students that they learn all these possible meson types by heart. If students take pleasure in lots of different types, then perhaps they would be better off studying botany. On the other hand, the number of possible variants is again limited by the four rules. I content myself with the portrayal of the π-mesons. These are the simplest mesons in their ground state. The lightest and most symmetrical is the electrically neutral π^0 meson with charge and spin 0, followed by the π^{-1} meson with charge –1. Additionally, in Section 12.10, I will analyses the K^0-meson, which during its kaon decay apparently violates time symmetry.

The **π^0-meson** consists of a *u*- and an anti-*u*-quark, or of a *d*- and an anti-*d*-quark. If the quark and antiquark have opposing spins, then the meson's spin will be 0. If they have the same spin $\pm\frac{1}{2}$, then the meson's spin will accordingly be ± 1. Figure 47 (p. 306) shows a π^0 meson, comprising a *d*- and an anti-*d* quark. The quark's time runs forwards, like that of the observer, i. e., $\Rightarrow (\mapsto) \Rightarrow (\hookleftarrow) \Rightarrow (|\!\!\leftarrow)$, whereas that of the antiquark runs backwards, i. e., $\Leftarrow (\mapsto) \Leftarrow (\hookleftarrow) \Leftarrow (|\!\!\leftarrow)$. The rotational direction of the black quark-triangle is in this model, seen from the outside, i. e., by the external observer, clockwise through –120° per event. Therefore, the quark has an electrical charge of $-\frac{1}{3}$. The rotational direction of the red anti-quark-triangle, seen from the outside, is also clockwise through -120° per event. Since its time runs counter to that of the observer's, however, the rotation of the antiquark, seen from the observer, corresponds to an anti-clockwise rotation through +120° per event. As a result, the antiquark has an electrical charge of $+\frac{1}{3}$. Due to the opposing rotational directions of the quark and antiquark, the colour of the mesons remains neutral in every phase: green is offset by anti-green, blue by anti-blue and red by anti-red. After 3 phases the meson is always back in its original state. For the sake of clarity, the spins of the neutrinos are not represented in this drawing.

If the spins of the individual quarks as well as those of the mesons as a whole are as small as possible, then the following spin distribution results. The meson has spin 0. Since the quark carries a negative charge, its orbital spin is *R*. For the quark spin to become as small as possible and at the same time, the spin distribution in the quark to be as symmetrical as possible, the 4 neutrinos of the quark must carry the spins *rrll*. Thus, the quark's spin *Rrrll* is also $+\frac{1}{2}$. Similarly, the spins of the antiquarks must be the reverse, i.e., *Lllrr* = $-\frac{1}{2}$.

Naturally, in the meson two neutrinos are part of both the quark and the antiquark. Thus, the meson's spin becomes *RrllLlrr* = 0 and it remains 0 with every event. If the spin distribution of the meson is such that the total spin becomes 1, then its mass is somewhat greater than that of the π^0-meson and that meson is called the rho meson (ρ^0).

One must first get used to the fact that the two different quark and antiquark times can run in opposite directions in the same meson. That is difficult to imagine. The external observer, who only directly observes the time direction of his own inner clock, sees the meson more as portrayed in Figure 47b (p. 307). Here the antiquark's time runs in same direction as that of the quark, but the orbital spin of the antiquark is reversed. Figures 47b and 47a are physically equivalent.

The $\boldsymbol{\pi^{-1}}$**-meson** is composed of a *d*- and an anti-*u*-quark (Figure 48, p. 308). The *d*-quark has the same characteristics as in the π^0-meson. In the red antiquark, time runs backwards. The rotation of the red triangle of the anti-*u*-quark per event is +240°. Since, however, its time runs backwards, the rotation as seen by the observer is -240° and accordingly, the electrical charge of the antiquark is $-\frac{2}{3}$. The π^{-1}-meson also can be portrayed in such a way that the times in the quark and the antiquark run in the same direction as the observer's inner clock (Figure 48b, p. 309). The π^{-1}-meson is without colour in every phase. Its electrical charge is $-\frac{1}{3}$ $-\frac{2}{3} = -1$. The orbital spins of both quarks of the π^{-1}-meson are *R*. If the meson has a total spin of 0, its spin distribution is thus *RllrRllr*.

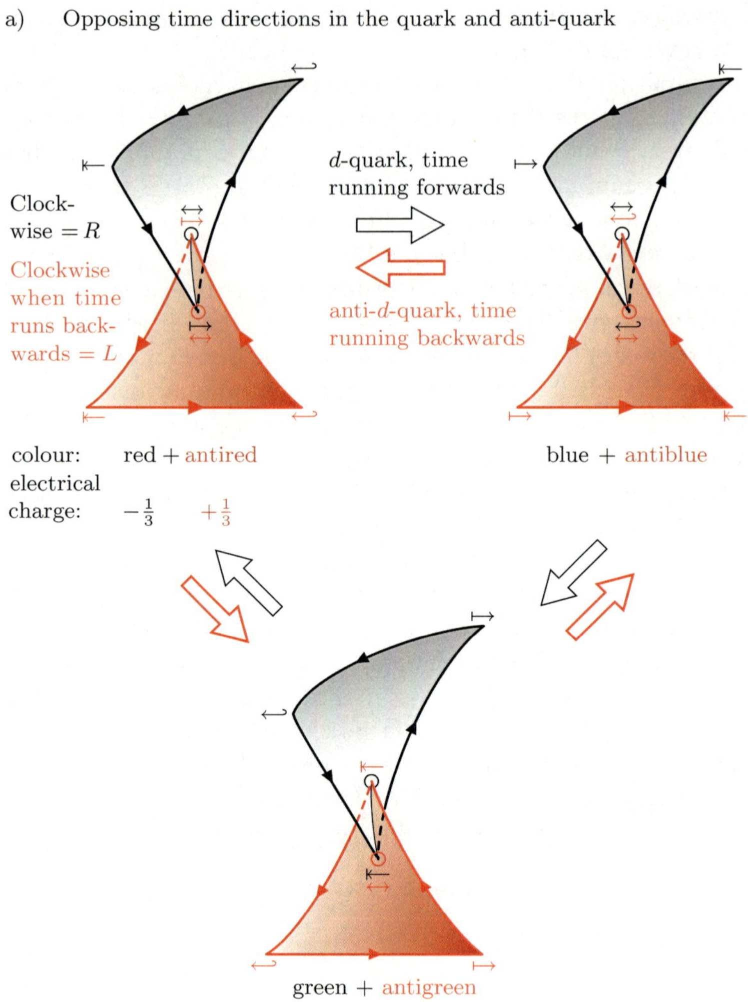

***Figure 47.** π⁰-meson consisting of a d- and an anti-d-quark, part a): a) Opposing time directions in the quark and anti-quark*

The black quark, seen from the outside, rotates clockwise through -120° per event and thus has a charge of $-\frac{1}{3}$. With the red anti-quark, time runs backwards and the sequence of its neutrino states is opposed to the orbital spin.

b) Same time direction in the quark and anti-quark. This is equivalent to a reversal of the orbital spin during the forwards-running time of the anti-quark.

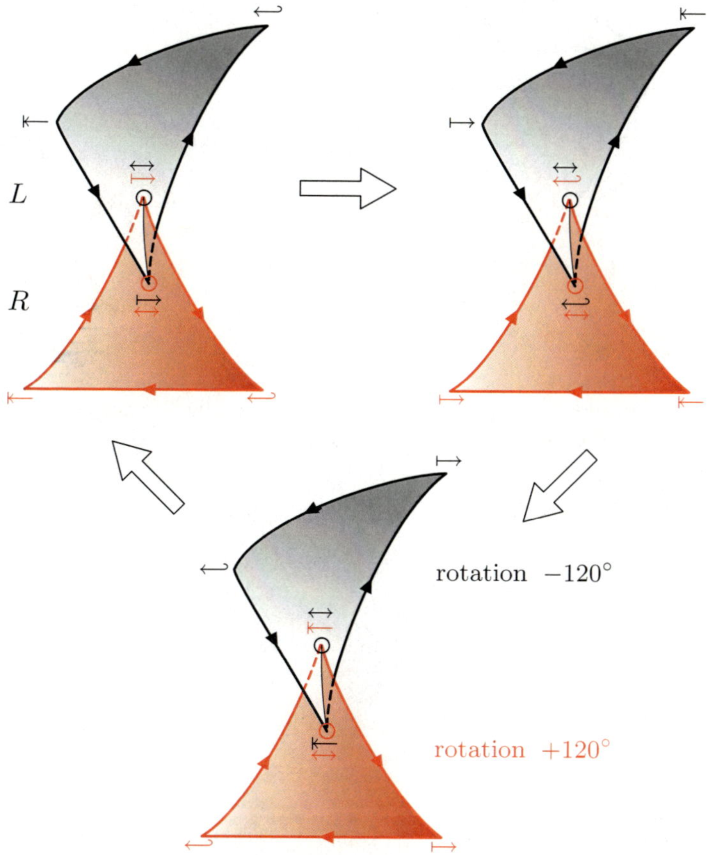

Figure 47. π⁰-meson consisting of a d- and an anti-d-quark, part b):
b) Same time direction in the quark and anti-quark. This is equivalent to a reversal of the orbital spin during the forwards-running time of the anti-quark. Following three events, the meson is back in its initial state. Its electrical charge is $-\frac{1}{3} + \frac{1}{3} = 0$ *and it is without colour.*

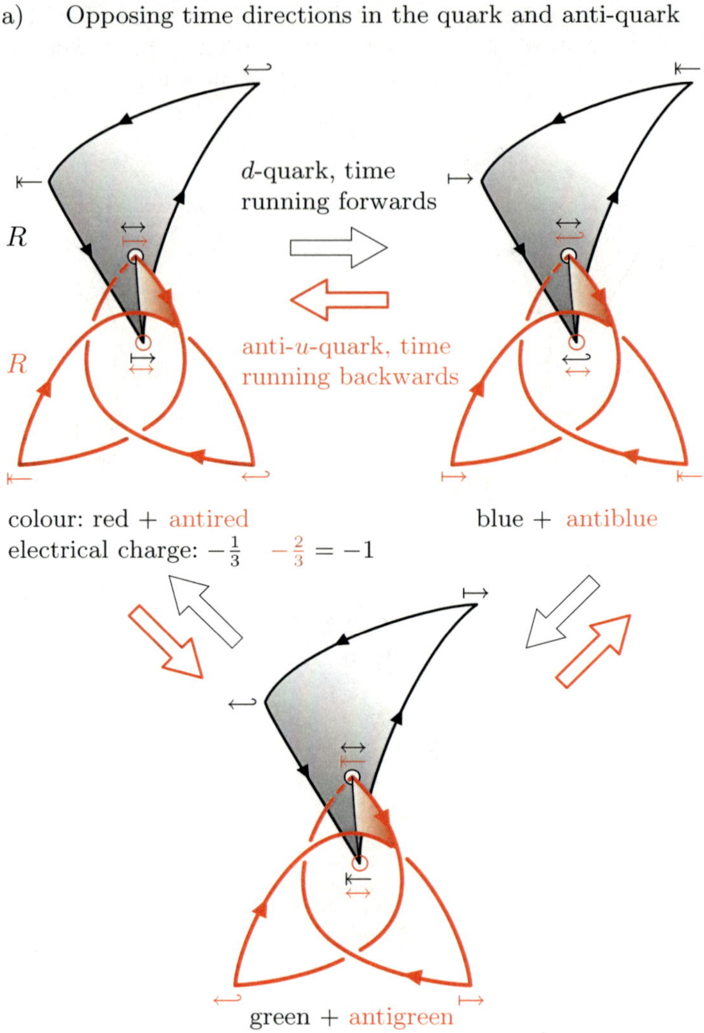

Figure 48. π⁻¹-meson consisting of a d- and an anti-u-quark, part a)
a.) Opposing time directions in the quark and anti-quark
The black d-quark rotates clockwise through -120° per event. Its electrical charge is $-\frac{1}{3}$. With the red u-quark, time runs backwards and the sequence of its neutrino states is opposed to the orbital spin. The rotation per event is -240° and its electrical charge is $-\frac{2}{3}$

b) Same time direction in the quark and anti-quark. This is equivalent to a reversal of the orbital spin during the forwards-running time of the anti-quark.

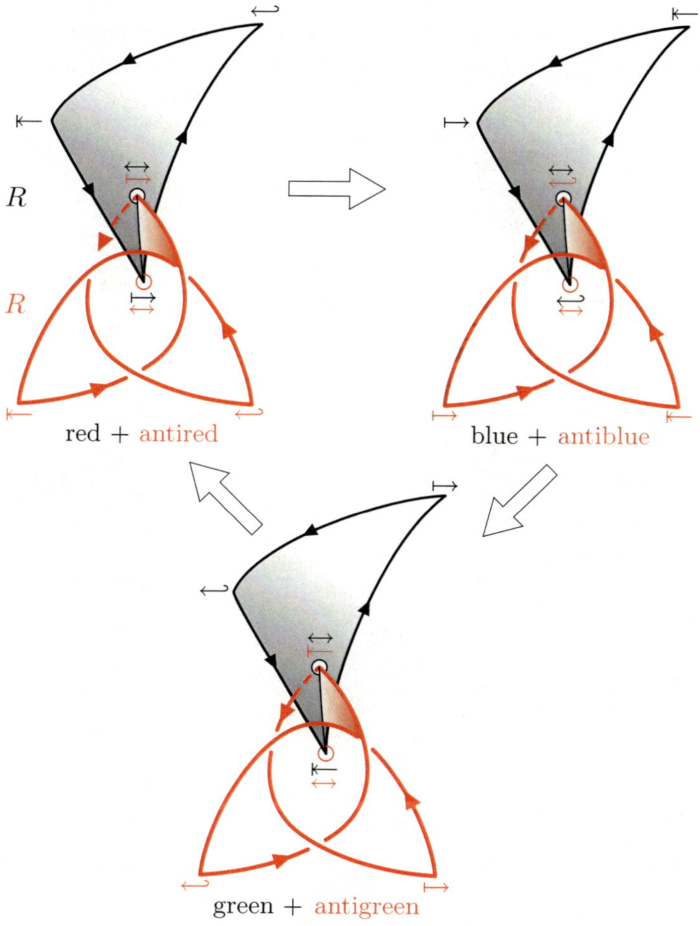

Figure 48. π^{-1}-meson consisting of a d- and an anti-u-quark, part b)
c) Same time directions in the quark and anti-quark. This is equivalent to a reversal of the orbital spin during the forwards-running time of the anti-quark.

As an illustration, yet another meson which is forbidden is briefly described, the so-called diquark, consisting of two d-quarks (Figure 49, p. 310). It cannot exist in isolation, since the colours of the quark are not offset in all phases by the anti-colours of an antiquark. Thus, the motions of the neutrinos in both quark and antiquark become so asymmetrical, that the meson already disintegrates after the first event.

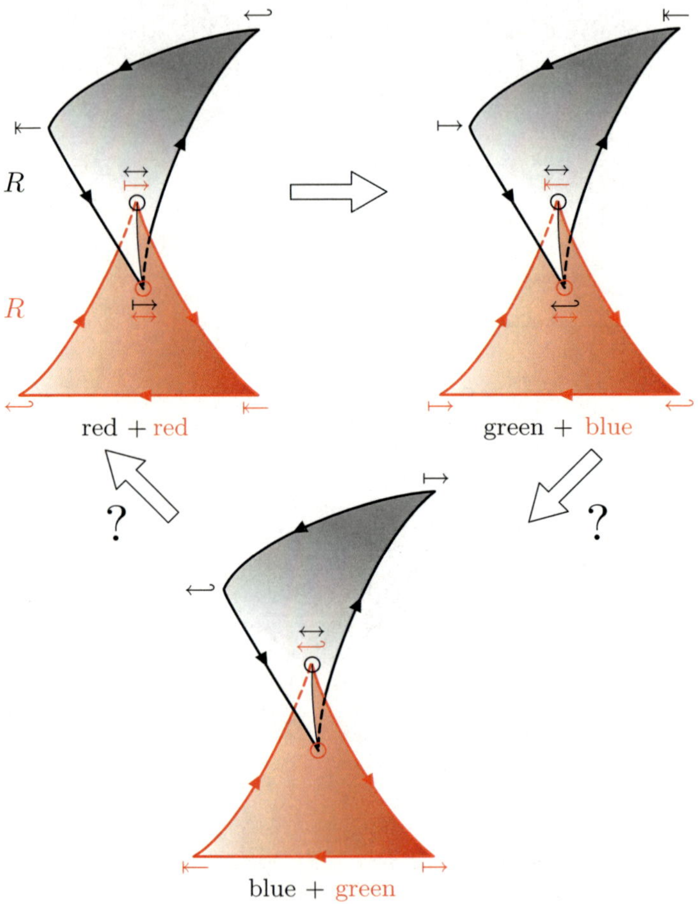

Figure 49. Forbidden diquark composed of two d-quarks: *A meson, comprised of two quarks, cannot exist. The relative motion of the neutrinos in the two quarks is so asymmetrical, that the meson disintegrates after the first event.*

12.6 Baryons

Baryons consist of three quarks or three antiquarks. In every phase, each of the three quarks is in the state red, blue or green, or anti-red, anti-blue or anti- green. Baryons are thus always colourless. Their colour charge is zero. Since the number of the quarks per baryon is larger than with the mesons, they are somewhat heavier than these, and there are even more baryon types than meson types. The lightest baryons are the proton, consisting of two *u*- and a *d*-quark with a half-life of more than $9 \cdot 10^{32}$ years [Klapdor-Kleingrothaus et al. (1997) p. 29] and the neutron, consisting of one *u*- and two *d*-quarks with a half-life of approximately one quarter of an hour. The proton has an electrical charge of +1 and the neutron is electrically neutral. Both have a spin of $\pm\frac{1}{2}$. There are also baryons, however, with other integral electrical charges and half-integer spins, for example the Δ^{++} with a charge of +2 and a spin of $\pm\frac{3}{2}$, which consists of three *u*-quarks. Moreover, particles comprising five quarks, the pentaquark, have been found [Wick (2004)]. They are made up of two *u*-, two *d*- and an anti-*s*-quark. It is unclear whether the pentaquarks concern the combination of a baryon with a meson, or the combination of two diquarks and one (anti)quark. Diquarks are energetically favourable pairings of two quarks or two antiquarks. They are not, however, to be regarded as colour neutral and neither therefore, as independent particles.

I limit myself here to describing the very stable proton (Figure 50, p. 312). Compared with the mesons, the baryons are simpler to imagine insofar as within a particle, there is only one time direction. This makes it possible for the individual neutrinos of the three entangled quarks to change at an event from one quark to the next. With the proton this is possible in the way that the proton after each event, one can also say in each phase, assumes a very similar state once more. That is probably the reason why the proton is much more stable than the most stable meson. The 3 quarks change thereby their colour with each event in turn, so that the proton as a whole always remains colourless. The angles of rotation of the quark triangles can be, as in the mesons, $\pm 120°$ or $\pm 240°$ with the resulting charges of $\pm\frac{1}{3}$ or $\pm\frac{2}{3}$. It is, however, not possible to unite quarks and antiquarks in the same baryon, since otherwise, the particle would no longer remain colourless with each event.

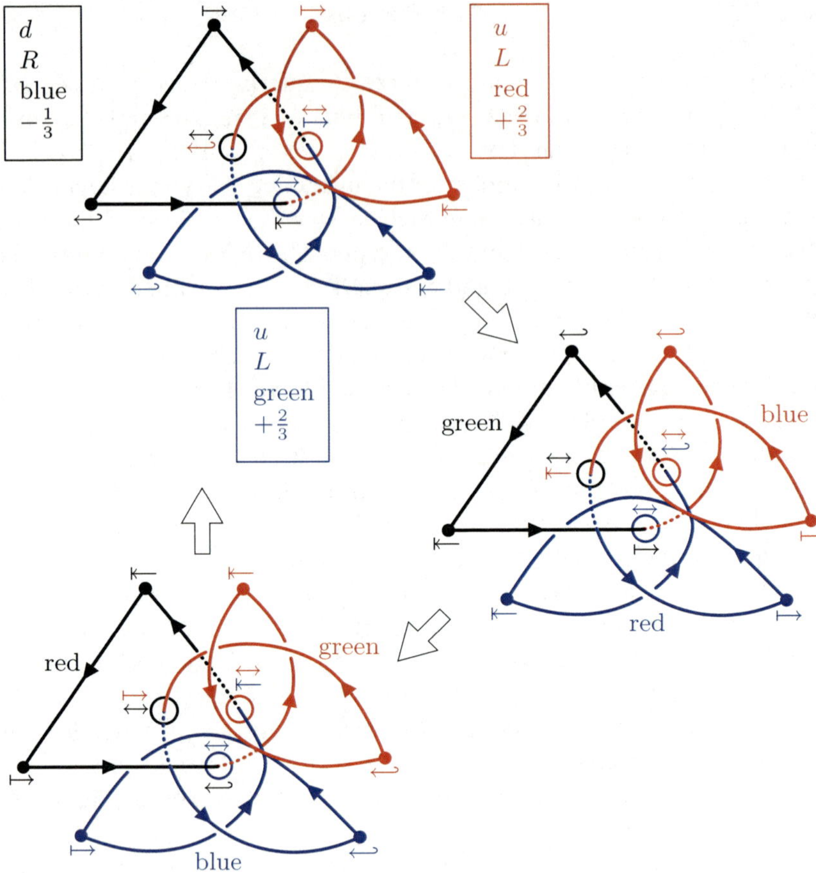

Figure 50. Proton: *The proton consists of two u-quarks and one d-quark, which change colour in turn in every phase. With every event moreover, 3 of the 9 neutrinos in total change the quark. The proton as a whole is colourless and has an electrical charge of* $\frac{2}{3} + \frac{2}{3} - \frac{1}{3} = 1$.

In conclusion, it is to be stressed once more that these illustrations of mesons and nucleons concern mathematical models which should by no means be over-interpreted. Indeed, the rather complicated Figure 50 showing the protons with the three entangled quarks and the somewhat confusing motions of the individual neutrinos may be quite pretty as a qualitative illustration, the proton in nature, however, is certainly rather different. Perhaps someone will one day come up with a more plausible representation.

12.7 Gluons

If one disregards the very weak force of gravitation, then the forces between two quarks come about quite similarly to those between the two charged leptons, i. e., as consequence of the relative movement of neutrinos. It is therefore no further surprise that the two forces can be described with a unified theory, the *G*rand *U*nified *T*heory GUT [Taylor (1989)]. The comparison of Figures 38 and 47 illustrates the analogy between the two kinds of interaction: particles, whose internal clocks run in opposing directions, attract each other. It is new with the quarks that because of their entanglement, the forces are only effective over short distances, and that this entanglement can be stable only if the events take place in three-time, whereby the particles become symmetrical and colourless. The colours, or rather, the **colour charges**, play the role in these forces which was played by the electrical charges in the electrical forces. The interaction between colour charges is called **strong interaction**. As is the case with electrical charges, identical charges repel each other. The theory of colour charges is called quantum chromodynamics (QCD) in contrast to quantum electrodynamics (QED). In the QED, I have called the difference between two states of an electrically charged pair of particles a photon. The **analogous difference between two states of a pair of quarks in the QCD is called a gluon**. The gluons are already contained implicitly in Figures 47–50.

The spin of the gluons is integral; they are thus bosons. Since in the QCD there are not only two different, opposing charges as in the QED, but three colours and anti-colours, there are, as a difference between colour states, eight possible variants of gluons [Georgi (1989b) pp. 425 ff.]. Furthermore, the gluons differ from the uncharged photons in that they carry a colour charge themselves (and at the same time an anti-colour charge [Moriyasu (1983) pp. 122 ff.]). Only so can the colour of a quark change during the interaction. As a consequence of that, there should be a QCD interaction between two colour-carrying gluons, indeed new particles consisting of two gluons, the so-called glueballs [Moriyasu (1983) pp. 71 f.]. It is assumed that about half of the proton mass is due to the quarks, the other half from the gluon energy [Close (1989) pp. 396 ff.]. As is the case with gravitons and photons, there is a real and a virtual variant with similar characteristics with gluons as well. Virtual gluons can transmit neither information nor energy, but only an action and a momentum.

Experiments show that gluons are only effective at distances of approximately 10^{-29} cm. This is not surprising, since two quarks must be entangled in order for the interaction to be possible, so the distances between the neutrinos must be very short.

12.8 Rules for the formation and decay of particles

All fermions discussed up to now are comprised of neutrinos. The mesons also, as bosons, are nothing other than a combination of neutrinos. For the structure and the interactions of all these particles there are certain rules which I already discussed. There are rules for the transformation of particles as well. They are transformed, according to chirality theory, by the exchange of individual points or whole neutrinos. Usually, physicists call this exchange of components interaction as well. I will come to the exchange of individual points in Section 14.3, which deals with open questions, because this aspect of chirality theory is very speculative. The exchange of individual neutrinos, however, is experimentally well-founded [Schmitz (1997)]. It is called – somewhat unfortunately – weak interaction. Weak, because on the one hand, the interaction is effective only at distances of up to about 10^{-16} cm and on the other hand, because the interaction does not work continuously like the strong or the electromagnetic interactions and gravitation, instead the exchange of neutrinos in each case is a unique, not periodic event. As a rule, but not always, new kinds of particle develop in the process. It can also be that after the weak interaction, there are more neutrinos than before, which complicates the understanding of this kind of interaction.

Since neutrinos possess energy, and the energy of the total system must remain constant from the perspective of one and the same observer (see Section 10.10), the question arises as to where the energy for the additional neutrinos comes from and which rules apply to the formation of these neutrinos. As an energy source, the kinetic energy from the relative motion of the particles before the interaction or the orbital energy, i. e., the frequency of the neutrinos from the periodic motion within a particle come under consideration. If the kinetic energy of the relative particle motion before the collision is large, then whole showers of new particles can be formed from it, most of which rapidly disintegrate. If the orbital energy is sufficient, then a particle can disintegrate of its own accord, without a collision with an external object. This occurs during radioactive decay, for example the β-decay of the neutron (n^0) into a proton (p^+), an electron (e^-) and an antineutrino (anti-ν_e), where as many as four additional neutrinos in the shape of an electron and an antineutrino are formed. The observer sees the relative translation movement of the particles before the impact as a non-periodic sequence of events. If it comes to an abrupt end with the collision, then the event frequency of the translation changes into a frequency of periodic events, i. e., into rotations with spin. These are the new neutrinos. The momentum, as well as the angular momentum and the spin of the total system, must be conserved during the interaction (see Section 10.11). With the spin, that even applies separately to the spin *R/L* of the neutrino's orbit, i. e., the electrical charge, as well as to the spin *r/l* of the neutrinos themselves, because the two types of spin belong to black holes on different levels, i. e., a different order.

Thus, the mandatory rules for the formation of new particles have been given: all particle transformations are permitted, during which the energy, the momentum, the angular momentum, the spin and the electrical charge are entirely conserved. A further rule is that each system should be as symmetrical as possible. The more asymmetrical a particle is, the more inclined it is to disintegrate, since its internal motion is less periodic. This is the reason for the high, yet probably not absolute stability of proton and neutron, although these are not black holes.

12.9 Weak interactions

With the strong interactions, neutrinos are periodically exchanged between quarks within a quark group. Since the neutrino density of the universe is estimated at approximately 500 neutrinos per cm^3 [Fritzsch

(2000) p. 78], the question arises as to whether these free neutrinos can also be exchanged with neutrinos from the inside of quarks or of leptons as well, and what effect this would have on the particles. It must be taken into account in each case that a particle is equivalent to its antiparticle in the opposite time direction. The time reversal, from the observer's perspective, turns the sign of spin, electrical charge and colour charge of a particle, as explained in Section 12.3, into its opposite. In other words, a neutrino which moves into the future is equivalent to an antineutrino from the past. A neutrino, which is emitted during a particle decay is equivalent to an antineutrino which is absorbed with the same decay. An electron with spin $+\frac{1}{2}$ in forward running time is equivalent to a positron (the anti-electron) with spin $-\frac{1}{2}$ in backwards running time. Certainly, neutrinos in leptons and quarks can only be exchanged if the neutrinos come close enough to the particle when the distances are short. Experiments show that at approximately 10^{-16} cm this is the case, which corresponds to approximately the diameter of a proton.

A simple example of such a neutrino exchange is the transformation of the electron (e^-) into a muon (μ^-), the so-called „inverse μ-decay" (Figure 51a). For this, an electron neutrino (ν_e) in the electron must be exchanged for a muon neutrino (ν_μ) that comes from the outside. The ν_μ is absorbed, the ν_e emitted, the e^- thus becomes a μ^-. Physicists are used to expressing all interactions with the help of individual „particles" which transmit the interaction. For gravitation it is the graviton, for electromagnetism the photon and for strong interactions, the gluon. The transmission of the weak interaction is done via the so-called **vector boson *W*.** It is composed, according to chirality theory, of an absorbed and an emitted neutrino, which in coming and going, fly past each other. Due to their relative motion, they can, as a pair of neutrinos, also carry an electrical charge 1. The electrical charge must be integral, because only thus it is renormalisable and thereby observable. The charge is transported in a specific direction, in the above example as W^+ from ν_μ to e^-. Depending upon the kind of neutrinos and according to whether it concerns emission or absorption, W has a spin of ± 1 or 0.

The μ-decay can also run in reverse (Figure 51b). Here a W^+ is likewise transmitted. If the analogous interaction occurs with antiparticles, however, then a W^- is transmitted (Figure 51c). Here the μ^- disintegrates spontaneously, i. e., without colliding with another particle. That is possible because the energy of the μ^- is sufficient for this decay process, i. e., for the formation of an additional neutrino and an antineutrino.

It can be, however, that it comes to an elastic collision between e^- and ν, an impact upon which no particles are transformed (Figure 51d). One can then say that a W^+ and a W^- were transmitted simultaneously. The pertinent interaction particle $W^+{+}W^-$ is called a neutral vector boson Z^0. Its electrical charge is 0. According to classical theory, the vector bosons (similar to mesons) develop from one quark and an antiquark, namely the W^- from d + anti-u, the W^+ from u + anti-d, the Z^0 from d + anti-d or u + anti-u [Schmitz (1997) p. 44]. The effect of the Z^0 consists of the fact that e^- and ν repel each other without being transformed in the process. Elastic collisions are also possible between neutrinos and antineutrinos (Figure 51e) [Moriyasu (1983) p. 114]. The elastic collision of two neutrinos can be described mathematically in this way, but one can wonder about the extent to which such a complicated representation of this simple process between two neutrinos makes sense. Is the model really still a good representation of nature? The two leptons and two vector bosons in the transitional state are virtual. Virtual in this context means that they are fundamentally not observable, because they exist for too short a time. It is probable that they disintegrate before they can arrive back in their initial state after few internal events. Thus, it is not to be doubted that an elastic collision between two neutrinos is in principle possible, even if (to my knowledge) it has never been observed. Probably one must simply imagine that the two neutrinos, with their total of eight points, come so close that their space inversions overlap. Since no stable new particle can emerge from this, the two neutrinos separate rapidly after a relatively chaotic transitional state, whereby moment and spin must naturally be maintained.

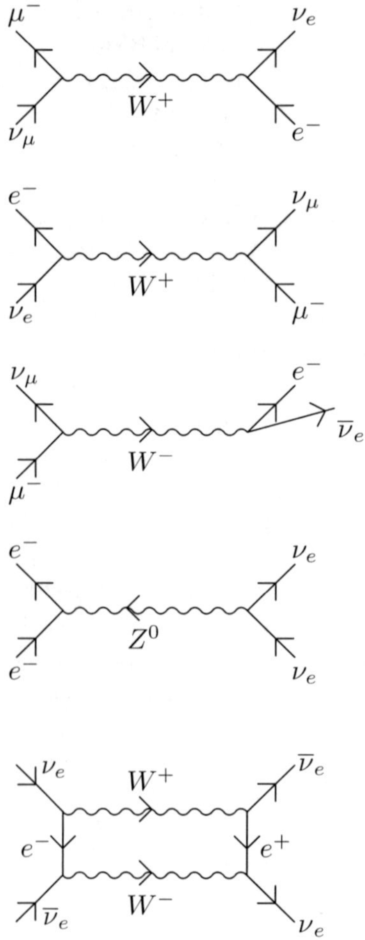

a) inverse μ-decay — Transformation of a e^- into a μ^- due to a collision between ν_μ and e^-

b) μ^--decay due to a collision between ν_e and μ^-

c) Spontaneous μ^--decay

d) elastic collision between e^- und ν_e. Mathematically $Z^0 = W^+ + W^-$. The arrow of the Z^0 signifies that a momentum is transmitted between the two particles.

e) elastic collision between a neutrino and an antineutrino

Figure 51. Weak interactions between leptons

Because the exchanged neutrinos are quite far apart during the weak interaction, which corresponds to a high event frequency, the mass of the *W* bosons becomes relatively large, i.e., corresponding to approximately 260 proton masses. The mass of the Z^0 bosons is somewhat larger still.

With quarks, similar weak interactions to those with leptons are possible [Schmitz (1997) pp. 43 f. and 72 ff.]. An example of this is the radioactive β-decay of the neutron (n^0) in a proton (p^+), an electron (e^-) and an antineutrino (Figures 52a and b, p. 320):

$$n^0 \Rightarrow p^+ + e^- + \text{anti-}\nu_e \text{ or } n^0 + \nu_e \Rightarrow p^+ + e^-.$$

The β-decay of the neutron, upon closer examination, is the transformation of a d-quark into a u-quark:

$$d^{-\frac{1}{3}} \Rightarrow u^{+\frac{2}{3}} + e^- + \text{anti-}\nu_e \text{ or } d^{-\frac{1}{3}} + \nu_e \Rightarrow u^{+\frac{2}{3}} + e^-.$$

Weak interactions cause the decay of the different pions (π^+, π^- and π^0) (Figures 52c, d and e).

As already shown in Section 4.7 in Figures 11–13 and repeated in Section 11.3.8, space parity is violated during the weak interaction, that is, particles with spin $+\frac{1}{2}$ (= right) react here completely differently from those with spin $-\frac{1}{2}$ (= left). The following particles cannot participate at all in weak interactions:

$$e^-_{\text{right}}\ ,\ e^+_{\text{left}}\ ,\ u^{-\frac{2}{3}}_{\text{right}}\ ,\ \text{anti-}u^{-\frac{2}{3}}_{\text{left}}\ ,\ d^{-\frac{1}{3}}_{\text{right}}\ ,\ \text{anti-}d^{+\frac{1}{3}}_{\text{left}}\ .$$

These are all particles with spin $+\frac{1}{2}$ and all antiparticles with spin $-\frac{1}{2}$. They are the so-called singlets. Here the spin distribution inside the particles is symmetrical, namely *Rrrll* or *Lllrr.*

The four doublets on the other hand, which produce weak interactions, are:

$$\begin{matrix} v_{\text{left}} & , & u^{+\frac{2}{3}}_{\text{left}} & , & e^+_{\text{right}} & , & \text{anti-}d^{+\frac{1}{3}}_{\text{right}} \\ e^-_{\text{left}} & , & d^{-\frac{1}{3}}_{\text{left}} & , & \text{anti-}v_{\text{right}} & , & \text{anti-}u^{-\frac{2}{3}}_{\text{right}}\ . \end{matrix}$$

The difference between the singlet and the doublet particles is that the singlet particles are substantially more symmetrical with respect to the spin distribution of their components, the neutrinos. The neutrino or the antineutrino on their own are completely asymmetrical spatially with their spins and occur therefore only in doublets. In the particles e^-_{left} , $d^{-\frac{2}{3}}_{\text{left}}$ and $u^{+\frac{2}{3}}_{\text{left}}$ spin distribution is *Rrlll.* In the antiparticles e^+_{right} , anti-$d^{+\frac{1}{3}}_{\text{right}}$ and anti-$u^{-\frac{2}{3}}_{\text{right}}$ spin distribution is *Llrrr.* One notes that with all doublets, the upper particle has an electric charge, which is greater than that of the lower by 1. As a consequence, during a weak interaction with emission

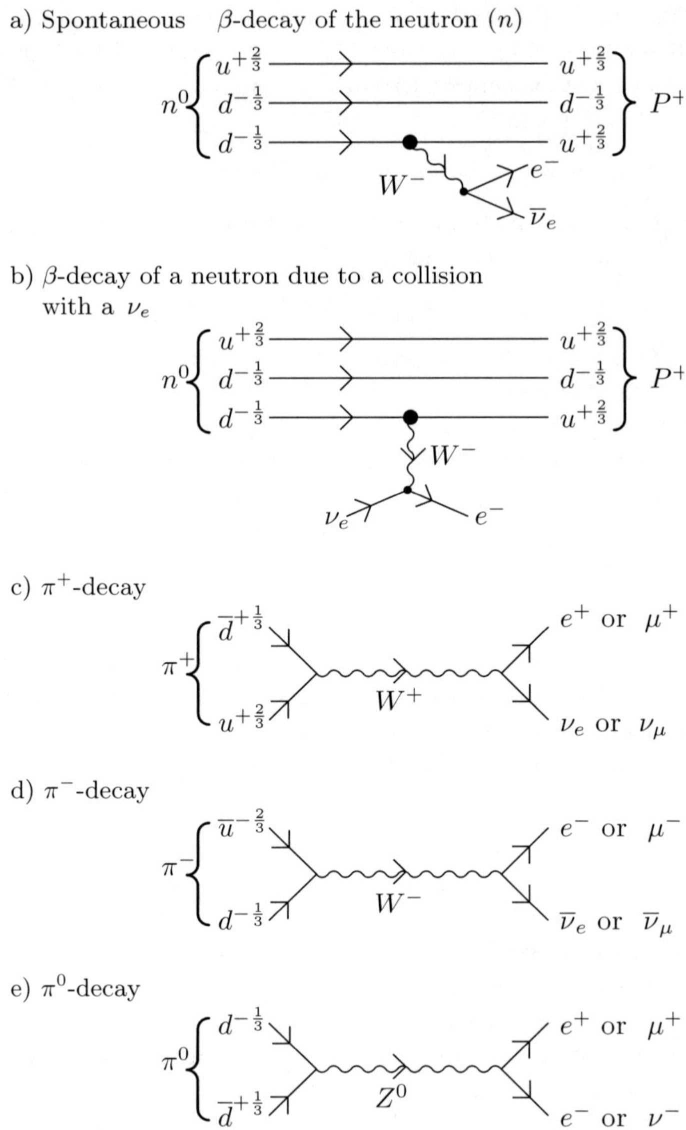

Figure 52. Weak interactions with quarks

of a W^-, the lower particle of the doublet changes into the upper; however, with the emission of a W^+, the upper becomes the lower, whereas all particles remain unchanged with the emission of a Z^0.

What, then, is the explanation for the violation of space parity and why does this only arise during weak interactions? The fundamental difference between the weak and all other interactions is that the weak interaction is a unique occurrence, for example the decay of a particle or the unique exchange of two neutrinos. With all other interactions, however, a continuous sequence of periodic events is involved. During the weak interactions the states of the particle before and after the decay can be distinguished clearly, whereby the time direction gains a prominent significance. This distinction is not only made by an external observer, but an internal observer can also differentiate the individual particles according to their state before and after the event. With all other interactions the objects involved do not change their state at all, or they return to their initial state again after a few events, so the time direction does not play an important role. It can no longer be said afterwards whether a particle has gotten into this state in a time which runs forwards or backwards. Only an external observer who can see all interacting particles ascertains a slight temporal change in the total system, because the distances between the particles change due to the interaction. As soon as the arrow of time plays a role during an interaction, however, it depends on whether one is dealing with particles in – seen from the observer's perspective – forwards-running time or with antiparticles in backwards-running time. Since through time reversal spin and electrical charge are also reversed, particles with positive spin or positive charge behave differently during weak interactions from those with negative spin or negative charge. Not until a particle's spin and electrical charge are both reversed does the particle react analogously to how it did before, simply in the opposite time-flow.

I have already stated in the discussion of electrical and the strong interactions that particles with the same direction of rotation repel each other, whereas those with opposing rotational direction attract each other. This applies – independently of the gravitation – apparently to neutrinos as well, at least if they are very close together. It means that two neutrinos with opposite spin „harmonise" or interlink like cogs, and thus form pairs if they are sufficiently close, which is the case on the inside of leptons and quarks. The electron with positive spin and the structure *Rrrll* can form two such pairs of neutrinos (*rl*) in its interior. In the electron with negative spin and the structure *Rrlll*, however, there is only one *r*-neutrino for a pair of neutrinos (*rl*), while two *l*-neutrinos remain and are available for the weak interaction. If an *l*-neutrino from the outside makes it into the inside of a lepton or a quark, then its periodic events are disturbed, indeed made impossible, since it only needs four neutrinos. The surplus *l*-neutrino will therefore immediately displace one of the two unpaired *l*-neutrinos, whereby the new particle becomes capable of periodic events again and thus stable.

Depending on the place in which, the direction in which and the velocity with which the absorbed neutrino penetrates the lepton or the quark, its angle of rotation per event, i. e., the electrical charge of the particle concerned can also change thereby. If with the collision there is still sufficient kinetic energy available, then additional neutrinos, whole leptons or quarks can develop.

That is not possible with the symmetrical leptons and quarks of the singlets with the spin distribution *Lllrr* or *Rrrll*. All neutrinos are paired and do not feel the weak interaction.

12.10 Kaon decay and time symmetry

It was already pointed out during the discussion of the arrow of time in Section 11.3.8 that the decay of the electrically neutral Kaons K^0 possibly violates time symmetry. It could be that a reaction $A \rightarrow B$ is not, or not completely, reversible. The K^0, however, is not a clearly defined particle with a clear structure. It consists of a combination of K^0 and anti-K^0, which are constantly transformed into one another by the exchange of virtual vector bosons. A particle is virtual according to Section 9.6 if it transfers an action to an object which cannot be observed by this object. So, the K^0 does not notice that it becomes an anti-K^0. It could only determine this by a comparison with another particle or antiparticle. Kaons are produced by making protons collide with antiprotons, a method by which theoretically, exactly as many K^0 should develop as anti-K^0. The K^0 consists of a *d*-quark and an anti-*s*-quark, the anti-K^0, however, of an anti-*d*-quark and an *s*-quark. According to chirality theory, with the reaction $K^0 \rightarrow$ anti-K^0, the ν_μ of the anti-s-quark is exchanged for a ν_e of the d-quark. Thus, the d-quark becomes an s-quark, the anti-s-quark becomes an anti-d-quark and the K^0 becomes an anti-K^0 (Figure 53). The apparent violation of the time symmetry has been indirectly deduced in different experiments since 1964, among other things from the observation that the reaction $K^0 \rightarrow$ anti-K^0 seems to be 0.66 % less probable than the reaction anti-$K^0 \rightarrow K^0$ [Wolschin (1999)].

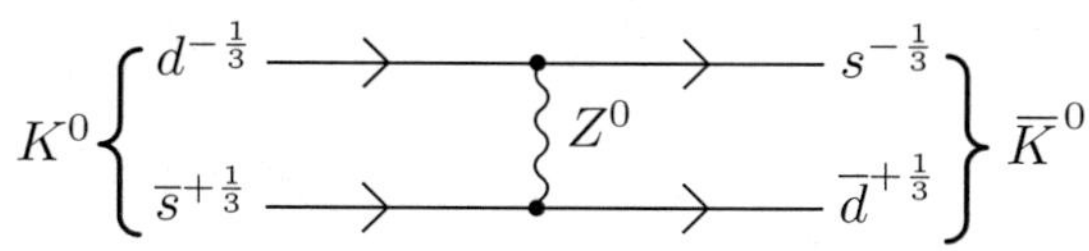

Figure 53. Oscillation of the K⁰-meson: *In this reaction inside the K^0, the Z^0 signifies the exchange of the ν_μ in the anti-s for a ν_e in the d, by which the K^0 becomes an anti-K^0.*

In Figures 54a) and b), two of the numerous possible kaon decays relevant to our discussion are portrayed. The K^0 decays into an observable π^- and an e^+, the anti-K^0 into a π^+ and an e^-. The neutrinos emitted in the process are not detectable. The relative frequency of K^0 and anti-K^0 can be inferred from the observable decay products. It can be seen from this that K^0 is clearly somewhat more frequent than anti-K^0. From this it is concluded that the reaction $K^0 \leftrightarrows$ anti-K^0 is not quite symmetrical.

Still, no allowance has been made yet for the spin distribution within K^0 and anti-K^0. The spin of the K^0 is always 0, i. e., one of the two quarks has spin $+\frac{1}{2}$, the other has spin $-\frac{1}{2}$. It remains open, however, in which of the two quarks the spin is $+\frac{1}{2}$ and in which it is $-\frac{1}{2}$. In the K^0, the d-quark spin can be $+\frac{1}{2}$. and the anti-s-quark spin can be $-\frac{1}{2}$. or vice versa. However, kaons only can participate in the weak interaction if the quark spin is $-\frac{1}{2}$ and the antiquark spin is $+\frac{1}{2}$, since only here are there no paired neutrinos. But even with this spin distribution between the two quarks there are still two different sub-variants which are portrayed in Figures 55a) and b). In 55a) we are dealing with an anti-ν_μ in 55b) with a ν_μ. With a) the anti-K^0 is the more probable and thus more frequent state, whereas with b) this is the K^0, indeed again due to the neutrino pairs formation, which is not possible with K^0 in a). and anti-K^0 in b). Nevertheless, the complicated situation is completely symmetrical. If, in all the particles

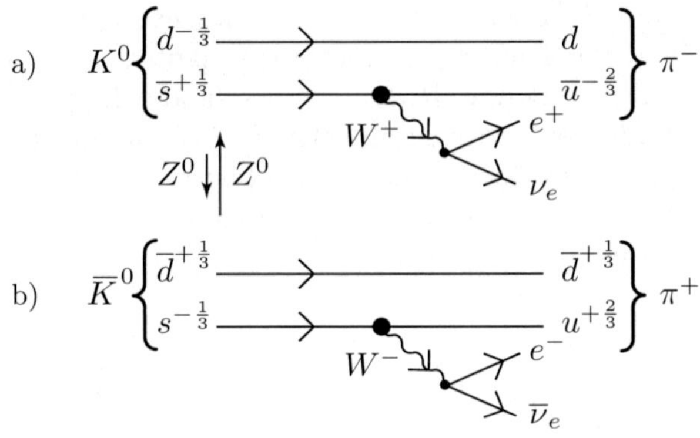

Figure 54. Violation of the time symmetry with K^0-decay: *a) is observed more frequently than b), from which the inference is made that anti-$K^0 \rightarrow K^0$ occurs more frequently than $K^0 \rightarrow$ anti-K^0. If no allowance is made for the spin distribution in K^0 and anti-K^0, the time symmetry would be violated during the reaction $K^0 \leftrightarrows$ anti-K^0.*

involved, one replaces spin $+\frac{1}{2}$ with spin $-\frac{1}{2}$ and at the same time, positive charges with negative, then K^0 turns to anti-K^0 and vice versa. It is not yet evident why the K^0-decay is observed somewhat more frequently than the anti-K^0-decay.

The violation of the time symmetry only shows up if one includes the real observer with his inner clock. This clock consists of matter, not anti-matter and therefore runs in the same time direction as the particles. The time direction of the antiparticles is the opposite of this direction. As soon as the time direction plays a role during an interaction, however, as it does with the kaon decay, the total situation becomes asymmetrical. With the kaon decay, three particles develop. It is improbable that in the reverse process these are simultaneously in the same location in order to form a kaon again. Past and future are fundamentally different, so for the kaon, the arrow of time is quite relevant. Thus, it is plausible that in the anti-K^0, the decay of the *s*-quark, whose time runs into the same direction as that of the observer, is perceived somewhat differently by this observer than the decay of an anti-*s*-quark in the K^0. However, the asymmetry is only simply, and to some extent, directly demonstrable in cases like the K^0, which spontaneously transforms into an anti-K^0. Thus, the kaon de-cay is not actually asymmetrical in itself, but only a chiral observer's

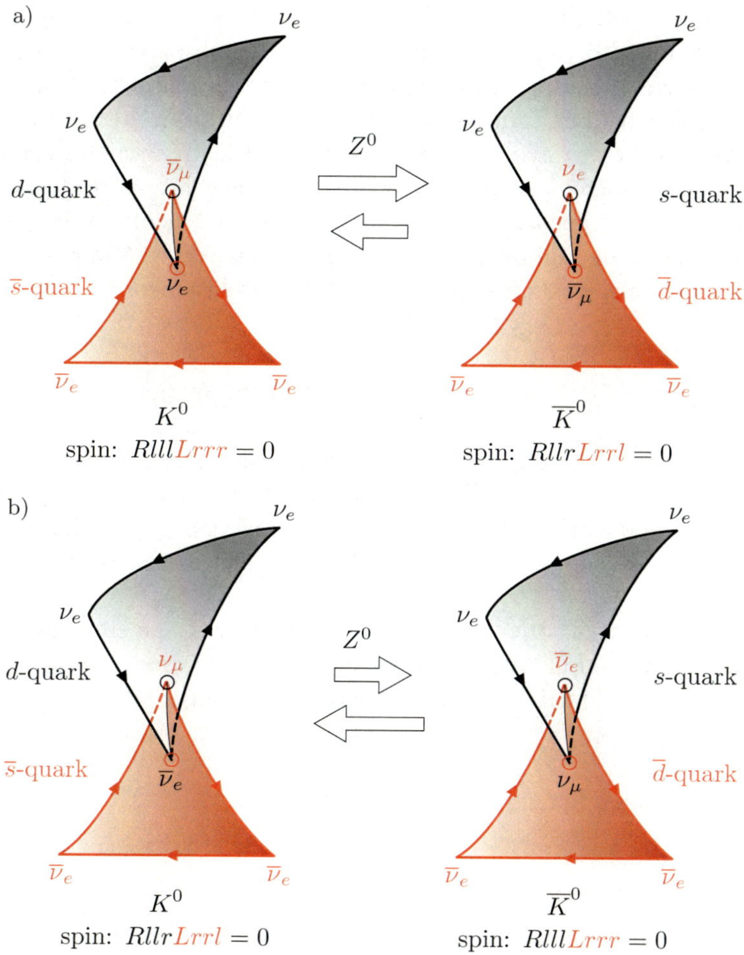

Figure 55. K^0-meson and anti-K^0-meson: *The neutrinos (ν) are levorotatory and the antineutrinos ($\bar{\nu}$) are dextrorotatory. The spin of the (black) quark is Rrlll = $-\frac{1}{2}$, the spin of the (red) anti-quark is Llrrr = $+\frac{1}{2}$. Nevertheless, the spin structures of the K^0 and the anti-K^0 in a) and b) are different. a) involves an anti-$\bar{\nu}_u$ whereas b) involves ^a ν_u. With a) the anti-K^0 with the anti-$\bar{\nu}_u$ is more stable, whereas with b), the K^0 with the ν_u is more stable because here the spin distribution enables a pair formation between the two neutrinos in each of the two quarks.*
Not shown are the kaons with spin distribution Rrrll and Lllrr, since due to the neutrino pairing, these quarks do not participate in the weak interaction.

perception of it. The arrow of time of the observer's clock points in the universe from the Big Bang into the future. An observer outside of our universe sees the universe as simply a black hole. For him, the time in the black hole would be completely symmetrical, but from the outside, he is unable to observe any events at all, such as the kaon decay for example. Presumably, such violations of the time symmetry are, in principle, observable with other decays also. The half-life of a neutron should, for example, deviate slightly from that of an antineutron with opposite spin. That would also be an experimental test of chirality theory.

12.11 Comparison of the four interactions

In the chaos of the Big Bang there was no order and thus also no time. The order resulted gradually from a series of symmetry breakings. With each symmetry breaking, a new kind of interaction occurred, up to the current state of the universe. Table 2 gives an overview of these interactions.

The nature of the four known interactions, namely gravitation, strong, weak and electromagnetic interactions, is compared in Table 3. The basis for all interactions is the axiom of chirality.

Table 2: Development of the four interactions

Chaos of points (Big Bang): no interactions

1. Symmetry breaking: formation of four-point spaces (neutrinos) as mini black holes

Theory of everything (TOE): quantum gravitation, attraction only

2. Symmetry breaking: distinction between

orderless neutrino paths	*and ordered neutrino paths. Black holes in the back hole with opposing eigentimes*
Gravitation Interaction between masses	Grand Unified Theory (GUT) Interaction between charges

3. Symmetry breaking: distinction between

entangled quarks	*and free electrons*
strong interaction between colour charges	electroweak interaction between (free or bound) particles

4. Symmetry breaking: distinction between

non-pair-ordered neutrion path	*and pair-ordered neutrion path*
electrical interaction between electrically charged particles without an exchange of neutrinos	weak interactions: exchange of (free or bound) neutrinos

Table 3: Characteristics of the four interactions

Interaction	Gravitation	Strong Interaction	Weak Interaction	Electrical Interaction
Structure of interaction	2 tetra-hedrons each of 4 points	2 entangled tetrahedrons each of 4 neutrinos, i. e., total of 6 neutrinos	Exchange of 2 (free or bound) neutrinos	2 free tetra-hedrons each of 4 neutrinos
Action	Only attractive	Attractive or repulsive	Decay or repulsive	Attractive or repulsive
Fermions, be-tween which there is an in-teraction	Neutrinos, possibly as part of com-postite parti-cles	Quarks with colour charge (electrical charge not re-levant for in-teraction)	Asymmetrical particles (levorotatory particles or dextrorota-tory antipar-ticles)	Quarks and electrically charged lep-tons
Boson, which transmits the interaction	Graviton (spin 2, mass 0)	Gluon (spin 1, colour charge, mass 0)	Vector boson (spin 1, mass $>$ 0)	Photon (spin 1, mass 0)
Bosons, sub-jected to the interaction	All energy-bearing, i. e. not virtual	Gluon	Mesons	Mesons, charged vector bosons

Chapter 13

Comparison of Chirality Theory with other Theories

I would rather discover a single cause than become king of the Persians.
Democritus (ca. 460–370 BC)

13.1 Physical Theories

During the last 3000 years, humanity suggested many theories to describe the world by natural laws. Whole libraries have been written. It is impossible to present an extensive comparison of all these theories and to name all their discoverers. Nevertheless, it is possible to compare the philosophical and mathematical methods and notions which were used. Such comparisons have been published among many others by Aristotle (384–322 b.c./1987), Capra (1991), Davies (1989), Einstein and Infeld (1950), Feynman (1965), Finkelstein (1996), Greene (1999), Grünbaum (1973), Kuhn (1967), Penrose (2004), Rovelli (2010), Smolin (2001), 't Hooft (2015), Weizsäcker (1985), Whitehead (1936) and by the author Wehrli (2008). All of them have influenced my chirality theory. Although there are fundamental differences between chirality theory and all the other theories there might be useful analogies which could be used to further develop other modern theories. To find a metaphysical strategy for the development of a physical theory is a difficult task. How this can be done starting from a formal language has been described by Döring and Isham (2007). They also have inspired my work although I could not follow all their ideas.

Of course, the strategy of chirality theory has not yet been completed and there are many open questions as is also the case for all the other theories. But a few fundamental principles that differ between chirality theory and many other physical theories shall be specified in this section.

13.2 Notions

13.2.1 Subject

The subject who observes and describes the universe might be
- a transcendental consciousness
- a subset of the universe which is also governed by the laws of nature
- a single ego
- a subject which can communicate with other subjects thereby giving rise to objectivity.

In chirality theory, one and only one subject with consciousness, the ego, exists. Since another subject never can unequivocally be differentiated from an object, every observation is actually subjective, even when the subject is not conscious of that. Every physical observation is a single and subjective one because an event can only be observed and counted once by one observer. One single subject simplifies the philosophical and consequently the mathematical description. In Chapters 7–9 the subject is mostly considered to be transcendental and not determined by the laws of nature; in Chapters 10–12 the subject is a real subset of the universe.

13.2.2 Universe and Multiverse

A physical observation is a transfer of information from the universe to a subject. Physical means that the observed system is directly or indirectly observable by empirical perception.

In some theories, also a multiverse plays a role:
- The quantum multiverse may be a transcendental set of possible outcomes of measurements or solutions of wave functions.
- It may be a set of individually different universes for every subject.
- There might be other universes which originate from „bubbles“ or new black holes in our own universe. Thus, our universe might be the interior of such a bubble or black hole.
- The multiverse may consist of other universes outside of our own.

In chirality theory, there is one and only one universe. This simplifies the theory.

13.2.3 Perception

The flow of information from the universe to the subject is called empirical perception. The perception might change both the universe as a whole and the subject. In some theories

- only part of the universe, namely the perceived objects might be changed by the perception
- or the universe remains completely unchanged
- or the transcendental subject remains unchanged
- or the subject as part of the universe remains unchanged.
- With extra-sensory perception (ESP) the information flow is not empirical.

In chirality theory, both the subject and the observed universe change with every perception. But there might be a transcendental „presumed observer" corresponding to the ego who is not changed by an information transfer. He perceives nature *per se*, i.e. not only real but also other existing elements of nature such as e.g. single points.

13.2.4 From language to (finite) quantity value objects and category theory

The physicist describes his observations („*Anschauung*") of the system S in his daily language $\mathcal{L}(S)$. The meaning of the signals, words or symbols of that language can in principle never be defined precisely. This causes a lot of confusion e.g. when scientists use words such as to be (Be-ing), reality, existence, truth, space, time, nature or measurement. In most articles, such notions are not defined at all.

To apply a theory of type Φ the physicist then translates the set of observed events $\varepsilon_{\varphi,S}$ into a mathematical language with quantity value objects $R_{\varphi,S}$, i.e. numbers:

$$\varepsilon_{\varphi,S} \rightarrow R_{\varphi,S}.$$

Number language is the most precise mathematical language for physics. But numbers theory is rather abstract. Thus, there are enormous unsolved mathematical difficulties in all modern theories such as quantum theory, loop quantum gravity, spin network theory, (super)-string theory, cosmology. One of the most difficult problems is to apply in principle not perceivable mathematical infinities to the always finite perceptions.

In chirality theory, terms such as to be (Be-ing), reality, existence, truth, space, time, nature or measurement are defined. The first representation of the observed events might be a graphical one with arrows as it is used with great success in chemistry. The advantage of such a language is that it is less abstract, but illustrative and easier to understand. The disadvantage is that calculations are not yet possible. Therefore, a procedure is looked for by which the graphs and arrows are further translated into a language with quantity values as it is done in most other theories. This aim is probably best accomplished by category theory with its topos:

$$\tau_{\varphi}(S) \text{ is the topos applied to } S.$$

13.2.5 Order and chirality

No information transfer, no language and therefore no physical theory can exist without a certain order. Zero order is chaos. This precondition for every physics is often neglected. It is seldom precisely defined what order means, where it comes from and how it is represented. The minimal condition for every order is that there is a duality of things that can be differentiated although they are characterized the same way, e.g. before and after, left and right, clockwise and counter-clockwise, positive and negative, yes and no. In most theories, there is simply a time that is thought to exist or even to be real, although never anybody has observed a time *per se*, one in which nothing happens. This (theoretical or real) time has a time arrow as the basis of order. Other theoretical basic notions for the duality are spin and (qu)bit.

In chirality theory, an explicit axiom of order (or chirality) is formulated. In set theory it reads

$$(P_n, P_m) \neq (P_m, P_n).$$

When reading the axiom, the physicist as an observer must know that there is a basic difference between left and right or between before and after respectively. For that the physicist himself has to be chiral – even if he is a transcendental subject. The origin of order in physics is the direction of the information flow from object to subject. This is best expressed by an arrow („if …, then …") as in category theory.

13.2.6 Infinity?

In all physical theories, numbers play an important role. In modern theories, this is the case for all types of numbers: natural, rational, surds, imaginary, finite and infinite numbers. Thus, number axioms are required for all theories.

For chirality theory, only finite rational numbers are required because all measurements are described by numbers of counted events. Thus, there are only few number axioms, namely the simple Peano axioms for finite numbers. If surds, imaginary or infinite numbers were introduced, e.g. to describe waves or information, this would be part of the mathematical theory and not an aspect of real nature. In chirality theory, there is no infinity.

13.2.7 Continuum and separation of the universe into objects

Most theories do not describe the universe as a whole but rather a subset called object. The separation of the universe into subsets is always an artificial one and this might falsify the description of a perception. Often the universe is described as a continuum, although it is much more difficult to define precise limits between objects in a continuum than if the universe consists of discrete sets. Also, the separation between subject and object is always an artificial one and leads to philosophical problems mainly in quantum theory, even when the observer is not part of the universe.

This is the case also in chirality theory, but this theory should be suitable for the description of the universe as a whole as well as of the perception of single objects such as elementary particles or black holes. In chirality theory, there is no continuum at all.

13.2.8 Background time and space

In most theories, time (and often also space) exist as a background for all objects as a prerequisite of the theory. Space and time might be real or not, but they exist. They are usually continuous, although neither space nor time nor a continuum can be observed as such. This way locality plays an important role for most laws of nature, although there are non-local phenomena in physics such as the interaction between entangled particles. But there are a few theories such as spin network theory where the discrete spin network is finite and has no space-time background. But even that theory is local.

In chirality theory, such problems do not arise at all because neither space nor time nor any continuum are prerequisites.

13.2.9 Local and non-local phenomena

In the last 80 years there has been much discussion but little agreement about the interpretation of non-local phenomena [Bell (1964)]. A convincing model to describe non-local phenomena is missing [Stanford Encyclopedia of Philosophy (2017)].

In chirality theory information transfer is always local whereas action transfer is always non-local.

13.2.10 Ontological real and ontological basic objects

Objects in physics are subsets of the universe, e.g. galaxy super clusters, the sun, man, a measuring apparatus, atoms, photons or neutrinos. A modern definition of the term object may be: An object is a combination of quantities whose present values permit common predictions about these quantities (in the future). Scientists have looked for a basic object for thousands of years. They have suggested water (Thales), the atom (Democritus), elementary particles, space-time loops, strings, spins, qubits or an orthonormal basis of Hilbert space as the ontological basic object. Sometimes it is not clear whether the basis is something real or rather theoretical.

In chirality theory, a system S consists of subsets called objects. The basic ontological object is the point P. It is the smallest possible object. It is one, has no parts, cannot be divided, is identical with itself, has no internal properties and corresponds to a point in topology. The point exists but it cannot be empirically observed; so, it is not real and cannot be counted or measured. It seems that a physical theory without such a not observable entity is impossible. Every object S consists of $\{P_n\}$, where $n = 1, 2, 3, \ldots, n \neq \infty$. The model of the real and observable basic ontological object is the 4-point space.

13.2.11 Natural laws as relations between objects

The relation between different objects is expressed as a natural law. Such laws may define the relative position or state of the objects and the change of these positions and/or states. In most modern theories, this is accomplished by so called fields.

In chirality theory, all points are related to each other. The relation r is expressed by $r_{n,m} = (P_n, P_m)$.

In $\{P_n\}$ there are $n(n\text{-}1)$ relations r called distance without considering any relations of relations and so on. No fields are used in the model, only points with their relations to one another.

13.2.12 Superposition and individuality

In quantum theory, there are superpositions of states with vague ontological status.

In chirality theory, every single point is an individual ontological basic object: $P_n \neq P_m$. There are no superpositions. But the same P can be an element of many real ontological 4-point spaces at the same time and these spaces may overlap.

13.2.13 Events, states and the quantum h

As a rule, physical theories describe measurements as states of the observed system at a certain time. States may be probabilistic and they stay unchanged for a certain time. Most physicists consider states to be real aspects of nature. The Planck quantum h is the minimal possible angular momentum, the quantum of action, i. e., the measure of any change of a state.

In chirality theory, there are no existing real states. An event is not a state. Only events are observable and real. An event occurs when something changes whereas something other remains unchanged. Time is not required. An event occurs, when any P_m changes its relation to $P_{1,\ 2,\dots,\ i}$ by getting in a between relation to any of these other P. In chirality theory, the quantum h is introduced as the measure for the counting of events. If an event does not change the spin of an object, it is a periodic event and no information is transferred to the outside. If the event changes the spin of the object, a quantized information is transferred to the outside. Every such event changes the orientation of the observed object by a rational number $\mathbb{Q}$ of quanta h. This way only rational numbers are required in chirality theory.

13.2.14 The arrow of time

There are whole books written about the arrow of time. Most authors agree that one cause of the arrow is the second law of thermodynamics. But with the interpretation of modern theories such as special relativity theory (objects moving at light speed, twin paradox), quantum electrodynamics (electrons theoretically moving backward in time), quantum chromodynamics (antiparticles) it might be possible that time stands still or even flows backward, at least theoretically. And for the instantaneous (inter)action between entangled particles no time at all is required. The mathematical model for the description of the arrow of time is the set of the real numbers.

In chirality theory, it is the phase arrow $\Rightarrow$ of the axiom of chirality from which the arrow of time and the chirality of space are derived.

13.2.15 Elementary particles

Elementary particles are objects with well-defined intrinsic values which are multiple subsets of the universe. Some elementary particles are stable, others decay spontaneously. A particle may be composed of smaller particles. The terms well-defined, multiple, stable and spontaneously may differ between theories.

In chirality theory, an elementary particle is a (directly or indirectly) perceivable object S_{Pi}, consisting of a definite number of points and a perpetual internal process of periodic events that do not change S_{Pi} and its orientation.

13.2.16 3-dimensionality of classical space

In most theories, the 3-dimensionality of classical space is an axiom, but Weizsäcker (1986, pp. 379–412) with his Ur-theory and Finkelstein (1996, pp. 319–358) with his qubit-formalism with Pauli matrices independently discovered a theoretical explanation. In these theories, chirality is not derived from metaphysics but is introduced mathematically by the imaginary numbers. These are always chiral and have an arbitrary orientation.

In chirality theory particles are only possible as finite periodic processes with 3-dimensionality. At least 4 points in 4 different states relative to the other points are required to produce a periodic event. Such a 4-point space forms a 3-dimensional space. The 4-point space is the real ontological basic object in chirality theory whereas the single point itself exists but is not real. Classical space corresponds to the synthetic *a priori* space of Kant (1781/2003), the space required for every „*Anschauung*". This does not exclude a space of higher dimensionality in nature, but such a higher dimensional space cannot be observed or be a precondition of observation. Thus, in chirality theory there are $2 \cdot 3 = 6$ additional „concealed" space dimensions due to black mini-holes and black mini-holes within black mini-holes (see Section 11.1).

13.2.17 Information

Natural laws determine how information is changed in any system. The quantity of information is the number of bits of the system. One bit is a binary alternative of a yes/no-question. In modern physical theories, the quantity of information corresponds to the energy. Sometimes there is not a clear distinction made between the quantity of information of the system, i. e. the number of yes/no-questions, and the content of that information i. e. the answers to the yes/no-questions. According to special relativity theory information cannot travel faster than light. Many physicists think that this must also be the case for the velocity of any action transfer.

In chirality theory information is related to the number of independent orientations in S_{Pi}. With each periodic event of an object in S_{Pi} neither the content of the information, i. e. the orientation of any subset of 4 points, nor the quantity of the information changes. The energy of S_{Pi} remains the same.

But if an event is not periodic, then orientations of some 4-point-spaces are changed and some information and energy is thereby transferred. Every point is an element of all possible 4-point spaces of the universe containing this point, i. e. many of these spaces overlap. Also, in chirality theory, information cannot travel faster than light, but the non-local (inter)action without any information transfer is instantaneous or timeless with each event. If some of the 4 points of the 4-pointspaces which change their relative orientation are far away from each other, the action according to the axiom of chirality is non-local. According to the axiom of chirality, single points always travel from event to event with light speed. As long as no point has travelled from one subsystem to another one, no information is transferred to the other system. For the action transfer on the other hand this is not required because it is simply a consequence of the timeless change of the configuration and the relative states of 4 points according to the axiom of chirality (Figures 27 and 28). This difference between chirality theory and all the others probably solves the paradox of entangled particles and of the 2-slit-experiment, probably also that of the collapsing wave function. The bit- and qubit-formalisms of other information theories might be translated for their use in chirality theory.

13.2.18 Space-time, duration and distance

Duration and distance are important parameters of most physical theories. They are the parameters of time, space or space-time. The properties of these entities differ from theory to theory:

- Time and space may ontologically be real, unreal but existing or they are simple ideas.
- Time has one time arrow or two opposite time arrows.
- Space-time is an extended entity or not.
- Space-time is 4-dimensional or it has more than 4 partly concealed dimensions.
- Space-time might be a continuum or it consists of discrete elements.
- In most modern theories space-time is no more Euclidean but curved.
- Movements are either absolute or relative and subjective.
- With Immanuel Kant space is chiral.

Duration Δt is measured with clocks, distance Δl with a one-dimensional measure or as $c \cdot \Delta t = \Delta l$.

In chirality theory, space and time do not exist but they can be theoretically derived from the only entity event. Events always occur when a 4-point space inverts, that is when a point gets between 3 other points of the universe.

In chirality theory, the duration between two moments t_m and t_n of time is a counted number $\mathbb{Q}_{m,n}$ of periodic events. The basic measure is the intrinsic clock of the (real or presumed) observer. Duration is a rational number. If the observed system consists of more than 4 points, more than one 4-point space with its events has to be considered, and duration becomes not a natural but a rational number. The algorithm that determines the counting of the duration in such a system is not yet known. The time between two events in a single 4-point space is the shortest possible time which probably corresponds to the Planck time. It is defined to be 1 (counted event).

In chirality theory, the distance between two observed objects S_m and S_n is the counted number $\mathbb{Q}_{m,n}$ of non-periodic events that occur, when a single point travels by the shortest way from one object to the other. Distance is a rational number. If the observed system consists of more than 4 points, more than one 4-point space with its events has to be considered and duration becomes not a natural but a rational number. The algorithm that determines the counting of the distance in such a system is not yet known.

Chapter 14

Open Questions

> Understanding the world will never end for mankind with a final truth, rather it is an infinite process, for which we can imagine neither a beginning nor end, and which nevertheless moves ever onwards.
>
> Werner Heisenberg[62]

I can sum up this last chapter relatively briefly. First of all, questions are mostly shorter than answers, and secondly, I cannot, and do not at all want to, draw up a complete list of all important open questions.[63] I limit myself to a few cases and areas which appear particularly attractive to me or which are easily overlooked.

14.1 Mass

The acid test for any theory is the experiment. It can confirm the theory or disprove it. Are there experiments for chirality theory? Or is it only beautiful because it explains and connects things which were unexplainable and isolated up until now?

Why are the masses of particles the way they are? Presently there is no answer to this. Yes, it is actually not even known why certain particles have a mass and others have not. Naturally, one can sometimes estimate individual masses by extrapolation in the context of certain mathematical

[62] Werner Heisenberg: „Das Verstehen der Welt wird für die Menschheit nie mit einer endgültigen Wahrheit enden, sondern es ist ein unendlicher Prozess, von dem wir uns weder Anfang noch Ende denken können, und der doch immer voranschreitet.“ [Fischer (2002) p. 193]

[63] „With each scientific insight gained, new areas where knowledge is lacking emerge. Knowledge is a ball which floats in the universe of the ignorance.“ [Mittelstrass (2001) p. 125]

models before they have been measured. There is no actual physical explanation for it, however. To work around this, quantum physicists have postulated a very short-lived, electrically neutral „Higgs particle“, which is supposed primarily to have played a role shortly after the Big Bang, by spreading a scalar Higgs field which gave mass to other, massless particles. A scalar field lends a value or a size without a pertinent direction to a location in space-time. Masses do not have a direction. The Higgs particle has first been observed by CERN in Geneva [Biever (2012)]. Its mass is between 125 and 127 GeV/c^2. Since then, the particle has been shown to behave, interact, and decay in many of the ways predicted for Higgs particles by the Standard Model, as well as having even parity and zero spin, two fundamental attributes of a Higgs boson. This also means it is the first elementary scalar particle discovered in nature.

In chirality theory there is no place for a „particle“ which spreads a scalar field. But in this case, one needs a replacement for the explanation of the phenomena shortly after the Big Bang. This replacement might be a symmetry breaking, either the symmetry breaking which made the formation of neutrinos from the symmetrical chaos of the points after the Big Bang possible, or the next symmetry breaking which effectuated the formation of black holes in the black holes, i. e., of electrons (see Section 12.11, table 2). For a symmetry breaking to occur there must be a minimal density and a maximal temperature, i. e. a specific energy in a specific location. Possibly these are the conditions fulfilled by the Higgs.

In chirality theory, mass is a consequence of the periodic event frequencies. If one knows this frequency, then one knows the mass. It is the ratio of two numbers, the event number of the object and that of the observer's inner clock. For the electron neutrino, the ratio according to the definition of chirality is one. Smaller masses are not possible. In order to calculate the masses of other particles, for example of the muon neutrino, the theory would have to explain how the frequency changes if the (↔) point lies over a side instead of being inside the triangle formed by the remaining points in the four-point space. For the electron with its total of 16 points, the calculation is already complicated. Above all, chirality theory lacks a prescription as to how these frequencies are to be calculated. It is presumably a matter of a relatively simple geometrical formula whose application, however, can be complicated with particles consisting of many points. Furthermore, I assume that asymmetrical particles have rather a larger mass than similar, more symmetrical particles. The electron with spin $-\frac{1}{2}$ (spin distribution *Rrlll*) should be somewhat heavier therefore than an electron with spin $+\frac{1}{2}$ (spin distribution *Rrrll*). To my knowledge, up to now nobody has ever measured the precise mass of electrons with definite spin, although such a measurement should not be to difficult. Because of such symmetry considerations, neutrons with spin $+\frac{1}{2}$ and $-\frac{1}{2}$ should, moreover, have different half-lives. However, these effects are probably very small, just as gravitation is very minor in comparison to electromagnetic interaction.

14.2 Pair formation

Occasionally, things of the same kind have been known to repel each other and opposites to attract each other. That is already evident with the simple neutrinos: those with the same spin clash, but those with opposing spin form pairs. It has also been seen in particles with an electrical or a colour charge that like charges repel each other, whereas opposites attract. Chemists well know that with the electron clouds of atoms, electrons with opposite spin like to form pairs, even though all electrons repel each other due to their negative charge. There are even cases where the two electrons of the same pair lie several hundred atoms apart, namely the Cooper pairs in superconducting substances [Moriyasu (1983) pp. 93 ff.]. Furthermore, the double helix of the genotype is also a pair of opposite, complementary macromolecules. And finally, multicellular organisms, in all their complexity, are rather prone to pairing up. The oppositeness, in all these cases, directly or indirectly involves chirality and thus the direction of time.

I suspect that the famous Pauli exclusion principle can also be explained more intelligibly on the strength of chirality theory. The Pauli exclusion principle states that two fermions cannot be in the same quantum state or – somewhat more roughly expressed – be in the same place at the same time. This is not an issue with bosons. Pauli's explanation is mathematically abstract and difficult to comprehend. For Feynman, that means that we have obviously not really understood the basic principle [Feynman (1966b) p. 4–3].

I am aware that I have only been able to explain the generation of pairs in my theory relatively weakly, for example with the picture of the two cogs which do not jam only if they rotate in opposite directions (Figure 39). I leave it to the reader to look for a better argumentation and to weave further thoughts about the generation of pairs. It could well be worthwhile.

14.3 The structure of the vacuum

Space does not actually exist as a perceivable thing in itself. A space, which does not exist, can also have no characteristics. It cannot be empty, for example. Emptiness, i. e., a place where there is nothing, cannot exist [Aristotle (1987) Physics vol. IV, 7 and 8 (209b)]. What, then, is a vacuum? In chirality theory, space (Section 8.4) and location (Section 8.12.1) can be defined as terms, although they are not directly perceivable. They are a consequence of the axiom of chirality and the relation „between", which create structures in the completely symmetrical chaos of the points. The term „completely symmetrical" means that in the chaos there is no place, no direction in space and no time direction which is distinguished in any way from all the other places, and directions in space or time. That is, however, only possible if at least some of the points have formed stable, and thus perceivable, mini black holes in the form of four-point spaces and don't just wildly move about amongst themselves in chaotic disorder. Fields (Section 9.5) can then be defined which lend characteristics to the different locations in the space, even if there is nothing there at all. As soon as an object is in such a location and is moved relative to all other objects, the order of the universe changes in the process and thereby becomes perceivable. This, according to chirality theory, would also have to be the case when the object is just an individual, not directly perceivable point. Are there such lonely, free points in the universe? A series of indicators point towards this.

Physicists have long searched the universe in vain for „dark matter + dark energy", which might constitute up to 90 % of the total matter of the universe [Schmitz (1997) pp. 398 ff.], whereby dark matter alone is about 27 % [NASA (2015)]. The existence of such yet to be perceived matter is a product of the rotation curves of spiral galaxies and of the observed velocity distributions of individual galaxies within galaxy clusters. A strong theoretical argument for dark matter is additionally delivered by the inflationary cosmological models. According to these, at the very beginning, the universe went through a short phase of a fast, exponential expansion called inflation, which led to a flat, i. e., an almost exactly closed universe with a critical density $\Omega = 1$ [Penrose (2004) pp. 772–778]. The inflation model is able, in a natural way, to explain the large-scale homogeneity and isotropy of the cosmic background radiation, the formation of structures on smaller scales following the inflation phase, the closeness of the density Ω to the critical density – it lies between 0.999,999,999,999,99 and 1.000,000,000,000,01 – and the absence of magnetic monopoles [Guth (2002)]. There are numerous candidates for the dark matter with imaginative names and characteristics, none of which have been proven to exist to date [Klapdor-Kleingrothaus (1997) p. 287; Hawking et al. (1996b) pp. 75–103].

On the strength of chirality theory, it is conceivable that the dark matter, or some of it, is nothing other than individual, free points. Such points would contribute to the frequency or the mass of the universe, although in principle, they would not be perceivable because the events in which they are involved are not periodic. Such individual points are also attracted by gravitation. They should thus appear more frequently in the proximity of galaxies or within matter generally. Now it has been proven that neutrinos within the sun and the earth oscillate between electron-, muon- and tauon-neutrinos [Schmitz (1997) pp. 278 ff. and 351 ff.]. Neutrino detectors measure different frequencies for the sun's neutrinos during the day from those at night, since at night, some neutrinos on their way from the sun through the earth are converted into other neutrino types [McDonald (2003)]. That is only possible, according to the hypothesis of chirality theory, if a ($\leftrightarrow$) point of the four-point space is substituted. For this to happen, individual free points are probably required, and these should be more prevalent inside the earth than outside.

According to this hypothesis, the vacuum may be free of particles, but it is not really empty because it can very well contain non-perceivable individual points. These points form a chaotic plasma with chaotic, i. e., non-periodic events. The plasma can abstract further points from the neighbouring „vacuum“ by gravitation if its pressure is lower than that of the neighbouring „vacuum“. Such negative pressures are postulated for the time shortly after the Big Bang, where the points were very densely packed and where stable neutrinos had not yet been able to form [Guth et al. (1989) pp. 34 ff.]. According to inflation theory, the vacuum's negative pressure meant that the volume of the universe, i. e., the vacuum, expanded exponentially. It is important to be conscious of the fact that according to this model, it was not the universe with its matter which expanded during the inflation, but rather the space in which the universe was situated.

14.4 Symmetry breaking

With the chirality theory three new symmetry breakings are introduces to physics that are all more radical than the two familiar symmetry breakings, which in one instance separates the strong interaction from the electroweak, and in the other instance, the weak from the electromagnetic interaction. The first symmetry breaking is the Big Bang, where space turned into time and time turned into space and where a chaos of points emerged. With the chaos of the Big Bang, the symmetry of the newly developed universe became much greater, however, not less; it thus involves a negative symmetry breaking. Thus, with the Big Bang the entropy of the new universe was minimal but by the following inflation it increased very fast. Through the second symmetry breaking, the four-point spaces could form with periodic events. The smallest possible black holes developed, the neutrinos with their gravitational field. During the third symmetry breaking, these black mini-holes formed a new generation of black holes, the charged quark/lepton plasma. Then from this, through the further symmetry breakings recognized by physicists, first the quarks separated from the charged leptons and later the weak interactions separated from the electromagnetic interactions [Mainzer (2005)]. All following newly-developing orders can also be viewed as symmetry breakings. The baryons formed ordered atomic nuclei, the atomic nuclei joined with leptons to form atoms, the atoms to molecules, the molecules to crystals, later to viruses, the viruses to single-celled organisms, the single-celled organisms to multicellular, the multicellular to higher organisms, these in turn to people, people to systems of social, political and further groupings. With living creatures and societies, functional replaces structural symmetry breaking. Along another line of development, gases became liquids, liquids solids, solids stars, stars solar systems, solar systems galaxies, galaxies galaxy clusters and superclusters, and at some time or other, large, new black holes with a new chaos emerged, from which a new universe was able to develop. The development is schematically represented in Figure 56 on a logarithmic distance scale. A similar scale in the dimensions of time and temperature can be found with Klapdor-Kleingrothaus et al. (1997) p. 130.

In a certain sense, every measurement is also a symmetry breaking, i. e., a break of the time symmetry: up until measurement, nearly all laws of nature were time symmetrical. A measurement, then, calls for an irreversible time, however, because without it, no documents for values measured in the past could be provided [Atmanspacher et al. (2003) p. 315]. For the future there are no documents. What is common to all these symmetry breakings, with the exception perhaps of the Big Bang? Weizsäcker's (1992) p. 910, assumption was: symmetry means the separability of the examined object in question from the rest of the world. On the one hand, owing to a „binding energy", a new, additional level of order develops with each breaking; on the other hand, the entropy increases despite the new order. The additional order is always purchased with additional disorder and the additional information with an information loss. The order is thereby probably relative; whether the system appears orderly or not depends upon the location and state of the observer. It depends, for example, on whether he is inside or outside of a crystal or a black hole. Moreover, common to symmetry breakings is that the system as a whole cools down precipitously with each breaking. Since the Big Bang, the average temperature of the universe has continued to drop steadily.

The distances, at which the symmetry breakings become effective, are unexplained. Although the strong and the weak interactions are both consequences of a new order in the relative motion of neutrinos, their distances lie at around a factor of 10^{13} apart. The factor is one hundred times greater than the size difference between us humans and the entire universe!

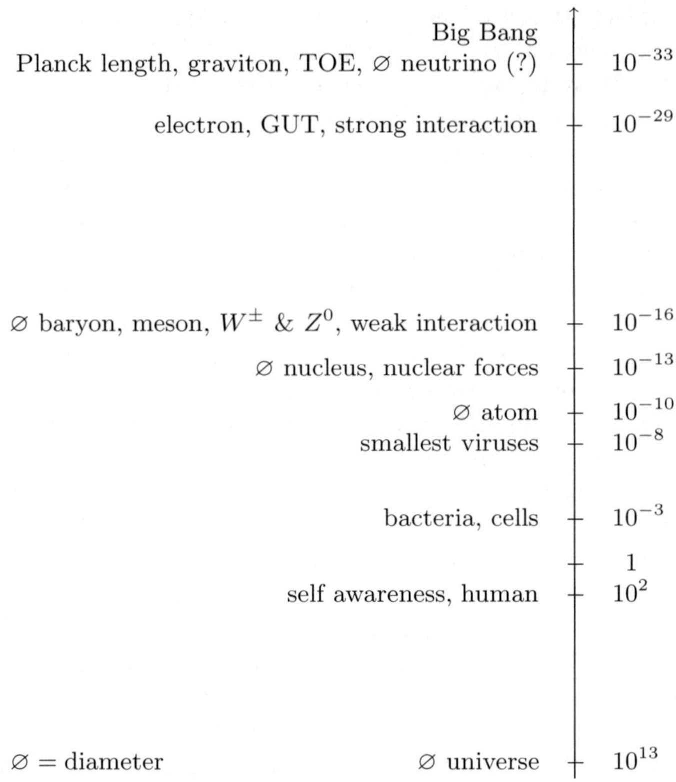

Figure 56. Distances with symmetry breaking *(logarithmic scale in centimeters)*

14.5 Cosmology and entropy

The considerations in the previous section indicate that the cosmos is not a universe, but a multiverse, because each new black hole is the beginning of a new universe for everyone who experiences the black hole from the inside. It is probable that in principle, different universes cannot communicate with one another because the information content is lost at the Schwarzschild radius. Mathematically the universe is defined as a set *U*, which fulfils five conditions that guarantee that all operations on elements of *U* lead again to elements of *U* [Mac Lane (1969/1998) pp. 22 f.]. This applies to the information *amount*, but not necessarily to the knowledge about the *content* of information. The quantity of information, i. e., the entropy, is one of the very few physical quantities which is completely independent of the frame of reference; perhaps even the only one in the whole cosmos [Møller (1969) p. 213]. With a computer also, one can know very well how large its memory capacity is and how many bits it has stored without knowing what the stored information actually consists of. One may know little, but not nothing. And the little possibly enables one to draw conclusions concerning the existence and size of the computer.

Seen from the outside, the boundary between outside and inside is a spherical plane, through which the quantity of the bits in the black hole transfer their action outward. The larger this surface is, the more bits of action can be transferred. Since the gravitational action of the mass is caused by the black hole and carried through its surface outward, the surface of the black hole at the Schwarzschild radius must be proportional to this mass. Theoretical considerations led to the result that the quantity of the bits, i. e., the amount of information of a black hole or its entropy, is

$$S = \frac{1}{4} \cdot \frac{A}{\hbar G}$$

whereby A is the surface of the black hole, $\hbar$ is Planck's constant divided by 2π and G is the consant of gravitation [Smolin (2001) pp. 90–178]. Hence it follows that the surface per bit of information is a fixed size, i. e., that there must be a smallest surface, or a smallest length and a smallest space volume, corresponding to the smallest possible amount of information, i. e., one bit. This surface's order of magnitude is a Planck length squared. The smallest possible object with one bit, according to chirality theory, is the neutrino, which can alternatively have a spin of $+\frac{1}{2}$ or $-\frac{1}{2}$. It is also the smallest possible black hole. The neutrino would therefore have a diameter of about a Planck length, unless it is even smaller still, because at these short distances, the physical preconditions for the consideration I present are perhaps, for theoretical reasons, not all valid. It is doubtful, however, that one can therefore assert that space is granular in the smallest unit volumes of a Planck length diameter, since according to chirality theory there is actually no such thing as space at all, and that which does not exist also cannot be granular. That which really exist are the events of the individual neutrinos and those are at most densely packed in the inside of a black hole, thus at a place we cannot observe at all. The pattern of an action transmission from the interior to the exterior of a black hole is thus

event in the black hole → information → event outside the black hole,

accordingly

volume → surface → volume.

One theory, which does most justice to this philosophy, was developed by Penrose (1998). It is called spin network theory and is not mathematically formulated in detail. The spin networks are not *in* space-time, they *cause* space-time. In so doing, they convey a causal relationship between events and change the space-time geometry with each information transfer [Smolin (2001) pp. 134–140, 172 and 219].

Perhaps it is even possible to draw conclusions about the existence and size of other universes without knowing what they concretely contain. One could perhaps draw such conclusions from the size and the mass of our universe, and the rate of its expansion due to *dark energy*, or from changes in these, or from the relationship of matter to antimatter. If we characterize the cosmos as the entirety of all that which appears not to be impossible to gain knowledge of in principle, then – in contrast to the universe – it must be „one". The science of the study of the cosmos is cosmology.

The cosmos might be a multiverse. According to this hypothesis, an information transfer between the inside and the outside of a universe which is interpreted as a black hole must cross the boundary of the black hole. Seen from the outside, this boundary is the Schwarzschild radius; seen from the inside, probably the Big Bang. At the Schwarzschild radius or at the speed of light, all clocks stand still, which means that cause and effect can no longer be differentiated. The information content is lost not, however, the knowledge about the amount of information, i.e., the action, for example gravitation. Just as the gravitation of a black hole is also perceivable outside of this hole, something about the entirety of the rest of the cosmos should be perceivable inside the black hole. What can we know about it? The *dark matter* – according to the model of chirality theory single points – enters the universe from outside, from the cosmos, since black holes attract also single points. Mass can be described and measured as a frequency $h \cdot \nu = m \cdot c^2$. When time at the Schwarzschild radius or with the Big Bang becomes space, then the mass, frequency and energy of the matter entering the black hole becomes an expansion of space inside the black hole. It is now the *dark energy*. Right after the Big Bang the expansion was extreme and caused the well-known inflation of the universe. With the inflation, not the particles but rather space itself expanded. This happens again with every point which falls into the black hole but since it doesn't happen regulary the expansion of the universe is not regular.

14.6 String theories

String theories, which are currently very popular and mathematically not simple, involve attempts to unify quantum theory and the general theory of relativity, i.e., something like quantum gravitation theories [Smolin (2001); Smolin (2004); Greene (1999)]. They require at least nine to ten spatial dimensions, of which six to seven are „concealed". These six are „rolled up" into objects of about a Planck length, i.e., about 10^{-33} cm in diameter and are not perceivable, because they are too small. In addition, there is a possibly complex time dimension. There are also versions, however, with up to 26 space dimensions and those with membranes or other multidimensional rolled up spaces in place of the one-dimensional strings.

The following aspects of the string theories are questionable in my opinion [Penrose (2004) pp. 869–933]: They operate with a space-time, which exists, although space and time as such can never be actually perceived. Moreover, this space is a continuum although a continuum cannot be perceived. Chirality in certain string theories plays no role, or at most, a subordinated one. The string theories are based on intrinsically fixed objects or atoms, on the strings without an internal structure. The strings, or the multidimensional „branes", can then vibrate, like the string of a violin. What is it that moves with the vibration? That could only be some parts of the strings, which, however, cannot exist at all according to string theory. The string theories in this form can probably never be a good model of observed nature. However, the number of nine space dimensions is remarkable. If one assumes that a black hole has three independent spatial dimensions, then one also arrives at nine spatial dimensions according to chirality theory: three in the neutrino, three in the lepton or quark, and three in the universe. The neutrino, lepton/quark and universe concern black holes or space curvatures on three different levels, which is why the nine spatial dimensions really are independent of each other. The six dimensions of the neutrinos and leptons would then be rolled up and not perceivable. Thus, it is quite possible that certain mathematical insights of the string theories can be carried over to chirality theory. Figures 47 to 50 of meson and proton models can also be understood as intertwined or knotted, closed strings. However, these are now no longer one-dimensional, vibrating strings, but finite sets of points, which change their configuration in relation to each other by periodic events. Hawking et al. (1996b) pp. 4 and 96, is (for different reasons) also of the opinion that string theory is overestimated.

Approaches related to string theory are loop theory, the spin network theory and the twistor theory. The **loop theory**, in contrast to string theory, is based on curved spaces which are described topologically. The individual loops form thereby a non-Euclidean spin network, in which a metric can be defined [Penrose (2004) pp. 941–946]. In the **spin network theory**, space as such does not actually exist (not even as a background), but it is the result of a discrete topological network of (spin) relationships between neighbouring objects. All probabilities for such spin relationships are countable and are therefore rational numbers [Penrose (2004) pp. 946–952; Rovelli (2010)]. The spin network theory, of all those theories known to me, probably comes closest to my theory of the four-point spaces. The **twistor theory** describes „twistors", which can be understood as punctiform photon states in a purely mathematical space. It is very abstract and has little to do with physics. Its advantage is that it is chiral (in Penrose's opinion, *too* chiral) and not local, as is suggested by the EPR experiment. Pieces of information are thus always stored and changed globally, never locally [Penrose (2004) pp. 958–1006].

14.7 Mathematics

Mathematics is neither absolute truth nor an end in itself. Every branch of mathematics presupposes a language, a logic and an axiomatic theory. Perhaps even a protologic. A protological principle could be that the logic may not change in the process of the mathematical activity, and that is already a principle which also presupposes a concept of time. Since there are innumerable languages, logics and axiomatic systems, mathematicians must each agree on one language, one logic and one axiomatic system before they can properly start work, with establishing proofs, for example. For the physicist, it is therefore a matter of finding a type of mathematics which can help him to describe his empirical observations in an as true-to-life manner as possible and nevertheless systematically. If theoretical physicists have been trying in vain for one hundred years to reconcile the theories of relativity and (quantum)electrodynamics, then that is because they have not yet found the appropriate mathematics. It is probably necessary to change fundamental principles of logic and axiomatic theory, to do without or reformulate them, before this century-old problem can be solved.

Conventional mathematics [Bishop (1985)] is badly suited to the description of empirical observations for the following reasons:

1. It takes set theory as its basis, i. e., is based on elements instead of on the relations between elements. Perceptions, however, are not things or objects, but relationships between things.
2. The geometrical space is derived from the numbers instead of the other way round. Geometrical arrangements and their change are perceived first of all. However, counting only takes place later.
3. Higher mathematics is infinite instead of finite. Infinity, or a continuum, is never perceivable, however.
4. Chirality is ignored to a large extent, although it is the basis of every perception.

Back in Chapters 2 to 7, I indicated in which direction conventional mathematics should be changed: the logical sets of $A \equiv A$ and $AB = BA$ can be dispensed with; the axiom of infinity likewise. But an axiom of chirality, or something like it, needs to be introduced. Thus, a completely new type of mathematics will emerge, in which the equals sign means either something else or where there is no more equals sign at all, where the concept of infinity is lacking and thus differentials and integrals as well. This is then no longer higher mathematics, but something simpler. It is clear that most mathematicians only do without the concept of infinity with some reluctance. After all, it is the concept of infinity that makes mathematics higher mathematics, and some mathematicians probably feel that it devalues their science to omit the higher part. Not a lot would then remain of the traditional mathematics we learned in school. We are dealing with something like a finite, chiral topology, in which there are probably – as in the axiom of chirality – arrows instead of the equals signs. Group theory, i.e., the theory of morphisms, would continue to be valid; at least to the extent it is finite. To my knowledge this type of mathematics does not yet exist, but topological theories have most certainly developed in this direction [Gilmore et al. (2002); Nakahara (2003)]. Mehlberg (1980) vol. I pp. 214–222 and vol. II pp. 137–145, and Bitbol (1993) pp. 91–100, take a different path. The entire relativistic space-time structure can be constructed on Mehlberg's axiomatic basis, with the help of the event concept and a relation between two objective events.

Jürgen Dümont (1999) has described a concept, along the lines of which such a mathematical system could be developed. In so doing, he relies particularly on the work of Shapiro (1997). Dispensing with infinity and continuum means that mathematics must be finite. *Finitism*, however, is mathematically unproblematic only in number theory and in those areas, which can be developed from number theory by finite combinatorial methods. Since the axiom of infinity of the conventional axiomatic theory is independent of all other axioms, dispensing with this axiom probably has similar consequences to dispensing with the parallel postulate in geometry [Suppes (1972) pp. 252 f.]. Now Dümont has developed a finite category theory as a new basic discipline. All important mathematical fields can be understood as categories. Thus, no field stands out from the others, not even set theory. The basic idea of category theory is *morphism*. Functional connections between different categories can be treated within the framework of category theory. In the process, epistemological and ontological questions arise. Since the existence of objects and morphisms is not brought up for discussion within the theory, one can ask whether morphisms are a consequence of the objects or the objects are a consequence of morphism. The two are equivalent. Since chirality theory is based on perception, thus on a morphism, the mathematical model would probably have to likewise describe the objects as a consequence of a morphism. Geometry can be well exploited by category theory, likewise *arithmetic*, in which the concept of well-order is often connected with the notion of processes, which advance in *time* in discrete, countable steps. In chirality theory, that would probably be the countable, progressive events. The formation of finite products and transformations with a given source and a target can then be described as elementary principles. Arithmetic is thus no longer a theory about particular objects, but about a particular *structure*, i. e., about that of progression. A structure is the abstract form of a system, which describes the relationships between objects. It is that which is common to different systems. Two structures are identical if they are isomorphic. Arithmetical truths never refer to the objects, but only ever to the role the objects play within the progression. Distinguishing a particular subset of the power set[64] of a given set defines a *topology*. The concept of topology is then understood in such a way that by the specification of a topology, a set is provided with a structure. Electron neutrinos e.g. with same spin are topologically identical and remain so also in the course of the internal events, which do not change their internal characteristics. External characteristics however, such as rotations or translations, which do not remain intact by an isomorphic transformation, are not structural characteristics of the neutrino. The morphisms of a category and not the objects represent the crucial „data“ for the category. And so, it should be technically possible, to completely dispense with objects, for example geometrical points, and to operate solely with arrows [Mac Lane (1969/1998) p. 9]. I have done precisely that with the axiom of chirality! Category theory is dynamic, intuitionist mathematics. In it, for example, the axiom *tertium non datur* no longer applies. Only if an ideal consciousness can produce the set of natural numbers as an actual infinite set, i. e., if it can be assumed that all natural numbers have already been constructed, does the axiom apply. It cannot apply, however, if the ideal consciousness is *not* able to

[64] The power set is the set of all subsets of a given set. The power set of a set of n elements contains 2^n elements.

effectively construct all natural numbers. This is because for a non-constructed number, it can neither apply, for example, that it is a prime number, nor that it isn't one. Its characteristics can only be determined if the number has been produced; therefore, classical logic does not apply.

I am not in a position to formulate such mathematics more precisely – that is for specialists to do. The procedure is quite normal. The transition from theoretical physicist to mathematician is not usually a smooth one. The relationship between the two is akin to that between the discoverer of new land and the farmer, who then follows, encloses and cultivates that land. The mathematical farmers want to settle everything very carefully and cleanly, and determine the exact boundaries of an idea, whereas the physical discoverers already love their ideas while they are still quite wild and untamed. Both have a tendency to believe that they have done the major part out of the work. In reality, however, both are needed; imaginative discoverers and scrupulously accurate mathematicians, as well as good communication between the two [Smolin (2001) p. 138]. Incidentally, I will never quite fully understand the mathematics I have suggested. Even Einstein is supposed to have once said that ever since mathematicians had begun to concern themselves with his theory of relativity, he no longer understood it.

14.8 Philosophy

Zurich is traditionally a philosophical desert. Many of the people who live here are proud of this. Philosophy is here, at best, a more or less smiled at, intellectual hobby. And the few people who regard philosophy highly are not exactly noted for philosophical creativity. As a pupil I missed out on something there, but deserts also have their advantages. One can pose questions in perfect liberty and naivety, without a wise or a knowing one coming along with his answers. To be sure, one is considered somewhat wacky when one expresses unusual thoughts out loud, but one can also keep them to oneself. That is actually an advantage because in this way, one is not sidetracked and doesn't waste any time. I came up with the greater part of my philosophy on my own and only discovered much later, with the reading of pre-Socratic philosophy, the old Indian and the Chinese, that these also posed the same questions and often gave very original answers to them, and sometimes similar answers to those I came up with myself.

I believe, not only for natural science, but in all areas of life, from human co-existence to politics to theology, it is advantageous for humans if they have an idea of the basic questions of philosophy, in particular of metaphysics. The questions thereby are always more important than the answers. They give us a feeling for the relativity of all knowledge and all truth.

In this, our society deludes itself a great deal and it is moving to witness again and again, how essentially intelligent people believe and do things without actually realizing why. They are rarely conscious of what is happening with them philosophically, (mass)psychologically and biologically. That would be a worthwhile area of research, also for mass psychologists. But which university has a chair for mass psychology? It is the basis for our living together, for our economic behaviour and for all wars.

As an illustration, I give five examples of basic philosophical questions of the pre-Socratic thinkers, none of whom, incidentally, lived in present-day Greece, orienting myself in the process by an early work of Nietzsche (1930b) [Diels (1903)].

Thales of Miletus (ca. 650–560 BC) also known as Thales the Milesian asked: Is there anything which is common to everything in existence? His answer was: Yes, everything is water. Naturally we know today that not everything is water. In the language of chirality theory, I would say: Everything is an event. Perhaps a modern physicist would answer: Everything is string, or perhaps put another way, everything is made of strings. And in 2500 years – or perhaps also somewhat earlier – scientists will smile in exactly the same way about the string idea as we do about the water of Thales. But the philosophically bold question remains unchanged and current: Is there a unity in that which exists, in all that is? That was the first natural-philosophical question ever asked. And the answer of most natural scientists is still: Yes!

Anaximander (610–546 BC) asked: What is transitory? His answer was: All which exists passes. Only the primordial „stuff", the *apeiron* – the indeterminate, is infinite. Multiplicity is made up of this uniform primordial matter. The *apeiron* in chirality theory would probably be the points.

Heraclitus (544–483 BC) asked: What can be said to „be"? His answer was: Nothing! There is only becoming; everything flows. Thereby he most likely based his conclusions consciously or unconsciously on the already centuries old Chinese tradition of the *I Ching*, the book of the change. Translated into chirality theory, this means: There are only events, since only they are perceivable. Furthermore, Heraclitus recognized that everything carries its opposite within itself at all times. That sounds very much like the discovery of chirality.

Parmenides of Elea in southern Italy (ca. 515–450 BC) asked: Are mind and matter two separate entities, two different worlds? His answer was: Only that which is constant and indestructible, is; all becoming is a deception. There is nevertheless the view that actually, quantum gravitation must also be a theory without time parameters, a quantum theory in which all quantum states are present in stationary form [Kiefer (2003)]. The senses are just illusion. Mind (*νους*[65]) and matter are separate entities. At first sight, this philosophy sounds rather absurd today. If one accepts, however, that space and time do not actually exist, but as in chirality theory, that they depend on the point of view and state of the observer, then becoming can indeed be understood as deception. In our modern language of physics, Parmenides would probably say something along the lines of: The universe is a compact and limited black hole, which my Self observes as a whole from the outside, about whose internal structure and changes, the „everything flows" of Heraclitus, there can be at most, assumptions. See also Popper (1998) pp. 141 ff. and 198 f. But the laws of nature always apply independently of this and these were probably a conceptual part of the „*νους*" of Parmenides. Moreover, Parmenides was convinced that there could not be infinity. In this point, he was ahead of most of today's physicists.

Anaxagoras (500–428 BC) asked: How does order come into the world? His answer was: Through the mind! It is a point of crystallization (focal point) of small, growing eddies which bring the order to the chaotic mixture of the primeval soup and innumerable different qualities. The mind does this without a goal, purpose or sense. Perhaps today one would say: Order is only such to the extent it is perceivable. That sounds very much like the second law of thermodynamics and almost resembles a black hole. Concerning the nature of time, Anaxagoras concluded that

[65] *νους* = nous = intellect, reason, knowledge + name + definition + portrayal of reality.

time as such does not actually exist but is a consequence of sensory perception. In this point he was ahead of most of today's physicists.

It would be fruitful, to once again seriously read and consider the questions of the ancients including the Chinese and Indians [Capra (1991)]. In Chinese and Indian thought the process is elementary, in West Asiatic or European thinking the actual is elementary. Couldn't the process be the actual? [Whitehead (1936) p. 11]. What do the ancient questions mean today? Perhaps a little courage is required for this. To my surprise, I recently even found with the ancient Maya a theory of cosmology, which modern natural scientists only in the 21st(!) century are slowly beginning to accept [Veneziano (2004)]: the universe developed from a chaos, in which there was no time. Time developed only as consequence of order [De la Garza (1998)]. The Maya also had modern ideas concerning the structure of the soul as the sum of intellect and insticts.

The Philosopher who advocated the ideas of my metaphysics to the greatest extent is, surprisingly, Gottfried Wilhelm Leibniz (1646–1716) [Holz (2013 a)]. Leibniz's „monad", as the simplest possible substance and atom of the soul, corresponds roughly to my concept of event and the 4-point space. The innumerable monads of the world all interact with each other according to a harmonious world law (theory of everything) through a dialectical, logical process with unlimited possibilities, which can lead to a flow of information within the one world (universe) [Holz (2013 b)]. Without knowledge of relativity and quantum theory, elementary particles and physical fields, Leibniz, as a monist, developed and philosophically founded his monad theory by pure thought, however without any significant reverberation in the natural sciences. I have clearly deviated from Leibniz in what he considers to be the fundamental concept of identity, as well as in ontological details. In its place I have introduced the concept of black holes. In this way I was able to elegantly circumvent most of Leibniz's aporias, however we are largely in agreement about the concept of God: He is not a person, but the best or only possible world law in itself. Science is everything that can lead to knowledge, including historical, legal, psychological, economic and sociological research. Only such a holistic education leads to happiness.

14.9 Free will

Nearly all humans, including scientists and philosophers, have been convinced for thousands of years that they have a free will. They are free to choose how they wish to behave and their reason enables them to make the best choice. Naturally they also often act in the heat of the moment, especially if there is little time for deliberation, if, for example, it concerns not burning one's fingers. But murders are always planned. One can also forgo a murder. Only those who have a choice can be held accountable for their decisions. And all societies since time immemorial have called to account those who did not adhere to the rules of that society. Humans are apparently, to a certain degree, free to act contrary to their own internal motivation as well as their emotions and instincts, be it a reasoned act or one on command. The free will is the capacity, according to certain rules, to determine one's own actions [Kant (1999)].

This simple truth contrasts with the insights of psychology and physics that force us to question free will [Lyre (2017)]. According to Wegner (2002) free will is an illusion which comes in three steps: First of all, our brain undertakes the planning of an action. Secondly, we become aware that we are thinking about the planning of the action and call that will. In the third step the action is carried out and we honestly, but erroneously, believe that our will caused the action. The second step limps at least a half second behind the first and it can happen, that the third step also takes place before the second [Blackmore (2003) pp. 61 ff. and 129 f.]. The tennis champion serves with a certain amount of conscious aforethought. His returns, however, all take place as pure reflex actions; otherwise the champion wouldn't be a champion for much longer. If one asks him about the game, then he is nevertheless honestly convinced that he played and reacted as he did from free will. Murders, on the other hand, function differently from tennis returns. They don't come about within fractions of a second but are planned and thereby conscious acts. They could thus certainly occur from free will. The more complicated an action is, the more likely it is conscious and can be triggered by free will. According to Penrose (1989) p. 411, one needs a consciousness for common sense, the judgement of truth, understanding and artistic appraisal. No consciousness is necessary for automatisms, mindless obedience, or programmed and algorithmic actions.

Dealing with the reservations within physics in relation to the free will is more difficult. Everything that exists in the universe is the fruit of coincidence and necessity. From this Democritus (ca. 460–370 BC) concluded that humans are subject to an absolute necessity. To the present day, hardly anything about Democritus' insight has changed from a physics point of view [Monod (1973)]. If consciousness with its free will is transcendental, then there is no problem with liberty in free will. The difficult question is then: What effect can a transcendental consciousness have on the empirically perceivable, real world, on that which exists and can be said to „be“? How does my Self make my finger bend on the trigger of the rifle? If consciousness is a function of the brain, however, then this obeys the laws of nature. As we have seen, these laws are deterministic; at least the probability of the future events is predetermined, quantum-mechanically and thermodynamically [Wechsler (1999) pp. 194–200]. So, the free will could, at most, intervene in the statistical probability of future events, and that can hardly satisfy anyone. In any case, that does not correspond to my feelings during an act of will. Penrose (1989) pp. 405–449, like many other thinkers, has considered this conflict. He, also, is yet to find a solution, but does not rule out the possibility that quantum mechanics might need emending. He believes that the mathematical system that the universe obeys is probably non-algorithmic, which means that the future is fundamentally non-predictable. Otherwise one could, in principle, predict what one would do next and then „decide“ to do something completely different. Practically, this would be a contradiction between the free will and the strong determinism of the theory. By introducing non-predictability to the theory, one can avoid this contradiction. However, even Penrose hopes that subtler rules can be found according to which the world functions. In this sense, Mainzer (2005) pp. 312 and 346 ff., has shown that finite networks also, such as the brain, the Internet or a Turing machine, due to their non-linearity and complexity, are in no way contrary to a free will since they do not answer questions by algorithmic means.

A subtler system could also be dualism, which has rather fallen out of favour in our materialistic culture (see Chapter 5). Determinism functions only in a purely temporal world, because only here is an order possible, whereby the cause comes before the effect. Actually, the concept of time is even more strongly relativised in chirality theory, however, than in the special theory of relativity. Perhaps a new understanding of the structure of time will help to build the long-sought bridge from the transcendental to the material world.

And sometimes it seems as if the free will is only possible outside of philosophy if one stops talking about it [Hersch (1974)].

14.10 Theology

What is going on in a wolf when it howls at the full moon? Are these the first religious stirrings? The Neanderthals were already burying their dead some 50,000 years ago in decorated graves. Neanderthals did not belong to the species of homo sapiens sapiens. Did they believe in an afterlife? From prehistoric times and to the present day, humans have had most varying conceptions of an afterlife and the gods. The gods, invented by humans after all, mostly have quite human characteristics. That can be taken so far that they even come to earth in human form or copulate with humans. Some humans, as did Spinoza and Einstein for example, believe that their god created the laws of nature and reveals himself in it to them [Weizsäcker (1999) pp. 254–263)]. There is, however, a contrary view: even the gods have to obey the laws of nature. The Greeks at the time of Hesiod believed this and this is still believed today by the Mormons, at least by those whom I have asked. The Buddhists have no answer to the question as to whether or not there is a God or gods. They say perhaps, and perhaps not [Blackmore (2003) pp. 401 ff.]. For them, it is an irrelevant question. In addition, there are atheists, whose answer to this question is: „There is no God" or „God is dead".

Despite these fundamentally different religious attitudes, there are two religious areas in which nearly all humans have always been in agreement:

First of all, everyone believes that there are laws, according to which the world functions. These laws apply generally and are immutable, except perhaps under the strictest conditions. We may not know where the laws come from, but one can also call their cause or their explanation God. That is purely a question of defining the term God. For Küng (2013) God is not a person. He is world immanent and world transcedental. He is everywhere and omni-present. He is not the idea of the Good as with Plato, no unmoved mover as with Aristotle, no inanimate

Oneness as with Plotin and He does not miraculously interfere with the history. No, God is the dynamic itself. This way God is about the same as the physicists' idea of a „theory of everything“ (TOE). The TOE is the basis of all laws of nature. It explains why and how the world changes and how forces cause this „movement“. The relation between force and movement is called dynamics. One can never know this final theory, but at most, believe that it must exist. Those who do not believe that there are such laws, that a God, so defined, exists, are nonviable. They will be steamrolled, because they do not believe in the law of steamrollers being dangerous. Seen in this way, humans are condemned to belief.

Secondly nearly all humans believe in an order of good and bad. They do not always agree on what is good and what is bad, and between good and bad there are also often conflicting aims. But the system of good and bad in itself is scarcely disputed. Most humans, of all religions and even atheists, can agree relatively easily on what is good and what is bad. Thus, there is the Universal Declaration of Human Rights, which has been adopted more or less unchanged into all national constitutions and a world ethos is generally accepted [Küng (2002)]. Primarily, opinions differ on the topic of force, where the boundary between good and bad is especially precarious. For some, this agreement comes as a surprise in view of all the wars stemming from religion. Since, however, all humans, all societies and all races want to survive, and since there is not an indefinite number of survival alternatives, it is actually quite clear that there must be certain rules for survival, about which reasonable people should be able to agree. There have been original thinkers who were of a different opinion, for example Nietzsche with his theory of the *Übermensch* (superman), but even Nietzsche would probably have been appalled, had he witnessed what Hitler did with his theory.

What has all that to do with physics? Chirality theory has at least three aspects which could be relevant to theology.

First of all, the question of communication, that is, between the gods, between humans, and between gods and humans arises in each religion. Communication involves information, knowledge, ignorance and the inability to know. Perhaps it would be interesting to rethink these concepts in light of chirality theory.

Secondly, the essence of religions is found in contrasts, for example God and humans, good and bad. The poles are always somehow in relation to each other. It is a similar situation with the principle of chirality: two things are alike and nevertheless opposite. Is the analogy coincidental?

Thirdly time, and thus causality, plays a fundamental role in chirality theory. The concept of time term is strongly relativised. Time is also fundamental in all religions. If time, on physical grounds, can stand still, reverse its direction or even turn into space, it is probably conceivable that it also does so within a religious context. There is a link here with communication, which is causal in conventional understanding and always takes place in time. Yet we have seen that at least the timeless, virtual bosons act without transmitting information. From a religious point of view, that is almost divine.

Bibliography

Abbott, B. P. et al. (LIGO Scientific Collaboration and Virgo Collaboration) (2016). Observation of Gravitational Waves from a Binary Black Hole Merger. Phys. Rev. Lett. 116 (6): 061102

Abramowicz, M. A. (1993). Black Holes and the Centrifugal Force Paradox. Scientific American March 1993, pp. 26–31

Abramowicz, M. A. and Prasanna, A. R. (1990). Centrifugal-force reversal near a Schwarzschild black hole (pp. 720–728). Mon. Not. R. astr. Soc. (1990) 245

Anderson, M. R. and Lemos, J. P. S. (1988). Mon. Not. R. astr. Soc. (1988) 233, p. 489

Aquinas, St. Thomas (1267–1273/1947). Summa Theologica. Benziger Bros. eds. www.nlx.com; Intelex {Past Masters: Aquinas: Collected Works}

Arasse, D. (1997). Leonard de Vinci: le rythme du monde (p. 18). Paris: Hazan

Aristotle (1984). On the Soul (418b); translated by J. A. Smith. In „www.nlx.com; Past Masters: Aristotle: Complete Works. J. Barnes ed.“. Princeton: Princeton University Press

Aristotle (1987). Physics. In „www.nlx.com; Past Masters: Aristotle: Complete Works. J. Barnes ed.“. Princeton: Princeton University Press

Atmanspacher, H., Primas, H. and Wertenschlag-Birkhäuser, E. eds (1995). Der Pauli-Jung-Dialog und seine Bedeutung für die moderne Wissenschaft. Berlin: Springer

Atmanspacher, H. (1995). Raum, Zeit und psychische Funktionen. In „Der Pauli-Jung-Dialog und seine Bedeutung für die moderne Wissenschaft (pp. 256–260). H. Atmanspacher, H. Primas and E. Wertenschlag-Birkhäuser eds“. Berlin: Springer

Atmanspacher, H. and Primas, H. (2003). Epistemic and Ontic Quantum Realities (pp. 301–321). In „Time, Quantum and Information. L. Castell and O. Ischebeck eds“. Berlin: Springer

Augustinus, A. (396–400 AD/1946). Confessiones book XI § III. In „The confessions of St. Augustine“. London: Dent

Barrow, J. D. and Webb, J. K. (2005). Inconstant Constants. Scientific American June 2005, pp. 32–39

Bauberger, S. (2003). Was ist die Welt? Zur philosophischen Interpretation der Physik (pp. 113 ff. and 136 ff.). Stuttgart: Kohlhammer

Baumgarten, H.-U. (1999). Erkenntnistheorie; Alber-Texte Philosophie vol. 3 (p. 21). K.-H. Lembeck ed. Freiburg im Breisgau: Alber

Bell, J. (1964). On the Einstein-Podolsky-Rosen paradox. Physics I (1964), pp. 195–200

Berkeley, G. (1710/1948–1957). On the Principles of Human Knowledge (vol. II, pp. v2. 41 ff.). In „www.nlx.com; Past Masters: Berkeley: Works. T. E. Jessop and A. A. Luce eds“. Princeton: Princeton University Press

Bertola, F. and Curie, U. eds (1993). The Anthropic Principle. Cambridge: Cambridge University Press

Biever, C. (2012). „It's a boson! But we need to know if it's the Higgs“. New Scientist 6. 7. 2012

Biggs, N. (1996). Discrete Mathematics. Oxford: Oxford University Press

Bishop, E. and Bridges, D. (1985). Constructive Analysis. Berlin: Springer

Bitbol, M. (1993). From the Anthropic Principle to the Subject Principle (pp. 91 ff.). In „The Anthropic Principle. F. Bertola, F. und U. Curie eds)“. Cambridge: Cambridge University Press

Blackmore, S. (2003). Consciousness. London: Hodder & Soughton

Böhme, G. (2000). Platons theoretische Philosophie. Stuttgart: Metzler

Bohr, N. (1960). Atoms and Human Knowledge (pp. 92 ff.). In „Atomic Physics and Human Knowledge“. New York: Wiley & Sons

Boole, G. (1847). The mathematical analysis of logics. Cambridge

Boyle, K. R. (1904). Lecture delivered before the Oxford University Junior Scientific Club on May 16, 1893. London: Clay

Brandes, J. (1995). Die relativistischen Paradoxien und Thesen zu Raum und Zeit. Karlsbad: Verlag relativistischer Interpretationen

Burger, P. (1998). Wittgensteinianismus. In „Vom Ersten und Letzten (p. 194). U. J. Wenzel ed.“ Frankfurt: Fischer

Canetti, E. (1960). Masse und Macht. Düsseldorf: Claassen

Capra, F. (1991). The Tao of Physics: an exploration of the parallels between modern physics and eastern mysticism. Boston: Shambhala

Carnap, R. (1928/1998). Der logische Aufbau der Welt. Hamburg: Meiner

Carnap, R. (1963). The Logical Syntax of Language (pp. 53 f.). In „The Philosophy of Rudolf Carnap. P. A. Schilpp ed.“. London: Cambridge University Press

Carter, B. (1993). The anthropic selection principle. In „The Anthropic Principle. F. Bertola and U. Curi eds“. Cambridge: Cambridge University Press

CERN (1979). Parity violation in real life. Link between left-right asymmetry in amino-acids and weak interactions (p. 115). CERN courier, May 1979

Chaitin, G. J. (1993). Randomness in Arithmetic and the Decline and Fall of Reductionism in Pure Mathematics (pp. 314–328). Bulletin of the European Association for Theoretical Computer Science 50 (June 1993)

Chalmers, D. (1996). The Conscious Mind. Oxford: Oxford University Press

Christenson, J. H., Cronin, J. W., Fitch, V. L. and Turlay, R. (1964). Phys. Rev. Letters 13 (1964) pp. 138–140

Clifford, Will (1989). The renaissance of general relativity. In "The New Physics (p. 27). P. Davies ed." Cambridge: Cambridge University Press

Close, F. E. (1978). New source of parity violation (p. 11). Nature 274, 6. Juli 1978

Close, F. E. (1989). The quark structure of matter. In „The New Physics. P. Davies ed.". Cambridge: Cambridge University Press

Collins, G. P. (2004). The Shapes of Space. Scientific American July 2004, p. 68

Comte, A. (1880). Einleitung in die Positive Philosophie (pp. 37 f.). Leipzig: Fues

Confucius (551–479 BC/1972). Gespräche (p. 102). Düsseldorf: Diederichs

Conradt, Rüdiger (1973). Empirische Erkenntnislehre (p. 154). Frankfurt am Main: Akademische Verlagsgesellschaft

Cushing, J. T. (1990). Foundational Problems in Quantum Field Theory (p. 33). In „Philosophical Foundations of Quantum Field Theory. H. R. Brown and R. Harré eds". Oxford: Clarendon Press

Davies, P. (1989). The New Physics. Cambridge: Cambridge University Press

Dawkins, R. (1976). The selfish gene. Oxford: Oxford University Press

Dawson, J. W. (1997). Logical Dilemmas: The Life and Work of Kurt Gödel. Wellesley Mass.: Peters

De la Garza, M. (1998). Die heiligen Mächte des mayanischen Universums (pp. 101 ff.). In „Maya. Die klassische Periode". München: Hirmer.

Descartes, R. (1619/2006). Méditations touchant la première philosophie. Méditation quatrième: Du vrai et du faux. www.nlx.com; Intelex {Past Masters: Descartes: Œuvres complètes}. A. Gambay et al. eds. Toronto: University of Toronto

Descartes, R. (1641/1960). Meditationes de prima philosophia. In „L. Gäbe ed., Meditationen über die Grundlagen der Philosophie". Hamburg: Meiner

Dewey, J. (1929/1958). Experience and Nature (pp. 166–297). New York: Dover

Diels, H. ed. (1903). Die Fragmente der Vorsokratiker. Berlin: Weidmann

Dirac, P. A. M. (1930). A Theory of Electrons and Protons (pp. 360–365). Proc. R. Soc. Lond. A. Royal Society Publishing. 126 (801)

Dirac, P. A. M. (1977). Annahmen und Voreingenommenheit in der Physik. Naturwissenschaftliche Rundschau 30 (1977) p. 429

Domotor, Z. (1973). Causal Models and Space-Time Geometries (pp. 3–55). In „Space, Time and Geometry. P. Suppes ed.". Dordrecht: Reidel

Döring, A. and Isham, C. J. (2007). I.–IV. A Topos Foundation for Theories of Physics. https://arxiv.org/abs/quant-ph/0703060. Accessed 21. December 2016.

Dreiding, A. and Wirth, K. (1980). The Multiplex. A Generalization of Finite Ordered Point Sets in Oriented d-Dimensional Spaces. MATCH, Communications in Mathematical Chemistry 1980 (8) pp. 341–352

Drieschner, M. (1981). Einführung in die Naturphilosophie (pp. 85–90). Darmstadt: Wissenschaftliche Buchgesellschaft

Dsi, Dschuang (ca. 340 BC/1912). Südliches Blütenland, translated by R. Wilhelm. Düsseldorf: Diederichs

Duhem, P. (1914). La Théorie Physique: son objet – sa structure (pp. 35 f., 63–76). Paris: Rivivière

Dümont, J. (1999). Formal-ontologische Kategorien in der Mathematik. München: Utz

Dürr, H. P. (1968). Dynamik der Elementarteilchen. In „Atome, Kerne, Elementarteilchen (pp. 229 ff.). G. Süssmann and N. Fiebiger eds". Frankfurt am Main: Umschau

Dürr, H.-P. et al. (1997). Gott, der Mensch und die Wissenschaft. Augsburg: Pattloch

Dyson, F. J. (1958). Innovation in Physics. Scientific American Sept. 1958, pp. 74–82

Earman, J. (1973). Notes on the Causal Theory of Time. In „Space, Time and Geometry (pp. 72–84). P. Suppes ed.". Dordrecht: Reidel

Eccles, J. and Popper, K. (1977). The Self and its Brain. New York: Springer

Eigen, M. (1987). Stufen zum Leben. Die frühe Evolution im Visier der Molekularbiologie (pp. 55–63 and 113). München: Piper

Einstein, A. (1918). In „E. Scheibe (2006). Die Philosophie der Physiker (pp. 127–142)". München: Beck

Einstein, A. (1922). Über die spezielle und die allgemeine Relativitätstheorie. Braunschweig: Vieweg

Einstein, A. (1926/1969). Briefwechsel 1916–1925. Kommentiert von Max Born. München: Nymphenburger

Einstein, A. (1949/1983). In „Albert Einstein als Philosoph und Naturforscher: eine Auswahl. P. A. Schilpp ed.". Braunschweig: Vieweg

Einstein, A. (1950/1991). Zur Erniedrigung des wissenschaftlichen Menschen (p. 172). 42. Tagung der Societá Italiana per il Progresso delle Scienze. Lucca 1950. In „Mein Weltbild. C. Seelig ed.". Frankfurt: Ullstein

Einstein, A. (1956a). The Meaning of Relativity (pp. 157 f.). London: Methuen

Einstein, A. (1956b). In „Helle Zeit – Dunkle Zeit (p. 72). C. Seelig ed.". Zürich: Europa

Einstein, A. (1977). In „Kuznecov, B. G.: Einstein, Leben-Tod-Unsterblichkeit" (p. 267). Stuttgart: Birkhäuser

Einstein, A. (1979). Frontiers of Time (p. 431). In „Problems in the Foundations of Physics. J. A. Wheeler ed.". Amsterdam: North-Holland Publishing Company

Einstein, A. (1991a). Zur Methodik der theoretischen Physik. Antrittsrede vor der Preussischen Akademie der Wissenschaften (1930). In „Mein Weltbild (pp. 113 ff.). C. Seelig ed.“. Frankfurt: Ullstein

Einstein, A. (1991b). Maxwells Einfluss auf die Entwicklung des Physikalisch-Realen. In „Mein Weltbild (pp. 159–162). C. Seelig ed.“. Frankfurt am Main: Ullstein

Einstein, A. and Infeld, L. (1950). Die Evolution der Physik (p. 343). Wien: Zsolnay

Einstein, A., Podolsky, B. und Rosen, N. (1935). Can Quantum-Mechanical Description of Physical Reality Be Considered Complete? Phys. Rev. 47, pp. 777–780

Enz, C. P. (1995). Rationales und Irrationales im Leben Wolfgang Paulis (pp. 21 ff.). In „Der Pauli-Jung-Dialog und seine Bedeutung für die moderne Wissenschaft. H. Atmanspacher, H. Primas and E. Wertenschlag-Birkhäuser eds“. Berlin: Springer

Faissner, H. (1968). Schwache Wechselwirkungen und Neutrinos. In „Atome, Kerne, Elementarteilchen. G. Süssmann and N. Fiebiger eds“. Frankfurt am Main: Umschau

Feest, U. (2017). Philosophie der Psychologie. In „Philosophien der Einzelwissenschaften (pp. 475–509). S. Lohse and Thomas Reydon eds“. Hamburg: Felix Meiner

Feyerabend, P. K. (1961). Comments on Grünbaum's Law and Convention in Physical Theory (p. 160). In „Current Issues in the Philosophy of Science. H. Feigl and G. Maxwell eds“. New York: Holt, Rinehart and Winston

Feynman, R. P. (1965). The Character of Physical Law (p. 172). Cambridge, Mass.: MIT Press

Feynman, R. P. (1966a). Lectures on Physics (Vol. III, 1–10). Reading: Addison-Wesley Publ. Co

Feynman, R. P. (1966b). Quantum Mechanics (p. 17–7). Reading: Addison-Wesley Publ.Co

Feynman, R. P. (1985). QED – The Strange Theory of Light and Matter. Princeton: Princeton University Press

Finkelstein, D. R. (1969). Space-Time Code. Phys. Rev. 184 (Aug. 1969) pp. 1261–1271

Finkelstein, D. R. (1996). Quantum Relativity. A Synthesis of the Ideas of Einstein and Heisenberg. Berlin: Springer

Fischer, E. P. (2002). Werner Heisenberg. Das selbstvergessene Genie (p. 193). München: Piper

Fischer, E. P. (2003). Die andere Bildung (pp. 397 f.). Frankfurt am Main: Ullstein

Fränkel, H. (1951). Dichtung und Philosophie des frühen Griechentums (pp. 452–473). New York: Am. Philological Association

Frege, G. (1918/1999). Der Gedanke (pp. 113, 132–135). In „Erkenntnistheorie. H. U. Baumgarten ed.“ Freiburg im Breisgau: Alber

Frege, G. (1993). Logische Untersuchungen (pp. 30–53). Göttingen: Vandenhoek & Ruprecht

Fritzsch, H. (1996). Die verbogene Raum-Zeit. München: Piper

Furger, A. (1997). Das Bild der Seele. Zürich: NZZ

Gadamer, H.-G. (1965). Wahrheit und Methode (pp. 277 and 289). Tübingen: Mohr

Gardner, M. (1964). The Ambidextrous Universe. New York: Basic Books

Geiger, M. (1966). Die Wirklichkeit der Wissenschaften und die Metaphysik (pp. 180–183). Hildesheim: Olms

Genz, H. (1994). Die Entdeckung des Nichts. Hamburg: Rowohlt

Genz, H. (2002). Wie die Naturgesetze Wirklichkeit schaffen. München: Hanser

Georgi, H. M. (1989a). Effective quantum field theories. In „The New Physics. P. Davies ed.“. Cambridge: Cambridge University Press

Georgi, H. M. (1989b). Grand unified theories. In „The New Physics. P. Davies ed.“. Cambridge: Cambridge University Press

Gibbs, W. (1998). Scientific American August 1998, pp. 9f

Gibson, J. J. (1979). The Ecological Approach to Visual Perception. Boston: Houshton Mifflin

Gilmore, R. and Lefranc, M. (2003). The Topology of Chaos. New York: Wiley

Gloy, K. (2004). Wahrheitstheorien (pp. 147 and 191 ff.). Tübingen: Francke

Gödel, K. (1983). Eine Bemerkung über die Beziehungen zwischen der Relativitätstheorie und der idealistischen Philosophie (pp. 225–231). In „Albert Einstein als Philosoph und Naturwissenschafter. P. A. Schilpp ed.“. Braunschweig: Vieweg

Gödel, Kurt (1931). Über formal unentscheidbare Sätze der Principia Mathematica und verwandter Systeme. Monatshefte für Mathematik und Physik 38 (1931) pp. 173ff

Goethe, J. W. (1797/1921). Faust, Prolog im Himmel (p. 241). Leipzig: Insel

Goethe, J. W. (1823–28/1921). Zahme Xenien VII (vol. 15, p. 386). Leipzig: Insel

Goethe, J. W. (1920). Gingo Biloba. In „West-Östlicher Divan. vol. II, p. 701“. Leipzig: Insel

Goldman, T., Hughes, R. J. and Nieto, M. M. (1988). Schwerkraft und Antimaterie. Spektrum der Wissenschaft, Mai 1988

Görnitz, T. (1999). Quanten sind anders. Berlin: Spektrum

Gould, J. L. and Grant Gould, C. (1994). Animal Mind. New York: Freeman

Greene, B. (1999). The Elegant Universe. Superstrings, Hidden Dimensions, and the Quest for the Ultimate Theory. New York: Norton

Grünbaum, A. (1973). Spatial and Temporal Congruence in Physics: A Critical Comparison of the Conceptions of Newton, Riemann, Poincaré, Eddington, Bridgman, Russell and Whitehead. In „Philosophical Problems of Space and Time". Dordrecht: Reidel

Guggenheim, K. (1952). Alles in Allem (p. 169). Zürich: Artemis

Guth, A. (2002). Die Geburt des Kosmos aus dem Nichts. Die Theorie des inflationären Universums (pp. 52 ff. and 271–304). München: Droemer

Guth, A. and Steinhardt, P. (1989). The inflationary universe (pp. 34–60). In „The New Physics. P. Davies ed.". Cambridge: Cambridge University Press

Habermas, Jürgen (1988). Nachmetaphysisches Denken. Frankfurt am Main: Suhrkamp

Hampe, M. (1998). Alfred North Whitehead. München: Beck

Han, B.-C. (1999). Martin Heidegger, eine Einführung. München: Fink

Hasenfratz, H.-P. (1986). Die Seele. Einführung in ein religiöses Grundphänomen (pp. 44 ff.). Zürich: Theologischer Verlag

Hawking, S. (1988). A Brief History of Time: From the Big Bang to Black Holes (p. 57). New York: bantam

Hawking, S. (1996b). Quantum Cosmology. In „The Nature of Space and Time. S. Hawking, and R. Penrose, Roger eds". New Jersey: Princeton University Press)

Hawking, S. (2001). The Universe in a Nutshell. New York: Bantam

Hawking, S. and Penrose, R. (1996a). The Nature of Space and Time. New Jersey: Princeton University Press

Heidegger, M. (1946). Brief über den Humanismus. Wikipedia

Heidegger, M. (1949). Was ist Metaphysik? Frankfurt: Klostermann

Heidegger, M. (1961). Nietzsche (vol. II, p. 455). Pfulligen: Neske

Heidegger, M. (1977). Sein und Zeit. Tübingen: Niemeyer

Heidegger, M. (1997). Besinnung. In „Gesamtausgabe vol. 66, p. 159". Frankfurt am Main: Klostermann

Heintel, E. (1988). Der Wienerkreis und die Dialektik der Erfahrungen; in „Gesammelte Werke vol. I, (p. 77–113)". Stuttgart: Frommann-Holzboog

Heisenberg, W. (1971). Der Teil und das Ganze. München: Piper

Hersch, J. (1974). Die Unfähigkeit Freiheit zu ertragen (p. 11). Zürich: Benziger

Hilbert, D. (1926). Über das Unendliche (pp. 161–190). Mathematische Annalen 95

Hinzen, W. (2017). Philosophie der Linguistik. In „Philosophien der Einzelwissenschaften (pp. 511–541). S. Lohse and Thomas Reydon eds". Hamburg: Felix Meiner

Hofmannsthal, H. von (1943). Der Rosenkavalier. Komödie für Musik (p. 49). London: Boosey & Hawkes

Hofstadter, D. (1979). Gödel, Escher, Bach: an Eternal Golden Braid. New York: Basic Books

Hofstadter, D. R. (1996). Die Fargonauten: über Analogie und Kreativität (pp. 349 ff.). Stuttgart: Klett-Cotta

Hölling, J. (1971a). Notizen zum Entwurf einer Erkenntnistheorie. In „Realismus und Relativität". München: Wilhelm Fink

Hölling, J. (1971b). Die Irreversibilität der Zeit und das Problem des absoluten Zeitbeginns (pp. 36–41). In „Realismus und Relativität". München: Fink) 1971

Holz, H.H. (2013 a). Leibniz. Darmstadt: Wissenschaftliche Buchgesellschaft.

Holz, H.H. (2013 b). Monadologie. In Kleine Schriften zur Metaphysik. H.H. Holz (Ed.), (pp. 439ff). Darmstadt: Wissenschaftliche Buchgesellschaft.

Hombach, D. (1994). Vom Quark zum Urknall. Die Selbstorganisation der modernen Wissenschaft (pp. 92–117). Grafrath: Boer

't Hooft, G. (2015). The Cellular Automaton Interpretation of Quantum Mechanics. https://arxiv.org/abs/1405.1548. Accessed 21. December 2016

Huber, G. (1995). Zur kategorialen Unterscheidung von „rational" und „irrational" (pp. 9–19). In „Der Pauli-Jung-Dialog. H. Atmanspacher, H. Primas and E. Wertenschlag-Birkhäuser eds". Berlin: Springer

Humboldt, W. von (1903). Schriften zur Sprachphilosophie. In „Gesammelte Schriften, vol. III, A. Filtner and K. Giel eds (1960–81)". Stuttgart: Gotta

Hume, D. (1739/1978). A Treatise of Human Nature (pp. 1–274). L. Selby-Bigge ed., Oxforf: Clarendon

Janich, P. (1997). Das Mass der Dinge. Protophysik von Raum, Zeit und Materie (p. 290). Frankfurt am Main: Suhrkamp

Janich, P. (2000). Was ist Erkenntnis? (p. 101). München: Beck

Jauch, J. M. (1973). Die Wirklichkeit der Quanten. München: Hanser)

Johnson, G. (2006). The Inelegant Universe. Scientific American September 2006, pp. 90f

Jung, C. G. (1967). Die Dynamik des Unbewussten. Zürich: Rascher

Jung, C. G. and Pauli, W. (1952). Naturerklärung und Psyche (pp. 3–107). Zürich: Rascher

Kamitz, R. (1980). Metaphysik und Wissenschaft (pp. 223 ff.). In „Handbuch wissenschaftlicher Begriffe. Speck, J. ed.". Göttingen: Vandenhoek & Ruprecht

Kant, I. (1768/1960). Von dem ersten Grunde des Unterschiedes der Gegenden im Raume. Wöchentliche Königsbergsche Frag- und Anzeigungs-Nachrichten (Nr. 6–8, 6., 13. und 20. Februar 1768). In „Werkausgabe Immanuel Kant. W. Weishedel ed. (pp. 991–1000)". Berlin: Suhrkamp

Kant, I. (1781/2003). Kritik der reinen Vernunft (A154–A166, pp. 253–264). J. Timmermann ed. Hamburg: Felix Meiner.

Kant, I. (1783/2001). Prolegomena zu einer jeden künftigen Metaphysik. Hamburg: Meiner

Kant, I. (1786/1957). Metaphysische Anfangsgründe der Naturwissenschaft. W. Weischedel ed., Wiesbaden: Insel

Kant, I. (1910). Gesammelte Schriften (vol. III, pp. 283 ff.). Berlin: Duncker & Humblot

Kant, I. (1999). Definition des Begriffes Wille, nach P. Prechtl. In „Philosophie Lexikon, p. 662". Stuttgart: Metzler) 1999

Kather, R. (2003). Was ist Leben? Philosophische Positionen und Perspektiven. Darmstadt: Wissenschaftliche Buchgesellschaft

Kelvin, R. (1893/1904). Boyle Lecture delivered before the Oxford University Junior Scientific Club on May 16, 1893. London: Clay

Kepler, J. (1595/1619). Harmonices Mundi. Linz: I. Planc

Kepler, J. (1937). Apologia. In „Gesammelte Werke. W. von Dyck, M. Caspar and F. Hammer eds". München: Beck

Kiefer, C. (2003). On the Interpretation of Quantum Theory – from Copenhagen to the Present Day (p. 297). In „Time, Quantum and Information. L. Castell and O. Ischebeck eds". Berlin: Springer

Kitcher, P. (1983). The Nature of Mathematical Knowledge. Oxford: Oxford University Press

Klapdor-Kleingrothaus, H. V. and Zuber, K. (1997). Teilchenastrophysik (p. 250). Stuttgart: Teubner

Klein, H.-D. (1984). Metaphysik. Eine Einführung (p. 21). Wien: Literas

Koch, A. F. (2004). Subjekt und Natur. Zur Rolle des ‚Ich Denke' bei Descartes und Kant, III. Raum und Zeit als Formen der Anschauung (pp. 83–122). Paderborn: mentis

Kölbel, M. (2002). Truth without Objectivity. London: Routledge

Krantz, D., Luce, R. D., Suppes, P. and Twersky, A. (1971). Foundations of Measurement. Vol. 1. New York: Academic Press

Kreisel, G. (1980). Modell. In „Handbuch wissenschaftstheoretischer Begriffe (p. 437). J. Speck ed.". Göttingen: Vandenhoek & Ruprecht

Kritias (2000). Fragment 3. In „Würfelt Gottt? (p. 123). A. Benz and S. Vollenweider eds". Düsseldorf: Patmos

Kuhlmann, M. (2017). Allgemeine Wissenschaftsphilosophie und die Philosophien der Einzelwissenschaften. In „Philosophien der Einzelwissenschaften (pp. 17–50). S. Lohse and Thomas Reydon eds". Hamburg: Felix Meiner

Kuhn, T. S. (1967). The structure of scientific revolutions. Chicago: Chicago University Press

Küng, H. (2002). Wozu Weltethos? Religion und Ethik in Zeiten der Globalisierung. Freiburg im Breisgau: Herder

Küng, H. (2013). Was bleibt. Kerngedanken. München: Piper

Kuznecov, B. G. (1977). Einstein, Leben-Tod-Unsterblichkeit (p. 267). Stuttgart: Birkhäuser

Langthaler, R. (1997). Nachmetaphysisches Denken? Berlin: (Duncker & Humblot)

Laotse (ca. 600 BC/1978). Tao te king, translated by R. Wilhelm. Düsseldorf: Diederichs

Lee, T. D. and Yang, C. N. (1956). Question of Parity Conservation in Weak Interactions. Phys. Rev. 104 (1956) 254

Leibniz, G. W. (1704/1961). Neue Abhandlungen über den menschlichen Verstand. W. Engelhardt and H. Holz eds. Frankfurt am Main: Insel

Leibniz, G. W. (1904). Initia rerum mathematicarum metaphysica. In „Hauptschriften zur Grundlegung der Philosophie, vol. 1. E. Cassirer ed.". Leipzig: Verlag der Dürr'schen Buchhandlung

Leibniz, G. W. and Clarke, S. (1715–16/1990). Der Briefwechsel zwischen Leibniz und Clarke 1715/1716. Dellian, E. ed.; Hamburg: Meiner

Lévy-Leblond, J.-M. (1979). The Importance of Being (a) Constant (pp. 237–263). In „Problems in the Foundations of Physics. G. Toraldo di Francia, ed.". Amsterdam: North Holland Publ. Co.

Lewis, D. (1998). Universals and Properties (pp. 163–229, especially pp. 173 ff.). In „Contemporary Readings in the Foundation of Metaphysics. S. Laurence and C. Macdonald eds". Oxford: Blackwell

Lieu, R. and Hillman, L. W. (2003). The Phase Coherence of Light from Extragalactic Sources: Direct Evidence against First-Order Planck-Scale Fluctuations in Time and Space. Astrophys. J. Letters 585 (2003) L77

Lineweaver, C. H. and Davis, T. M. (2005). Misconceptions about the Big Bang. Scientific American March 2005, pp. 24–33

Lohse, S. and Greve, J. (2017). Philosophie der Soziologie. In „Philosophien der Einzelwissenschaften (pp. 543–582). S. Lohse and Thomas Reydon eds". Hamburg: Felix Meiner

Luminet, J. P., Starkmann, G. D. and Weeks, J. R. (1999). Is Space Finite? Scientific American April 1999, pp. 68ff

Lyre, H. (2003). Weizsäcker's Reconstruction of Physics: Yesterday, Today, Tomorrow (pp. 374–383). In „Time, Quantum and Information. L. Castell and O. Ischebeck, eds". Berlin: Springer

Lyre, H. (2017). Philosophie der Neurowissenschaften. In „Philosophien der Einzelwissenschaften (pp. 319–353). S. Lohse and Thomas Reydon eds". Hamburg: Felix Meiner

Mac Lane, S. (1969/1998). Categories for the Working Mathematician. New York: Springer

Mach, E. (1889). Die Mechanik, in ihrer Entwicklung historisch-kritisch dargestellt (pp. 213 ff. and 477). Leipzig: Brockhaus

Maddox, J. (1998). What remains to be discovered. London: Free Press

Mainzer, K. (2005). Symmetry and Complexity. New Jersey: World Scientific

Maxwell, J. C. (1881). Treatise on Electricity and Magnetism (2 vols.). Oxford: Clarendon Press

McDonald, A. B., Klein, J. R. and Wark, D. L. (2003). Solving the Solar Neutrino Problem. Scientific American April 2003, pp. 22–30

McGinn, C. (1996). Bewusstsein und Raum (pp. 183–200). In „Bewusstsein. Beiträge aus der Gegenwartsphilosophie. T. Metzinger ed.“. Paderborn: mentis

McLarty, C. (1992). Elementary Categories, Elementary Toposes. Oxford: Clarendon Press.

Mehlberg (1980). Time, causality and the quantum theory. Dordrecht: Reidel

Metzinger, T. (2003). Being No One: The Self-Model Theory of Subjectivity (p. 589). Cambridge: MIT Press

Minsky, M. L. (1968). Matter, Mind and Models. In „Semantic Information Processing. M. Minsky ed.“. Cambridge. Mass.

Mittelstaedt, P. (1972). Logik und Quantenlogik. In „Philosophische Probleme der modernen Physik (pp. 166–207)“. Mannheim: Bibliographisches Institut

Mittelstaedt, P. (1972). Philosophische Probleme der modernen Physik. Mannheim: Bibliographisches Institut

Mittelstaedt, P. (1979). Quantum Logic. In „Problems in the Foundations of Physics. G. Toraldo di Francia ed. (pp. 264 ff.)“. Amsterdam: North Holland

Mittelstrass, J. (2001). Wissen und Grenzen. Frankfurt am Main: Suhrkamp

Möbius, A. F. (1827). Der barycentrische Calcul (p. 184). Leipzig: Barth

Møller, C. (1969). The Theory of Relativity. Oxford: Clarendon

Monod, J. (1973). Le hasard et la nécessité: Essai sur la philosophie naturelle de la biologie moderne (pp. 115 ff.). Paris: Seuil

Moriyasu, K. (1983). An Elementary Primer for Gauge Theory. Singapore: World Sci. Publ. Co

Morrison, M. (2000). Unifying Scientific Theories (p. 255). Cambridge: Cambridge University Press

Müller, S. (2004). Programm für eine neue Wissenschaftstheorie. Würzburg: Königshausen & Neumann

Musser, G. (1998). A massive discovery. The weight of neutrinos. Scientific American August 1998, pp. 9f

Musser, G. (2003). Frozen Stars. Scientific American July 2003, pp. 11f

Musser, G. (2004). Was Einstein Right? Scientific American Sept. 2004, p. 72

Nagel, T. (2003). Relativismus und Vernunft (pp. 107–130). In Wissen zwischen Entdeckung und Konstruktion. M. Vogel and L. Wingert eds. Frankfurt am Main: Suhrkamp

Nakahara, M. (2003). Geometry, Topology and Physics. Bristol: Institute of Physics

NASA (2015). Dark Matter, Dark Energy. NASA Science 5 June 2015

Neumann, J. von (1932). Mathematical Foundations of Quantum Mechanics (pp. 417–445). Princeton: Princeton University Press

Newton, I. (1973). Newton as data fudger. Scientific American 228 (April 1973) p. 44

Nicolis, G. (1989). Physics of far-from-equilibrium systems and self-organisation (pp. 316–347). In „The New Physics. P. Davies. ed.". Cambridge: Cambridge University Press

Nietzsche, F. (1930a). Also sprach Zarathustra (vol. IV, p. 32). Leipzig: Kröner. English translation: Project Gutenberg etext of *Thus Spake Zarthustra*; translated by Thomas Common http://www.gutenberg.org/dirs/etext99/spzar10.txt

Nietzsche, F. (1930b). Die Philosophie im tragischen Zeitalter (pp. 257–338). Leipzig: Körner

Northrop, F. S. C. (1928). The Theory of Relativity and the First Principles of Science. The Journal of Philosophy 25, 16 (pp. 421–435)

Pauli, W. (1919). Merkurperihelbewegung und Strahlenablenkung in Weyls Gravitationstheorie. Verhandlungen der Deutschen Physikalischen Gesellschaft 21, pp. 742–750

Pauli, W. (1955). Niels Bohr and the development of physics: essays dedicated to Niels Bohr on the occasion of his seventieth birthday (pp. 30–51). London: Pergamon

Pauli, W. (1961). In „Aufsätze und Vorträge über Physik und Erkenntnistheorie (p. 156). W. Westphal ed.". Braunschweig: Vieweg

Pauli, W. (1992). In „Wolfgang Pauli und C. G. Jung. Ein Briefwechsel 1932–1958 (p. 54). C. A. Meyer ed". Berlin: Springer

Pauli, W. (1995). Die Klavierstunde. Eine aktive Phantasie über das Unbewusste (pp. 317ff.). In „Der Pauli-Jung-Dialog. H. Atmanspacher, H. Primas and E. Wertenschlag-Birkhäuser eds". Berlin: Springer

Peirce, C. S. (1903/1997). Pragmatism as a Principle and Method of Right Thinking. Albany: State University of New York Press

Penrose, R. (1989). The Emperor's New Mind Concerning Computers, Minds and Laws of Physics. New York: Oxford University Press

Penrose, R. (1994). Shadows of the Mind. Oxford: Oxford University Press

Penrose, R. (1996). The Twistor View of Spacetime (pp. 105–120). In „The Nature of Space and Time. S. Hawking and R. Penrose eds". New Jersey: Princeton University Press

Penrose, R. (2004). The Road to Reality. London: Cape

Piaget, J. (1970). Genetic Epistemology. New York: Columbia University Press

Planck, M. (1949). Scientific Autobiography and Other Papers. Westport, Conn.: Greenwoodpress
Plato (1953). Parmenides (pp. 126 ff.). In „The dialogues of Plato, translated by B. Jowett ed.". Oxford: Oxford University Press
Plato (1957). Sophistes 248c–e. In „The dialogues of Plato. B. Jowett ed." Oxford: Oxford University Press
Plato (1961–1963a). Theaitetos 159e. London: Heinemann
Plato (1961–1963b). Timaios. London: Heinemann
Plato (1961–1963c). Philebos 16c, ff. London: Heinemann
Platon (1981). Theätet (p. 160). E. Martens ed. Stuttgart: Reclam
Poincaré, H. (1898/2003). Das Mass der Zeit (pp. 19–32). In „Wert der Wissenschaft". Berlin: XENOMOS
Popper, K. (1959). The Logic of Scientific Discovery. London: Hutchinson
Popper, K. (1972). Objective Knowledge (pp. 44 ff., 264 ff.). Oxford: Clarendon
Popper, K. (1979). Objective Knowledge (pp. 153–190). Oxford: Clarendon
Popper, K. (1998). The World of Parmenides – Essays on the Presocratic Enlightenment. London: Routledge
Popper, K. and Eccles, J. (1977). The Self and its Brain. New York: Springer
Poser, H. (2001). Wissenschaftstheorie. Eine philosophische Einführung (p. 107). Stuttgart: Reclam
Primas, H. (1995). Über dunkle Aspekte der Naturwissenschaft (pp. 205–238). In „Der Pauli-Jung-Dialog. H. Atmanspacher, H. Primas, Hans and E. Wertenschlag-Birkhäuser eds". Berlin: Springer
Putnam, H. (1981). Reason, truth and history (p. 55). Cambridge: Cambridge University Press
Putnam, H. (1998a). Philosophy of Logic. In „Contemporary Readings in the Foundations of Metaphysics (pp. 404–434). S. Laurence, and C. Macdonald eds". Oxford: Blackwell
Putnam, H. (1998b). On Properties (pp. 148–162). In „Contemporary Readings in the Foundation of Metaphysics. S. Laurence and C. Macdonald eds". Oxford: Blackwell
Putnam, H. (2003). Ethik: In den Strömungen Kurs halten. In „Wissen zwischen Entdeckung und Konstruktion (pp. 288 ff.). M. Vogel, and L. Wingert eds". Frankfurt am Main: Suhrkamp
Quine, W. V. O. (1963). Set Theory and its Logic. Cambridge: Harvard University Press
Quine, W. V. O. (1969). Ontological Relativity and other Essays (pp. 26–29). New York: Columbia University Press
Rae, A. (2004). Is it all in the mind? (p. 72). In „Quantum Physics. Illusion or Reality?". Cambridge: Cambridge University Press

Regge, T. (1961). Relativity without coordinates. Nuovo Cimento 1961, 19 (pp. 558–571)

Reidemeister, K. (1957). Über den Unterschied der Gegenden im Raum (pp. 53–69). In „Raum und Zahl." Berlin: Springer

Reinhardt, F. and Soeder, H. (1974). dtv-Atlas zur Mathematik. München: dtv

Rhine, J. B. and Pratt, J. G. (1957). Parapsychology (pp. 45–65). Oxford: Blackwell

Richardson, M. (1998). The Penguin Book of Firsts. London: Penguin

Rikken, G. L. J. A. and Raupach, E. (2000). Enantioselective magnetochiral photochemistry. Nature 405 (2000) pp. 932–935

Rogler, G. (1996). Die hermeneutische Logik von Hans Lipps und die Begründbarkeit wissenschaftlicher Erkenntnis (p. 161). Würzburg: Ergon

Rorty, R. (2001). Wider den Dogmatismus. In Neue Zürcher Zeitung (p. 79). Zürich 4.1.2001

Roth, G. (2003). Aus Sicht des Gehirns (pp. 128 ff.). Frankfurt am Main: Suhrkamp

Rovelli, C. (2010). Quantum Gravity (p. 415). Cambridge: Cambridge University Press.

Russell, B. (1903/1956). The Principles of Mathematics. London: Allen & Unwin

Russell, B. (1919). Introduction to Mathematical Philosophy (ed. 11, pp. 194–206). London: Allen and Unwin

Russell, B. (1921/1991). Merkmale mentaler Phänomene (pp. 149–159). In „Prozess, Gefühl und Raum-Zeit. M. Hampe and H. Maassen eds". Frankfurt am Main: Suhrkamp

Russell, B. (1925/1958). The ABC of Relativity. London: Allen & Unwin

Russell, B. and Whitehead, A. N. (1935). Principia Mathematica (pp. 85–324). London: Cambridge University Press

Rýzl, M. (1992). Parapsychologie (pp. 203–219). Genf: Ariston

Saller, H. (2003). The Operational Structure of Spacetime (pp. 385–395). In „Time, Quantum and Information. L. Castell and O. Ischebeck eds". Berlin: Springer

Santayana, G. (1940). Scepticism (pp. 515–518). In „The Philosophy of George Santayana. P. A. Schilpp ed.". Evanston: Northwestern University

Sayre, K. M. (1980). Leib-Seele-Problem (pp. 368–372). In „Handbuch wissenschaftstheoretischer Begriffe. J. Speck ed.". Göttingen: Vandenhoek & Ruprecht

Schaff, A. (1980). Objektivität (p. 462). In „Handbuch wissenschaftstheoretischer Begriffe. J. Speck ed.". Göttingen: Vandenhoek & Ruprecht

Scheibe, E. (1994). Struktur und Theorie in der Physik (pp. 101–120). In „Philosophie und Physik der Raum-Zeit. J. Audretsch and K. Mainzer eds". Mannheim: (BI-Wiss.-Verlag)

Schiller, F. (1945). Sprüche des Konfuzius (Gedichte, 4. Buch). Basel: Birkhäuser

Schlüter, A. (1993). Der wachsende Kosmos und die Realität der Quanten. Nova Acta Leopoldina NF 69, Nr. 285 (1993) pp. 127–135. Cited from „T. Görnitz (1999). Quanten sind anders: die verborgene Einheit der Welt (pp. 206 ff.)". Heidelberg: Spektrum

Schmitz, N. (1997). Neutrinophysik. Stuttgart: Teubner

Schmutzer, E. (1972). Symmetrien und Erhaltungssätze der Physik (pp. 156–162). Berlin: Akademie-Verlag

Searle, J. (1997). The Mystery of Consciousness (pp. 120 and 212). New York: New York Review of Books

Sexl, R. and H. (1990). Weisse Zwerge – Schwarze Löcher. Braunschweig: Vieweg

Sexl, R. and Schmidt, H. K. (1987). Raum-Zeit-Relativität. Braunschweig: Vieweg

Shapiro, R. (2007). A Simpler Origin for Life. Scientific American June 2007 (pp. 24–31)

Shapiro, S. (1997). Philosophy of Mathematics. Structure and Ontology. Oxford: Oxford University Press

Singh, S. (1997). Fermat's Last Theorem. The Story of a Riddle that Confounded the World's Greatest Minds for 358 Years (pp. 157–164). London: Forth Estate

Smolin, L. (2001). Three Roads to Quantum Gravity. London: Phoenix

Smolin, L. (2004). Atoms of Space and Time. Sci. American January 2004, pp. 56–65

Spierling, V. (1990). Kleine Geschichte der Philosophie (pp. 166–170). München: Piper

Spinoza, B. (1677/1990). Die Ethik. Stuttgart: Reclam

Stanford Encyclopedia of Philosophy (2008). https://plato.stanford.edu/archives/win2008/entries/qm-action-distance/

Steiner, R. (1894/1992). Die Philosophie der Freiheit. Grundzüge einer modernen Weltanschauung. Seelische Beobachtungsresultate nach naturwissenschaftlicher Methode. Dornach: Rudolf Steiner

Stent, G. S. (2002). Paradoxes of Free Will. Philadelphia: Trans. Am. Phil. Soc. 92

Sternberg, S. (1997). Group theory and physics. Cambridge: Cambridge University Press

Suppes, P. (1972). Independence of the Axiom of Choice and the Generalized Continuum Hypothesis. In „Axiomatic Set Theory. P. Suppes ed.". New York: Dover

Suppes, P. (1973). Time and Geometry. Dordrecht: Reidel

Suppes, P. (1980). Messung. In „Handbuch wissenschaftstheoretischer Begriffe (pp. 415–423) with numerous further references. J. Speck, ed.". Göttingen: Vandenhoek & Ruprecht

Tarski, A. (1959). What is Elementary Geometry? In „The Axiomatic Method. L. Henkin, P. Suppes and A. Tarski eds“. Amsterdam: North Holland

Taylor, J. (1989). Gauge theories in particle physics (pp. 458 ff.). In „The New Physics. P. Davies ed.“. Cambridge: Cambridge University Press

Tryon, E. P. (1973). Is the Universe a Vacuum Fluctuation? Nature vol. 246, pp. 396f

Turing, A. M. (1948/1992). Intelligent Machinery (pp. 107–127). In „Collected Works of A. M. Turing: Mechanical Intelligence. D. C. Ince ed.“. Amsterdam: North-Holland

Uhlenbeck, G. E. and Goudsmit, S. (1926). Spinning Electrons and the Structure of Spectra. Nature 117 (1926)

van Erkelens, H. (1995). Kommentare zur „Klavierstunde“ (pp. 336 f.). In „Der Pauli-Jung-Dialog. H. Atmanspacher, H. Primas, and E. Wertenschlag-Birkhäuser eds“. Berlin: Springer

Veneziano, G. (2004). The myth of the Beginning of Time. Scientific American May 2004, pp. 30–39

von Franz, M. L. (1970). Zahl und Zeit. Psychologische Überlegungen zu einer Annäherung von Tiefenpsychologie und Physik (p. 106). Stuttgart: Klett

von Franz, M. L. (1988). Symbole des Unus Mundus (p. 75). In „Psyche und Materie. Einsiedeln: Daimon“

von Franz, M. L. (1995). Reflexionen zum „Ring *i*“ (p. 331). In „Der Pauli Jung-Dialog. H. Atmanspacher, H. Primas and E. Wertenschlag-Birkhäuser eds“. Berlin: Springer

von Salis, J. R. (1975). Grenzüberschreitungen 1. Teil (p. 260). Zürich: Orell Füssli

von Salis, J. R. (1979). Grenzüberschreitungen, 2. Teil (p. 257). Zürich: Orell Füssli

Wagner, M. (1998). Gruppentheoretische Methoden in der Physik. Braunschweig: Vieweg

Walker, A. G. (1959). Axioms for Cosmology. In „The Axiomatic Method (pp. 308–321). L. Henkin, Ü. Suppes and A. Tarski eds“. Amsterdam: North Holland

Walter, S. (2017). Philosophie der Kognitionswissenschaft. In „Philosophien der Einzelwissenschaften (pp. 441–471). S. Lohse and Thomas Reydon eds“. Hamburg: Felix Meiner

Ward, P. D. (2006). Impact from the Deep. Scientific American October 2006, pp. 43–49

Wechsler, D. (1999). Die Quantenphilosophie des Bewusstseins. Die Leib-Seele-Diskussion des 20. Jahrhunderts und die moderne Naturwissenschaft. Neuried: Ars Una

Wegner, D. M. (2002). The Illusion of Conscious Will. Cambridge: MIT Press

Wehrli, H. (2008). Metaphysics. Chirality as the Basic Principle of Physics. www.hanswehrli.ch. Zurich: Wehrli

Wehrli, P. (2017). Das Universum, das Ich und der liebe Gott (pp. 93–126). Alsdorf: NIBE

Weinberg, S. (1977). The Search for Unity: Notes for a History of Quantum Field Theory (pp. 17–35). Cambridge: Daedalus 106 (4) 1977

Weinberg, S. (1999). Scientific American December 1999, p. 36

Weizsäcker, C. F. von (1971a). Die Einheit der Natur. München: Hanser

Weizsäcker, C. F. von (1971b). Voraussetzungen naturwissenschaftlichen Denkens. München: Herderbücherei

Weizsäcker, C. F. von (1972). Sprache als Information. In „Voraussetzungen des naturwissenschaftlichen Denkens (pp. 106 ff.)". Freiburg im Breisgau: Herderbücherei

Weizsäcker, C. F. von (1986). Aufbau der Physik. München: Hanser

Weizsäcker, C. F. von (1999). Grosse Physiker. München: Hanser

Wenzel, Uwe Justus ed. (1998). Vom Ersten und Letzten. Frankfurt: Fischer

Weyl, H. (1923). Raum, Zeit, Materie (p. 5). Berlin: Springer

Whitehead, A. N. (1919). Time Space and Material: Are they, and if in what Sense, the Ultimate Data of Science? (pp. 44–57). In „Proceedings of the Aristotelian Society, Supplement 2"

Whitehead, A. N. (1925/1939). Science and the Modern World (pp. 103 ff.). New York: Macmillan

Whitehead, A. N. (1936). Process and Reality. An Essay in Cosmology. New York: Macmillan

Wick, H. (2004). Neue Teilchen im Puzzle der Materie. Pentaquarks versprechen tiefere Einblicke in die starke Kraft. Neue Zürcher Zeitung 21. 7. 2004 (p. 9)

Wiehl, R. (2000). Subjektivität und System. Frankfurt am Main: Suhrkamp

Wigner, E. (1970). Remarks on the Mind-Body Question (pp. 171–184). In „Symmetries and Reflections". Cambridge: M.I.T.-Press

Wigner, E. P. (1950). The limits of science. Proceedings of the American Philosophical Society 94 (1950) 424

Wigner, E. P. (1973). Epistemological perspective on quantum theory (pp. 369–385). In „Contemporary Research in the Foundations and Philosophy of Quantum Theory. C. A. Hooker ed.". Dordrecht: Reidel Publ.

Wilhelm, R. (1912/1923). I Ging, das Buch der Wandlungen. Düsseldorf: Diederichs

Wilhelm, R. (1978). Laotse. Tao te king. Düsseldorf: Diederichs

Wilholt, T. (2017). Philosophie der Mathematik. In „Philosophien der Einzelwissenschaften (pp. 53–75). S. Lohse and Thomas Reydon eds". Hamburg: Felix Meiner

Will, C. (1989). The renaissance of general relativity (p. 27). In „The New Physics. P. Davies ed.". Cambridge: Cambridge University Press

Wingert, L. (2003). Wissen zwischen Entdeckung und Konstruktion (p. 219). Frankfurt am Main: Suhrkamp

Wittgenstein, L. (1963/1994). Tractatus logico-philosophicus. Zur Axiomatik der Raumzeit-Theorie (pp. 121–144). In „Philosophie und Physik der Raum-Zeit. J. Audretsch und K. Mainzer eds". Mannheim: BI-Wiss.-Verlag

Wittgenstein, L. (1969). Über Gewissheit. Oxford: Basil Blackwell

Wolschin, G. (1999). Kaonen spüren den Zeitpfeil (p. 14 ff.). Spektrum der Wissenschaft April 1999

Wu, C.-S. et al. (1957). Experimental test of parity conservation in β-decay. Phys. Rev. 105 (1957) 1413–1415

Wuketits, F. M. (1983). Evolutionäre Erkenntnistheorie (pp. 11 ff.). In „Die Evolution des Denkens. K. Lorenz and F. Wuketits eds". München: Piper

Wüthrich, C. (2017). Philosophie der Physik. In „Philosophien der Einzelwissenschaften (pp. 201–228). S. Lohse and Thomas Reydon eds". Hamburg: Felix Meiner

Zermelo, E. (1908a). Neuer Beweis für die Möglichkeit einer Wohlordnung (pp. 107–128). Mathematische Annalen 65

Zermelo, E. (1908b). Untersuchungen über die Grundlagen der Mengenlehre I (pp. 261–281). Mathematische Annalen 65

Zimmer, Heinrich (1994). Philosophie und Religion Indiens (especially p. 223). Frankfurt am Main: Suhrkamp

Zwart, P. J. (1973). The Flow of Time. In „Space, Time and Geometry. P. Suppes ed.". Dordrecht: Reidel

Index

FSC
www.fsc.org
MIX
Papier | Fördert gute Waldnutzung
FSC® C083411

Kolibri 360 GmbH
Ferdinand-Jühlke-Straße 7
99095 Erfurt, Deutschland
produktsicherheit@kolibri360.de